THE IMPORTANT THING IN SCIENCE

NEW FACTS AS TO DISCOVER NEW V

SIR WILLIAM LAWRENCE BRAGG

WE CANNOT SOLVE OUR PROBLEMS WITH THE SAME THINKING WE
USED WHEN WE CREATED THEM.
ALBERT EINSTEIN

IF YOU WANT TO IMPRESS PEOPLE, MAKE THINGS COMPLICATED. IF
YOU WANT TO HELP PEOPLE, KEEP IT SIMPLE.
FRANK KERN

JAKOB SCHWICHTENBERG

NO-NONSENSE CLASSICAL MECHANICS

NO-NONSENSE BOOKS

no-nonsense books

First printing, September 2020

Copyright © 2020 Jakob Schwichtenberg
With illustrations by Corina Wieber

All rights reserved. No part of this publication may be reproduced, stored in, or introduced into a retrieval system, or transmitted in any form or by any means (electronic, mechanical, photocopying, recording, or otherwise) without prior written permission.

UNIQUE ID: D34675F9D37C0C61E47091459FAF3575282B25EDFB2275C645477FC995F583AC
Each copy of No-Nonsense Classical Mechanics has a unique ID which helps to prevent illegal sharing.

BOOK EDITION: 1.65

Dedicated to my parents

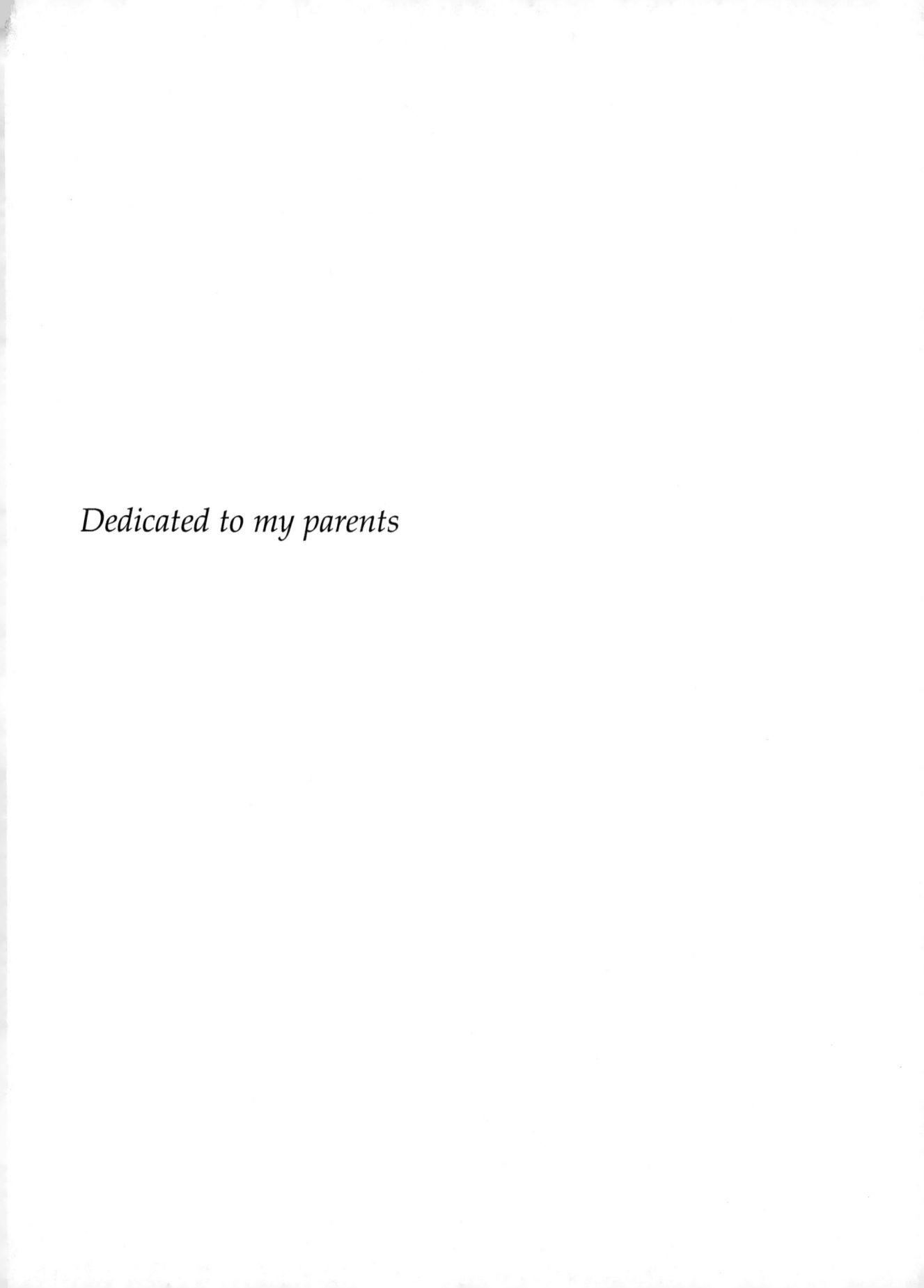

Preface

Classical mechanics is more than 300 years old and dozens, maybe even hundreds, of textbooks on the subject have already been written. So why another one?

First of all, this book focuses solely on the fundamental aspects of classical mechanics.[1] This narrow focus allows us to discuss all of the important concepts several times from various perspectives.

In contrast, most other classical mechanics textbooks try to do a lot at once. For example, it's not uncommon that in addition to the Newtonian, Lagrangian and Hamiltonian formulations, dozens of applications, edge cases, advanced topics, historical developments or even biographies of the most important contributors are discussed. I think this is problematic because, as the saying goes, if you try to be good at everything, you will not be great at anything.

So a clear advantage of the approach used in this book is that the reader has multiple chances to understand a given concept, while in a "normal" textbook the reader immediately has a problem when a passage is not understood perfectly.[2] A second advantage of our narrow focus is that it minimizes the risk of unnecessarily confusing the reader. Like all other fundamental theories, classical mechanics is, at its heart, quite simple. However, using it to describe complicated systems is far from easy and this is where most of the difficulties usually arise.[3]

[1] Applications are only discussed insofar as they help to deepen our understanding of the fundamental concepts and not as an end in themselves. In addition, there are already dozens of great books which discuss applications or other special topics in great detail. Some of the best ones are recommended in Chapter 13.

[2] In a "normal" textbook, each topic is only introduced once. As a result, later chapters become harder and harder to understand without a full understanding of all previous chapters. Moreover, it's easy to become discouraged when a few passages are not perfectly clear since you know that you need the knowledge to understand later chapters.

[3] Most of the difficulties are really mathematics problems, not physics problems anyway, e.g., solving a difficult integral or solving a given differential equation.

In summary, restricting ourselves to the fundamentals allows us to introduce classical mechanics as gently as possible.[4]

[4] While advanced applications are, of course, important, they are not essential to understand the fundamentals of classical mechanics. There are many great books which focus on specific applications. After you've developed a solid understanding of the fundamentals, it's far easier to learn more about those applications you're really interested in.

While this alone may already justify the publication of another classical mechanics textbook, there are a few other things which make this book different:

▷ Firstly, it wasn't written by a professor. As a result, this book is by no means an authoritative reference. Instead, this book is written like a casual conversation with a more experienced student who shares with you everything he wishes he had known earlier. I'm convinced that someone who has just recently learned the topic can explain it much better than someone who learned it decades ago. Many textbooks are hard to understand, not because the subject is difficult, but because the author can't remember what it's like to be a beginner[5].

[5] This is known as the "Curse of Knowledge."

▷ Secondly, this book is unique in that it contains lots of idiosyncratic hand-drawn illustrations. Usually, textbooks include very few pictures since drawing them is either a lot of work or expensive. However, drawing figures is only a lot of work if you are a perfectionist. The images in this book are not as pretty as the pictures in a typical textbook since I firmly believe that *lots* of non-perfect illustrations are much better than a few perfect ones. The goal of this book, after all, is to help you understand classical mechanics and not to win prizes for my pretty illustrations.

▷ Finally, my only goal with this book was to write the most student-friendly classical mechanics textbook and not, for example, to build my reputation. Too many books are unnecessarily complicated because if a book is hard to understand it makes the author appear smarter.[6] To give a concrete example, nothing in this book is assumed to be "obvious" or "easy to see". Moreover, calculations are done step-by-step and are annotated to help you understand faster.

[6] To quote C. Lanczos: "Many of the scientific treatises of today are formulated in a half-mystical language, as though to impress the reader with the uncomfortable feeling that he is in the permanent presence of a superman."

Without any further ado, let's begin. I hope you enjoy reading this book as much as I have enjoyed writing it.

Karlsruhe, June 2018　　　　　　　　*Jakob Schwichtenberg*

PS: I regularly incorporate reader feedback. So if you find an error or have any other kind of comment, I would appreciate an email to errors@jakobschwichtenberg.com.

Acknowledgments

Special thanks to Dr. Florian Colbatzky whose comments, ideas and corrections have made this book so much better. I also want to thank Michael Havrilla, Hywel Griffiths, Daniel Thomas Speckhard, Alex Huang, Eduard Sackinger, Mark Sacchetti, Ronnie Webb, Stephan Lange, Alexandre Felix, Luis Arias and Fabian Waetermans for reporting several typos and to Dena Russell and Jacob Ayres for carefully proofreading the manuscript.

Before we dive in, we need to talk about two things. First, a crucial question:

Why should you care about classical mechanics?

First of all, classical mechanics is still the state of the art when it comes to many problems. Modern theories like quantum mechanics or quantum field theory do not help us when we want to describe how a ball rolls down a ramp or how a rocket flies. While classical mechanics does not help us to describe the fundamental building blocks of nature, it's still the best theory of macroscopic objects that we have.[7]

[7] We will talk about the relationship of classical mechanics to modern theories like quantum mechanics and quantum field theory in Section 12.1.

This alone makes classical mechanics an invaluable tool in the toolbox of any competent physicist.

But even if you only care about truly fundamental aspects of physics, there are several reasons why learning classical mechanics makes sense:

▷ Firstly, classical mechanics is an ideal playground to learn many of the most important concepts which are used everywhere in modern physics. For example, the Lagrangian formalism is presently our best tool to explore new models of nature, and there is no better way to learn it than by studying how we can use it in classical mechanics.[8] In addition, Noether's theorem —a cornerstone of modern physics —can be understood in beautiful and natural terms using the framework of classical mechanics.

[8] To quote Roger Penrose: "*In modern attempts at fundamental physics, when some suggested new theory is put forward, it is almost invariably given in the form of some Lagrangian functional.*" [Penrose, 2016] This is discussed in more detail Section 12.4.

▷ Secondly, by discussing the various formulations of classical mechanics we can understand why there are usually multiple ways to describe a given system, how these alternative descriptions are related and why studying multiple formulations is often a smart thing to do. Understanding this aspect of modern physics is especially important if we want to think about new theories of fundamental physics.[9]

▷ Finally, classical mechanics provides an intuitive arena to study basic mathematical tools and basic mathematical arenas which we need all the time.

[9] To quote Paul Dirac: "*It is not always so that theories which are equivalent are equally good, because one of them may be more suitable than the other for future developments.*"

The second thing we need to talk about is the meaning of a few special symbols which we will use in the following chapters.

Notation

▷ Three dots in front of an equation ∴ mean "therefore", i.e., that this line follows directly from the previous one:

$$\omega = \frac{E}{\hbar}$$
$$\therefore \quad E = \hbar\omega \, .$$

This helps to make it clear that we are *not* dealing with a system of equations.

▷ Three horizontal lines $\equiv$ indicate that we are dealing with a definition.

▷ The symbol $\stackrel{!}{=}$ means "has to be", i.e., indicates that we are dealing with a condition.

▷ The most important equations, statements and results are highlighted like this:

$$\boxed{\frac{\partial L}{\partial q} - \frac{d}{dt}\left(\frac{\partial L}{\partial \dot{q}}\right) = 0} \qquad (1)$$

▷ $\frac{\partial f(x,y,z,t)}{\partial t}$ denotes the partial derivative with respect to t, while $\frac{df(x,y,z,t)}{dt}$ denotes the total derivative.[10]

[10] The difference between partial and total derivatives is discussed in Appendix A.3.

▷ A dot above a function denotes the derivative with respect to time $\dot{q}(t) \equiv \frac{dq(t)}{dt}$ and $\ddot{q}(t) \equiv \frac{d^2q(t)}{dt^2}$.

▷ To unclutter the notation, we often use q as a shorthand for all coordinates, i.e., $q \equiv (q1, ..., q2)$. So for example, instead of $f(x,y,z)$, we write $f(q)$.

That's it. We are ready to dive in (after a short look at the table of contents).

Contents

1 Bird's-Eye View of Classical Mechanics 17

    Part I What Everybody Ought to Know About Classical Mechanics

2 Fundamental Concepts 25
    2.1 Basic Quantities 26
        2.1.1 Mass 28
        2.1.2 Momentum and Angular Momentum 29
        2.1.3 Energy 32
    2.2 Equations of Motion 41
    2.3 Mathematical Arenas 49
        2.3.1 Configuration Space 50
        2.3.2 Phase Space 54

3 Newtonian Mechanics 59

4 Lagrangian Mechanics 67
    4.1 Action and the Lagrangian 73
    4.2 Variational Calculus 81
    4.3 The Euler-Lagrange Equation 86
        4.3.1 Meaning of the Euler-Lagrange Equation . . 90

5 Hamiltonian Mechanics 93
    5.1 Hamilton's Equations 95
        5.1.1 Meaning of Hamilton's Equations 104
    5.2 Hamilton's General Equation 107

6 Summary 113

Part II Essential Systems and Tools

7 Algorithms and Tools — 121
- 7.1 The Newtonian Algorithm 125
- 7.2 The Lagrangian Algorithm 127
  - 7.2.1 Constraints 130
  - 7.2.2 Point Transformations and Generalized Coordinates 134
  - 7.2.3 Gauge Transformations 141
- 7.3 The Hamiltonian Algorithm 148
  - 7.3.1 Canonical Transformations and Canonical Coordinates 152
  - 7.3.2 Canonical Point and Gauge Transformations 160
  - 7.3.3 Infinitesimal Canonical Transformation . . . 169
  - 7.3.4 Generating Functions 172

8 The Harmonic Oscillator — 181
- 8.1 Newtonian Description 184
- 8.2 Lagrangian Description 187
- 8.3 Hamiltonian Description 188
- 8.4 Solving the Equation of Motion 190
- 8.5 Solving the Harmonic Oscillator Using a Canonical Transformation 196

9 The Pendulum — 201
- 9.1 Newtonian Description 203
- 9.2 Lagrangian Description 207
- 9.3 Hamiltonian Description 211
- 9.4 Solving the Equation of Motion 213

Part III Get an Understanding of Classical Mechanics You Can Be Proud Of

10 Noether's Theorem — 223
- 10.1 Symmetries 224
  - 10.1.1 Symmetries of Physical Systems 224
- 10.2 Noether's Theorem Intuitively 228
- 10.3 Noether's Theorem in the Hamiltonian Formalism 232
  - 10.3.1 Noether's Extended Theorem 239
  - 10.3.2 Noether's Converse Theorem 241

    10.4 Noether's Theorem in the Lagrangian Formalism . 242
          10.4.1 Noether's Extended Theorem 245
    10.5 Summary 248

11 Additional Formulations of Classical Mechanics   **251**
    11.1 Hamilton-Jacobi Mechanics 252
          11.1.1 Meaning of Hamilton's Principal Function . 255
          11.1.2 Harmonic Oscillator 257
    11.2 Statistical Mechanics 261
          11.2.1 Probability Density 264
          11.2.2 Conservation of Phase Space Paths 267
          11.2.3 Liouville's Equation 274
          11.2.4 Liouville's Theorem 275
          11.2.5 Summary and Comments 280
    11.3 Koopman-von Neumann Mechanics 287
          11.3.1 Hilbert Space 287
          11.3.2 Koopman-von Neumann Equation 290

12 The Origins of Classical Mechanics   **295**
    12.1 The Cube of Physics 297
    12.2 The Origin of the Least Action Principle 300
          12.2.1 The Origin of the Classical Path 304
          12.2.2 The Origin of the Hamilton-Jacobi Equation 308
    12.3 The Origin of the Classical Lagrangian 311
          12.3.1 Special Relativity 313
          12.3.2 Time Dilation 314
          12.3.3 The Lagrangian of Special Relativity 318
          12.3.4 The Free Classical Lagrangian 320
          12.3.5 Understanding the Minus Sign 322
          12.3.6 General Relativity 323
    12.4 Lagrangians in Modern Physics 332

13 Further Reading Recommendations   **335**

    One Last Thing

Part IV Appendices

A Calculus   **343**
    A.1 Product Rule 344

A.2	Integration by Parts	345
A.3	Total and Partial Derivatives	345
A.4	Chain Rule	348

B The Legendre Transform — 351

C Lagrange Multipliers — 359

D Invariance, Covariance and Functional Form — 367

E Active vs. Passive Transformations and Symmetries vs. Redundancies — 373

F Taylor Expansion — 377

G Vector Calculus — 381
- G.1 The Dot Product — 381
- G.2 The Cross Product — 384

Bibliography — 387

Index — 391

1

Bird's-Eye View of Classical Mechanics

As mentioned in the preface, classical mechanics is, at its heart, quite simple. However, specific applications can be extremely complicated. For this reason it's easy to lose the forest for the trees. To prevent this, we start this book with a quick overview. Afterwards, we will talk about the various concepts in more detail and gradually refine our understanding until we are ready for concrete applications.

So don't worry if not everything is immediately clear in this chapter. Our goal is solely to get an overview and each idea mentioned here will be discussed later in more detail.

Now first of all, what is our goal in classical mechanics?

The short version is:[1]

> We want to describe how macroscopic objects behave.

[1] Macroscopic means big enough such that we don't need quantum mechanics or quantum field theory to describe it.

A bit more technically we can say that:

> We want to derive and solve the equations of motion for macroscopic systems.

One way to accomplish this is by using Newton's second law

$$\frac{d}{dt}\vec{p} = \vec{F}, \quad (1.1)$$

where $\vec{p}$ denotes the momentum of a given object and $\vec{F}$ is the total force acting on it.

But one thing which makes classical mechanics (and physics in general) extremely interesting is that physicists are quite imaginative. Since it is so hard to discover a new theory, it is often a huge step forward to find an alternative method to describe an existing theory. And in fact, for each theory there are different ways of how we can use it to describe a given situation.

This is possible because there are different mathematical arenas we can use as the stage on which we describe what happens. The easiest one is the **physical space** we live in, but there are also more abstract ones like **configuration space**, **phase space** and **Hilbert space**. Each of these mathematical arenas have particular advantages.[2]

[2] We will talk about these arenas in detail in Section 2.3.

The laws of classical mechanics were originally written down using vectors living in physical space. We can describe the behavior of these vectors by using Newton's second law (Eq. 1.1). Nowadays this is known as the **Newtonian formulation**.

[3] It's also possible to formulate classical mechanics in Hilbert space. This is known as the **Koopman-von Neumann formulation** and we will discuss it in Section 11.3. In contrast, quantum mechanics was *originally* formulated in Hilbert space. But it's equally possible to formulate it in phase space, configuration space or physical space.

But it's equally possible to describe macroscopic systems using configuration space or phase space. If we write down the laws of classical mechanics in configuration space, we end up with the **Lagrangian formulation** of classical mechanics. And if we use instead phase space, we end up with the **Hamiltonian formulation**.[3]

In general, we call the description of a given theory in a particular mathematical arena a **formulation** of the theory. So in other words, there are always different formulations of each theory.[4]

This is similar to how we can describe the number 1021 using the English word "one thousand twenty-one" or using the German word "Eintausendeinundzwanzig" or "MXXI " in Roman numerals or "1111111101" in the binary numbering system. Each of these descriptions has particular advantages depending on the problem at hand. For example, saying "1111111101" is extremely awkward in everyday life but essential if we want to do calculations using a computer.

Analogously, the Newtonian formulation of classical mechanics is extremely useful for simple systems because all we have to do is to specify a few vectors in physical space. But for more complicated systems involving constraints, the Lagrangian formalism is a much better choice. And the Hamiltonian formulation is awesome to understand the general structure of classical mechanics and to describe systems consisting of *lots* of objects.

[4] To define what a theory is all about, we need to use a specific formulation. But none of the formulations are more fundamental than the others. We can imagine a theory as something abstract living in "theory space". And to really investigate it, we need to map this abstract thing to something more tangible, i.e., to a set of rules acting on objects living in a specific mathematical arena. Each such map yields a different formulation. (See the diagram below.)

As mentioned above, in the Newtonian formulation, we use Newton's second law (Eq. 1.1) to derive the equations of motion.

In contrast, in the Lagrangian formulation, our first task is always to write down the appropriate **Lagrangian** for the system at hand

$$\boxed{L = T - V,} \qquad (1.2)$$

where T denotes the kinetic energy and V the potential energy.

As soon as we have a concrete Lagrangian, we can derive the corresponding equations of motion by using the **Euler-Lagrange** equation

$$\frac{\partial L}{\partial q} - \frac{d}{dt}\left(\frac{\partial L}{\partial \dot{q}}\right) = 0. \quad (1.3)$$

In the Hamiltonian formulation, we start by writing down the **Hamiltonian** for the system at hand

$$H = p\dot{q} - L \quad (1.4)$$

where p denotes the momentum, $\dot{q}$ the velocity and L, as before, the Lagrangian.[5]

[5] In mathematical terms, the connection between the Hamiltonian and Lagrangian is given by the **Legendre transform**. The Legendre transform is discussed in detail in Appendix B.

As soon as we have a concrete Hamiltonian, we can derive the equations of motion by using **Hamilton's equations**

$$\begin{aligned}\frac{dp}{dt} &= -\frac{\partial H}{\partial q} \\ \frac{dq}{dt} &= \frac{\partial H}{\partial p}.\end{aligned} \quad (1.5)$$

All of this is summarized by the following diagram:

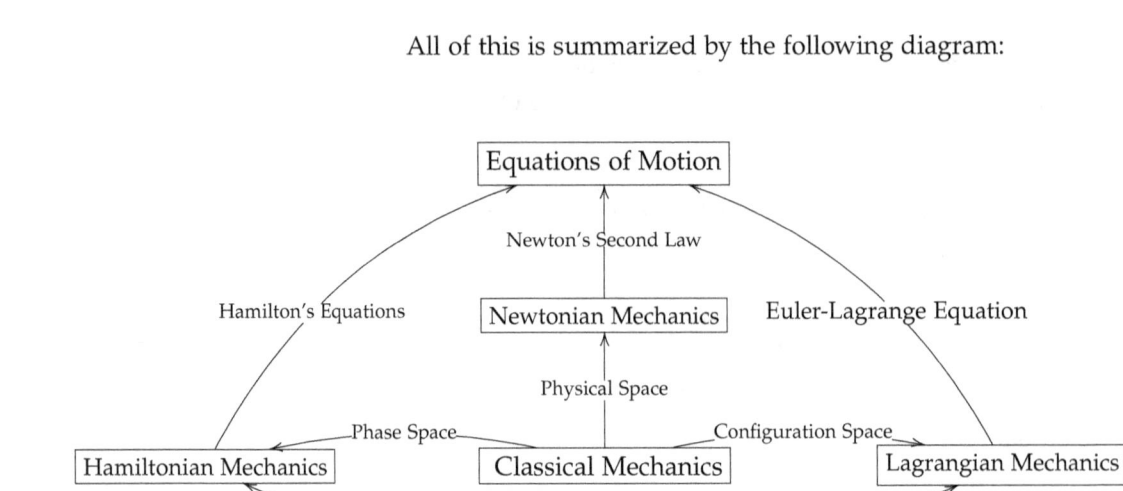

```
                    Equations of Motion
                            ↑
                     Newton's Second Law
                            |
    Hamilton's Equations    |    Euler-Lagrange Equation
                    Newtonian Mechanics
                            ↑
                      Physical Space
                            |
           Phase Space      |      Configuration Space
  Hamiltonian Mechanics  Classical Mechanics  Lagrangian Mechanics
                     Legendre Transform
```

Now after this quick overview, let's move on and discuss everything mentioned here in more detail.

Part I
What Everybody Ought to Know About Classical Mechanics

"The action principle turns out to be universally applicable in physics. All physical theories established since Newton may be formulated in terms of an action. The action formulation is also elegantly concise. The reader should understand that the entire physical world is described by one single action."

Anthony Zee

PS: You can discuss the content of Part I with other readers and give feedback at www.nononsensebooks.com/cm/bonus.

2
Fundamental Concepts

Before we can start talking about classical mechanics, we should talk about what exactly we want to describe.

In short, we want to describe macroscopic (big) objects and we characterize them by using labels like their position, velocity, acceleration, momentum, mass and energy. Since in this book we only care about fundamental aspects, we will treat all objects as if they were mass points. This means that we ignore all effects which arise as a result of the size of objects. There are many important effects which only arise for extended objects, but since each extended object can be considered as a collection of mass points, we will focus on these elementary building blocks.[1]

[1] Detailed discussions of extended objects can be found in most of the textbooks recommended in Section 13. But be warned that the motion of rigid bodies like a spinning top are among the most complicated things students usually have to learn.

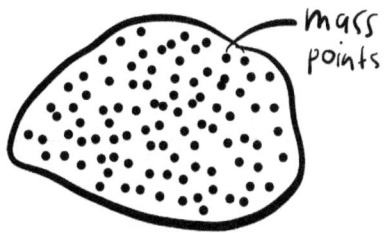

Our goal in classical mechanics is to find the correct equation of motion which allows us to calculate how objects move as time passes in each system. Formulated more technically, the solutions of the equation of motion describe the trajectories of the various objects within the system.

But especially for systems containing multiple objects, keeping track of the various trajectories can become quite cumbersome. For such systems, a better approach is to treat the system as one big thing. We can then describe the time evolution of the system as a path in either configuration space or phase space.

Next, let's talk about all these fundamental concepts one by one.[2]

[2] If you're already familiar with the notions mentioned above and don't need a refresher, feel free to skip this chapter. But at least make sure to skim Section 2.3 because a solid understanding of configuration space and phase space is essential for everything that follows.

2.1 Basic Quantities

Any object can be described using a few key quantities. For example, in classical mechanics our main goal is usually to calculate a function $x(t)$ which tells us for each moment in time t, at which **position** we can find a given object.

[3] The dimension of a space corresponds to the smallest number of coordinates which is sufficient to localize an object in the space. To describe the location of an object on a line, we only need one number which indicates the distance from the origin. Similarly, to describe an object which moves on a circle, we only need the angle φ. Usually objects are free to move in all three dimensions. But to keep the notation as simple as possible, we will usually consider the case where our object can only move freely in one dimension. This is the case, for example, for a bead on a wire.

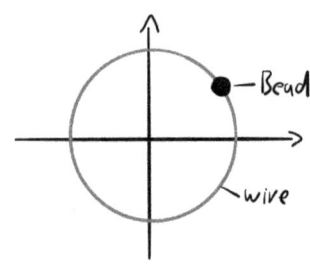

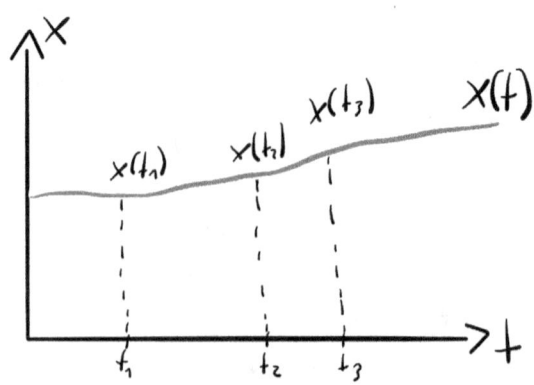

If our object moves in three dimensions, we need three such functions[3]

$$\vec{q}(t) = \begin{pmatrix} x(t) \\ y(t) \\ z(t) \end{pmatrix}. \qquad (2.1)$$

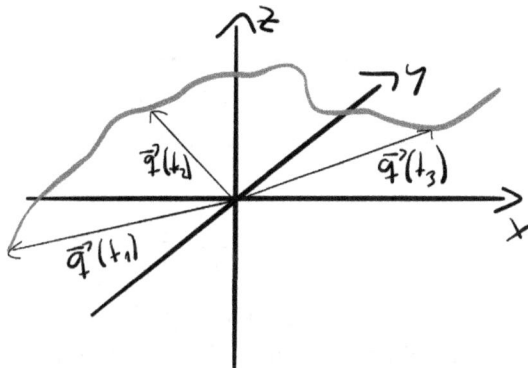

But there are also additional quantities we often use to describe objects. For example, we often want to know how quickly an object moves around. This is described by the **velocity** function $v(t)$ which is defined as the rate of change of the location:

$$v(t) = \frac{dx(t)}{dt}. \qquad (2.2)$$

In words, this means that $v(t)$ tells us exactly how quickly the position of the object changes as time passes. Since derivatives with respect to t are so common in physics, we introduce the shorthand notation

$$\dot{x}(t) \equiv \frac{dx(t)}{dt}. \qquad (2.3)$$

This means that whenever we write a function with a little dot on top of it, we mean its rate of change.

Again, if our object moves in three dimensions, we need three velocity functions

$$\vec{v}(t) = \begin{pmatrix} v_x(t) \\ v_y(t) \\ v_z(t) \end{pmatrix}. \qquad (2.4)$$

Using our shorthand notation, we can write this as

$$\dot{\vec{q}}(t) = \begin{pmatrix} \dot{x}(t) \\ \dot{y}(t) \\ \dot{z}(t) \end{pmatrix}. \tag{2.5}$$

[4] We will see later why we care about how quickly the velocity changes. But to spoil the surprise: the velocity of an object changes whenever a force acts on it. Describing how objects react when forces act on them is what classical mechanics (or really almost all of physics) is all about.

Sometimes, we not only care about how quickly a given object moves around but also about how quickly its velocity changes.[4] The rate of change of the velocity function is known as the **acceleration**

$$\ddot{x}(t) \equiv \frac{d^2 x(t)}{dt^2} = \frac{d}{dt}\frac{dx(t)}{dt} \equiv \frac{d}{dt}\dot{x}(t). \tag{2.6}$$

The relationship between the position $\vec{q}(t)$, velocity $\dot{\vec{q}}(t)$ and acceleration $\ddot{\vec{q}}(t)$ is illustrated in the following figure.

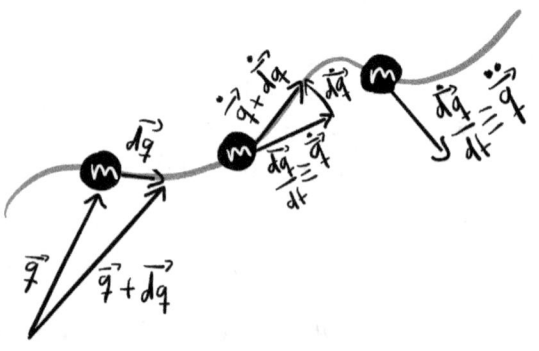

2.1.1 Mass

The most important additional property of objects in classical mechanics is their **mass**. Historically, even Newton only gave a hopelessly circular definition of mass as the object's volume times its density.[5] From a modern perspective, we can understand mass as the "charge" of gravity, analogous to how the electric charge is the charge of electromagnetic interactions. A large mass means that an object has a big gravitational impact, analogous to how an object with a large electric charge has a big electrodynamical impact on other electrically charged objects.

[5] This definition is circular because a (mass) density is defined as an object's mass divided by its volume.

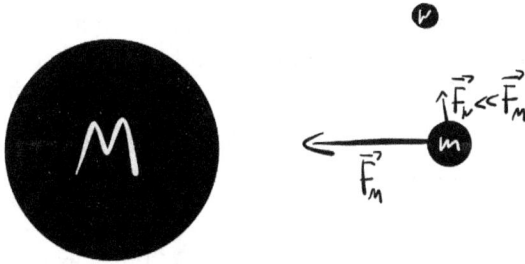

But the main reason why we care about mass in classical mechanics is that it encodes the resistance of an object to being moved or to having its velocity changed if it already moves with uniform velocity.[6] In other words, the mass of an object is a label we use to describe how difficult it is to change its velocity with a force.[7] For example, it's much easier to push a ball made of cork than a ball of equal size made of iron.

2.1.2 Momentum and Angular Momentum

There are three further basic quantities that we regularly use in classical mechanics: momentum, angular momentum and energy. These quantities are useful because they are conserved. This means that while everything within the system may change all the time (in particular, the locations and velocities of the objects), these quantities remain unchanged. In this sense, conserved quantities are like anchors which we can always cling to in an otherwise chaotic world.[8] However, take note that the momentum, angular momentum and energy of individual objects *can* change. Only the total momentum, total angular momentum and total energy within a closed system remain constant.[9]

One of the most beautiful aspects of classical mechanics is that we can actually understand the origin of conserved quantities.

[6] You might rightfully wonder how these two descriptions of mass fit together. Historically, physicists used the notions of "gravitational mass" and "inertial mass" because they thought that the two roles played by mass meant that they are actually two kinds of mass and that we need to consider them separately. However, the idea that inertial mass and gravitational mass are actually the same thing (which we now simply call the mass) was one of the key insights which led Einstein to his famous theory of general relativity. From this perspective, the resistance of an object to being moved is a result of the gravitational pull of all surrounding objects in the whole universe.

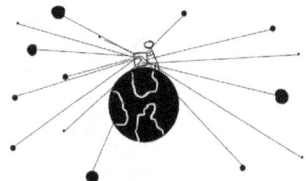

This idea is known as Mach's principle. However, take note that there are still lots of ongoing discussions about the validity of Mach's principle and how to formulate it properly.

[7] Intuitively, we can define a force as something which changes the velocity of objects.

[8] Mathematically, a conserved quantity is something with a vanishing rate of change, for example, $\frac{d}{dt}E = 0$.

[9] A closed system is sufficiently isolated such that we can ignore all effects from the outside world. In particular, no energy or momentum is leaking out from a closed system.

[10] To spoil the surprise: for each symmetry of a system, we get a conserved quantity. For example, whenever it makes no difference whether we perform an experiment today or tomorrow, the energy within the system is conserved. This is known as Noether's theorem.

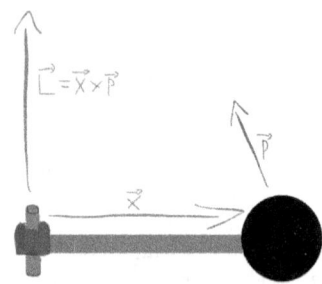

[11] If you're unfamiliar with the cross product, see Appendix G. We use the cross product because it allows us to multiply two vectors and get another vector as a result. This is what we need because to describe angular momentum, we need to know the axis about which an object is rotating (a direction) and the speed of the rotation (a magnitude). Angular momentum is, therefore, naturally represented by a vector. And it is the cross product that allows us to take two vectors, position and momentum, and combine them to obtain the angular momentum. In contrast, the dot product of two vectors $\vec{a} \cdot \vec{b}$ yields a number and therefore cannot encode any directional information.

[12] As mentioned above, this is known as Newton's second law and we will discuss it in more detail in Chapter 3.

[13] We can multiply by dt because dt simply means a little bit of t, i.e., a really short time interval.

In other words, we do not need to introduce momentum, angular momentum and energy ad hoc as additional quantities but we can actually *derive* them. In Chapter 10, we will talk about this in detail.[10] But for the moment it is sufficient to use the following rather rough definitions:

▷ **Momentum** is the velocity of an object times its mass
$$\vec{p}(t) \equiv m\dot{\vec{q}}(t). \tag{2.7}$$

▷ **Angular momentum** is defined as the cross product of the position vector and the momentum vector[11]
$$\vec{L}(t) = \vec{q}(t) \times \vec{p}(t) = m\vec{q}(t) \times \dot{\vec{q}}(t). \tag{2.8}$$

In intuitive terms, we can say that:

▷ **Momentum** is the total "oomph" an object has behind it. To understand how this interpretation comes about, we can use the fact that the rate of change of the momentum of an object is always equal to the force acting on it:[12]
$$\frac{dp}{dt} = F. \tag{2.9}$$

Therefore, the total change in momentum $\Delta p = p(t_f) - p(t_i)$ during some time interval $\Delta t = t_f - t_i$ is equal to the force F times the time interval Δt:[13]

$$\frac{dp}{dt} = F$$
↲ multiplication by dt and integrating
$$\int_{t_i}^{t_f} dp = \int_{t_i}^{t_f} F dt$$
↲ assuming the force is constant
$$\Delta p = F \Delta t.$$

In words, this means that the momentum of an object tells us how long it takes a given force F to stop it. An object with a large momentum is much harder to stop. Formulated differently, we need a much bigger force to stop it quickly.

Alternatively, we can consider collisions of two objects. The main point here is that the object with the larger momentum "wins".

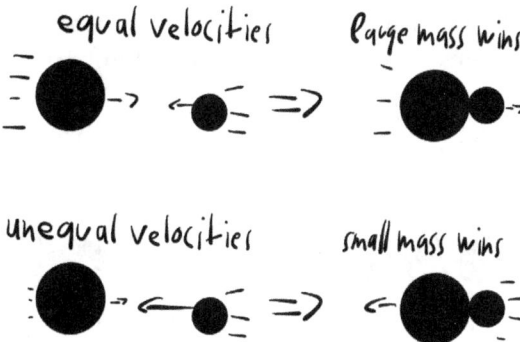

Winning in such a collision can either be achieved by having a large mass or having a large velocity and this is exactly what momentum encodes: $p = mv$.

▷ **Angular momentum** is the total "rotational oomph" an object has behind it. While (linear) momentum tells us how hard it is to stop an object using a specific force, angular momentum tells us how hard it is to stop it spinning. For example, it's certainly much harder to stop the rotation of a dumbbell made of iron than to stop the rotation of a dumbbell made of Styrofoam.

Similarly, it's much harder to stop a fast-spinning dumbbell than a slowly spinning one. A third (but a bit more subtle) aspect of angular momentum is that it's much harder to stop a long dumbbell.

And that's why angular momentum is proportional to mass × velocity × radius.[14]

As mentioned above, we are primarily interested in momentum

[14] Don't worry if the meaning and importance of angular momentum is not immediately clear. For the moment it is sufficient to keep in mind that angular momentum is a useful quantity whenever things are spinning around. In modern physics, angular momentum is *extremely* important because all elementary particles except for the Higgs boson carry some kind of internal angular momentum which we call spin. Using this property of elementary particles we can understand, for example, why matter is stable at all.

and angular momentum because they are conserved. To understand why conserved quantities are so helpful, let's consider a head-on collision of two objects. For simplicity, let's assume that the two objects stick together after the collision.[15]

[15] In more technical terms, this means that the collision is completely inelastic. An elastic collision is one in which both momentum and kinetic energy are conserved. In an inelastic collision, some of the kinetic energy is transferred to internal degrees of freedom of the objects. Parts of the object may crumble or bend, for example. For the collision we consider here, a maximum amount of kinetic energy is "lost" to internal degrees of freedom. (This is necessarily the case here because otherwise the two objects wouldn't stick together.) But take note that the total energy is always conserved. It's just that after the collision some of the energy is now present in different forms.

Then, solely using the fact that momentum is conserved, we can calculate the velocity that the combined objects have after the collision:

$$p_i^{system} = p_f^{system}$$

↺ definition of momentum

$$(m_1 v_1 + m_2 v_2) = (m_1 + m_2) v$$

↺ rearranging terms

$$v = \frac{m_1 v_1 + m_2 v_2}{m_1 + m_2}.$$

Next, let's talk about one additional conserved quantity which we use all the time in classical mechanics: energy.

2.1.3 Energy

Energy is a bit more tricky because it comes in different forms. At first, physicists believed that energy could be described by the formula[16]

[16] Take note that $\vec{\dot{q}}^2 = \vec{\dot{q}} \cdot \vec{\dot{q}}$, i.e., the dot product of the velocity vector with itself. If you're unfamiliar with the dot product, have a look at Appendix G.1.

$$T = \frac{1}{2} m \vec{\dot{q}}^2. \tag{2.10}$$

But it was quickly noted that this quantity is not always conserved. From a modern perspective, we say that energy comes in two forms: kinetic and potential. The formula in Eq. 2.10 only describes **kinetic energy** and is therefore incomplete. The total energy is always conserved and therefore, whenever the kinetic energy becomes smaller, it doesn't vanish but is converted into potential energy. Mathematically, this means that the **total energy** can be written as

$$E = T + V, \quad (2.11)$$

where V denotes the potential energy. The specific formula for the potential energy V always depends on the system at hand.

In general, kinetic energy is the energy associated with motion, while potential energy represents the energy which is "stored" in a physical system. Moreover, the total energy is always conserved. But while the total energy remains unchanged, the kinetic and potential parts of the total energy can change all the time.

Imagine, for example, a pendulum which swings back and forth. When it swings, it sweeps out an arc and then slows down as it comes closer to its highest point. At this highest point the pendulum does not move at all. So at this point, the energy is completely given in terms of potential energy. But after this brief moment of rest, the pendulum swings back again and therefore part of the total energy is then given in the form of kinetic energy. So as the pendulum swings, kinetic and potential energy are constantly converted into each other.

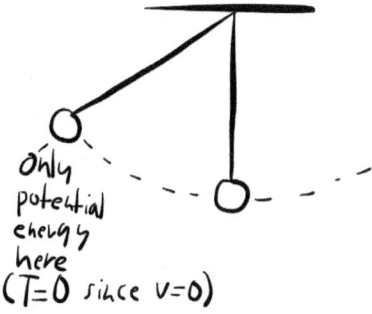

To understand the interplay between kinetic and potential energy a little better, let's assume that we are dealing with an object with constant mass and we notice that its kinetic energy is getting smaller.[17] This can happen whenever a force $\vec{F}$ acts on

[17] Take note that the mass is often constant but not always. For example, the mass of a rocket gets smaller because it burns fuel.

[18] This is, in fact, Newton's second law and we will talk about it in more detail in Chapter 3.

the object. We can define a force as something which causes the momentum of our object to change[18]

$$\vec{F} = \dot{\vec{p}}. \tag{2.12}$$

Using this, we can write the rate of change of the kinetic energy as follows:[19]

[19] Don't worry if some of the steps are not completely clear. We'll talk about everything mentioned here in more detail below. The sole goal in this section is to get a rough understanding for what we mean by potential energy and a potential.

$$\frac{dT}{dt} = \frac{d}{dt}\left(\frac{1}{2}m\dot{\vec{q}}^2\right)$$

↷ $\frac{d}{dt}\dot{\vec{q}}\cdot\dot{\vec{q}} = \ddot{\vec{q}}\cdot\dot{\vec{q}} + \dot{\vec{q}}\cdot\ddot{\vec{q}} = 2\ddot{\vec{q}}\cdot\dot{\vec{q}}$ (product rule)

$$= m\ddot{\vec{q}}\cdot\dot{\vec{q}}$$

↷ $\vec{p} = m\dot{\vec{q}} \Rightarrow \dot{\vec{p}} = m\ddot{\vec{q}}$

$$= \dot{\vec{p}}\cdot\dot{\vec{q}}$$

↷ Eq. 2.12

$$= \vec{F}\cdot\dot{\vec{q}}. \tag{2.13}$$

So the change in kinetic energy of the object, as it travels from some specific location q_i at t_i to another location q_f at t_f, is given by[20]

[20] We calculate this because changes in the kinetic energy mean that the potential energy must change. Therefore, we can learn something about the potential energy. Moreover, take note that

$$T(t_f) - T(t_i) = \int_{t_i}^{t_f} \frac{dT}{dt}\, dt$$

follows from the fundamental theorem of calculus

$$\int_a^b \frac{d}{dx} f(x)\, dx = f(b) - f(a).$$

$$T(t_f) - T(t_i) = \int_{t_i}^{t_f} \frac{dT}{dt}\, dt \underbrace{=}_{\text{Eq. 2.13}} \int_{t_i}^{t_f} dt\, \vec{F}\cdot\dot{\vec{q}}$$

↷ $\dot{\vec{q}} = \frac{d}{dt}\vec{q}$

$$= \int_{t_i}^{t_f} dt\, \vec{F}\cdot\frac{d}{dt}\vec{q}$$

↷ $\frac{dt}{dt} = 1$

$$= \int_{\vec{q}_i}^{\vec{q}_f} \vec{F}\cdot d\vec{q}. \tag{2.14}$$

This final expression is known as the **work** done by the force $\vec{F}$ as the object moves from q_i to q_f. What we've calculated here therefore means in words: the change in kinetic energy is equal to the work done.

Now the key idea is that we can imagine that (at least all fundamental) forces originate from an underlying potential[21]

[21] More precisely, forces for which this is the case are known as **conservative forces** because the work done by them as we move an object along a closed curve is zero, i.e., nothing gets lost or added. Conservative forces only depend on the location $\vec{q}$ and not on the velocity $\dot{\vec{q}}$, etc.

$$\vec{F} = -\nabla V(\vec{q}) \equiv -\begin{pmatrix}\frac{\partial}{\partial x}\\\frac{\partial}{\partial y}\\\frac{\partial}{\partial z}\end{pmatrix} V(\vec{q}) = -\begin{pmatrix}\frac{\partial V(\vec{q})}{\partial x}\\\frac{\partial V(\vec{q})}{\partial y}\\\frac{\partial V(\vec{q})}{\partial z}\end{pmatrix}. \tag{2.15}$$

For one-dimensional problems, the relationship between a force and the corresponding potential simplifies to

$$F = -\frac{\partial V(q)}{\partial q}. \qquad (2.16)$$

Using this, we can rewrite Eq. 2.14 as follows:[22]

$$T(t_f) - T(t_i) = \int_{\vec{q}_i}^{\vec{q}_f} \vec{F} \cdot d\vec{q} \qquad \text{this is Eq. 2.14}$$

$$\circlearrowright \text{ Eq. 2.15}$$

$$= \int_{\vec{q}_i}^{\vec{q}_f} \left(-\nabla V(\vec{q})\right) \cdot d\vec{q}$$

$$\circlearrowright \int_a^b f'(x)dx = f(b) - f(a)$$

$$= -\left(V(\vec{q}_f) - V(\vec{q}_i)\right). \qquad (2.17)$$

By rearranging the terms, we can therefore conclude that

$$T(t_f) + V(\vec{q}_f) = T(t_i) + V(\vec{q}_i). \qquad (2.18)$$

In words, this means that the sum of kinetic and potential energy at the initial moment in time (for which the object is at $\vec{q}_i$) is equal to the sum of kinetic and potential energy at the final moment in time (for which the object is at $\vec{q}_f$). We've therefore derived that changes in the kinetic energy can always be accounted for by changes in the potential energy. The total sum of kinetic and potential energy always stays the same. In other words, the total energy $E = T + V$ is conserved.[23]

Here's another perspective.

When we move an object attached to a spring away from its rest position, there will be a force which pushes the object back to this rest position.

[22] If you're confused by the steps here, it may be helpful to consider a one-dimensional problem. The calculation then reads

$$T(t_f) - T(t_i) = \int_{q_i}^{q_f} F dq$$

$$= \int_{q_i}^{q_f} \left(-\frac{\partial V(q)}{\partial q}\right) dq$$

$$= -\left(V(q_f) - V(q_i)\right).$$

[23] This is only the case for systems in which all forces can be written using a potential (Eq. 2.15). But since this is the case for all fundamental forces (gravity, electromagnetic force, weak nuclear force and strong nuclear force) and we only care about fundamental aspects, this is all we need to know.

One way to understand this is by describing the spring using the potential $V(x) = Cx^2$. This potential has its minimum at the rest position, and any movement away from it results in a force $F = -\frac{\partial V(x)}{\partial x} = -2Cx$ which pushes the object back to the rest position.

For concreteness, we are dealing with a spring described by the constant $C = 1 \frac{kg}{s^2}$.[24] When we pull the object attached to this spring to the $x = 2$ m location, there will be a force

$$F = -2Cx = -4 \frac{kg \cdot m}{s^2},$$

which is directed toward the equilibrium position at $x = 0$. And if we push the object in the opposite direction to $x = -2$ m, there will be a force

$$F = -2Cx = 4 \frac{kg \cdot m}{s^2},$$

which again points toward the equilibrium position.

[24] The constant C has units $\frac{kg}{s^2}$, where "kg" denotes kilograms and "s" denotes seconds, because then $F = -2Cx$ has the correct units of a force $\frac{kg \cdot m}{s^2}$.

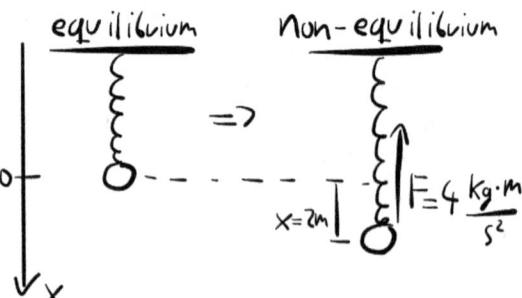

One final comment before we move on: we can't measure the potential or potential energy of an object directly. Instead, we can only measure the resulting force $F = -\frac{\partial V(x)}{\partial x}$. This follows because we can add an arbitrary constant $V(x) \to \tilde{V}(x) = V(x) + c$ without changing the resulting force:

$$V(x) \quad \Rightarrow \quad F(x) = -\frac{\partial V(x)}{\partial x}$$

$$\tilde{V}(x) = V(x) + c \quad \Rightarrow \quad \tilde{F}(x) = -\frac{\partial \tilde{V}(x)}{\partial x} = -\frac{\partial (V(x) + c)}{\partial x} = -\frac{\partial V(x)}{\partial x} = F$$

because $\frac{\partial c}{\partial x} = 0$. This means that we can shift the potential however we want.[25] In geometrical terms, we can say that the force remains unchanged because the slope remains unchanged by constant shifts of the potential:[26]

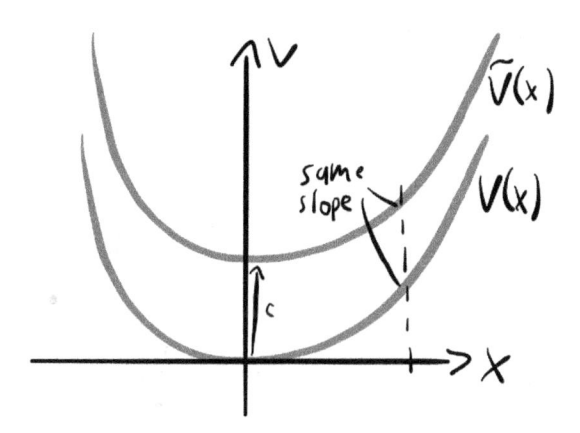

The freedom in how we define the potential for any system also implies that we can't measure the potential energy of an object directly. Instead, only differences between the potential energy of an object at one position and its potential energy at some reference position are physically important. We usually shift the potential such that any object at the equilibrium position has potential energy zero and then specify the potential energy at other positions relative to this position. For example, we usually shift the Earth's gravitational potential such that it is zero when the ball lies on the ground. We are then able to write the potential energy as $V(z) = mgz$, where z is the height above the ground.[27] But we could equally shift the potential such that it is non-zero at the ground, and then we would have to use the formula $\tilde{V}(z) = mgz + c$, where c is the value of the potential at the ground ($z = 0$). For example, we could shift the potential such that it is zero for $z = 2$ m. The potential energy would then read $\tilde{\tilde{V}}(z) = mgz - 2mg$ because then we get zero for $z = 2$ m: $\tilde{\tilde{V}}(2) = mg(2) - 2mg = 0$.

However, everything we can measure in experiments remains completely unaffected by such shifts of the potential. This follows because only forces play a role in our equations of motion

---

[25] Historically, physicists used this observation to conclude that potentials are solely convenient calculation devices which, however, aren't really real because we can't measure them directly. Working with a potential is much simpler than working with a force because a force is described by a vector function $\vec{F}(x)$ while a potential is an ordinary function $\phi(x)$. Moreover, the freedom to shift the potential often allows important calculatory simplifications. But from a modern perspective, potentials are more than purely mathematical tools. For example, the famous Aharonov-Bohm experiment demonstrated that particles can be affected in regions in which only the potential is non-zero but nothing else. In other words, it has been experimentally verified that potentials are indeed real. In some sense, potentials are really at the heart of modern physics and if you want to learn more about how we understand their role in nature, you might enjoy:

Jakob Schwichtenberg. *Physics from Finance*. No-Nonsense Books, Karlsruhe, Germany, 2019b. ISBN 978-1795882415

[26] The slope of a potential corresponds to the force since $F = -\frac{\partial V}{\partial q}$.

[27] Semantics: the potential here reads $\phi(z) = gz$ and an object of mass m which is located at z therefore has the potential energy $V(z) = mgz$.

(which we will discuss next), and all constant shifts leave the resulting forces unchanged:

$$V(z) = mgz \quad \Rightarrow \quad F = -\frac{\partial V(z)}{\partial z} = -mg$$

$$\tilde{V}(z) = mgz + c \quad \Rightarrow \quad \tilde{F} = -\frac{\partial \tilde{V}(z)}{\partial z} = -mg$$

$$\tilde{\tilde{V}}(z) = mgz - 2mg \quad \Rightarrow \quad \tilde{\tilde{F}} = -\frac{\partial \tilde{\tilde{V}}(z)}{\partial z} = -mg$$

To summarize:

▷ **Energy** is defined as the sum of an object's kinetic energy T and its potential energy V

$$E = T + V. \tag{2.19}$$

- The **kinetic energy** is defined as

$$T = \frac{1}{2}m\dot{\vec{q}}^2. \tag{2.20}$$

- There is no general formula for the **potential energy** because different systems are characterized by different potentials and, therefore, we need different formulas for the potential energy. But take note that usually the formula for potential energy of an object only depends on its location $\vec{q}(t)$ and not on its velocity $\dot{\vec{q}}(t)$ or acceleration $\ddot{\vec{q}}(t)$. For example, the potential energy in the Earth's gravitational field can be described by

$$V = mgz, \tag{2.21}$$

where z is the height above sea level, m the mass and $g \approx 9.81 \, \frac{m}{s^2}$, a constant which characterizes the Earth's gravitational field.

In intuitive terms, we can say that:

▷ Kinetic energy is a measure for how much is going on in our system.
  – If all objects are moving around quickly, the kinetic energy is high.
  – In contrast, if all objects are sitting still, the kinetic energy is zero.

▷ Potential energy is a measure for how much *could* happen, but isn't.
  – If all objects are placed on the top of a mountain, they have a large (gravitational) potential energy and they could start moving quickly by falling down. Currently, they are stationary. In this sense, their large potential energy is unrealized potential, which is why we use the word *potential* energy.
  – In contrast, if all objects are placed on the floor, they have zero potential energy since they can't fall any further down.

Finally, to demonstrate once more that conserved quantities are indeed useful, let's try to describe a ball which is thrown straight up into the air. The ball's total energy is

$$E = T + V$$
$$\quad\quad\quad\quad\quad\quad\quad\quad T = \frac{1}{2}m\dot{z}^2, V = mgz$$
$$= \frac{1}{2}m\dot{z}^2 + mgz$$
$$\quad\quad\quad\quad\quad\quad\quad\quad \text{definition of momentum } p = m\dot{z}$$
$$= \frac{p^2}{2m} + mgz.$$

For concreteness, let's assume that the ball's mass is 1 kg and we throw it with $\dot{z} = 2\,\frac{m}{s}$ starting from $z = 2$ m into the air. Therefore, the ball's initial momentum is $p = 2\,\frac{\text{kg·m}}{s}$. Moreover, the ball's energy at the moment we throw it is

$$\begin{aligned}
E &= \frac{1}{2m}p^2 + mgz & &\circlearrowright\ p = 2\ \frac{\text{kg}\cdot\text{m}}{\text{s}},\ z = 2\ \text{m} \\
&= \frac{1}{2m}\left(2\ \frac{\text{kg}\cdot\text{m}}{\text{s}}\right)^2 + mg\,(2\ \text{m}) & &\circlearrowright\ g \approx 9.81\ \frac{\text{m}}{\text{s}^2},\ m = 1\ \text{kg} \\
&= \frac{1}{2(1\ \text{kg})}\left(2\ \frac{\text{kg}\cdot\text{m}}{\text{s}}\right)^2 + (1\ \text{kg})(9.81\ \frac{\text{m}}{\text{s}^2})2\ \text{m} & &\circlearrowright\ \text{simplifying} \\
&= 2\ \frac{\text{kg}\cdot\text{m}^2}{\text{s}^2} + 19.62\ \frac{\text{kg}\cdot\text{m}^2}{\text{s}^2} \\
&= 21.62\ \frac{\text{kg}\cdot\text{m}^2}{\text{s}^2}\,. & & (2.22)
\end{aligned}$$

Using only these numbers and that the ball's energy is conserved, we can directly predict the ball's maximum height.[28] This is possible because when the ball reaches its maximum height, it will stop moving upward. This means that $\dot{z} = 0$ and therefore that the kinetic energy vanishes $T = \frac{1}{2}m\dot{z}^2 = 0$. But since the total energy is conserved, we know automatically that at this point, the ball's total energy must be in the form of potential energy:

[28] Maybe you wonder about the conservation of momentum. While the ball's momentum is not conserved (and even becomes zero at the point of maximum height), the total momentum of the system consisting of ball + Earth is conserved.

$$V_{\max} = mgz_{\max} \stackrel{!}{=} 21.62\ \frac{\text{kg}\cdot\text{m}^2}{\text{s}^2} = E$$

$\circlearrowright\ g \approx 9.81\ \frac{\text{m}}{\text{s}^2},\ m = 1\ \text{kg}$

$$(1\ \text{kg})(9.81\ \frac{\text{m}}{\text{s}^2})z_{\max} \stackrel{!}{=} 21.62\ \frac{\text{kg}\cdot\text{m}^2}{\text{s}^2}$$

$\circlearrowright$

$$z_{\max} \stackrel{!}{=} 2.204\ \text{m}\,.$$

This means that we can predict correctly how high the ball will fly solely by using the fact that energy is conserved.

Of course, there are additional properties of objects which can be important in specific systems like, for example, the electric charge. But since these are only important for specific systems, we will not discuss them any further here.[29]

Next, let's talk about how we can describe objects at all times (and not just at some extremal point like the point of maximal height).

[29] In particular, to describe the behavior of electrically charged objects, we need electrodynamics. If you want to learn about electrodynamics, you might enjoy

Jakob Schwichtenberg. *No-Nonsense Electrodynamics*. No-Nonsense Books, Karlsruhe, Germany, 2018a. ISBN 978-1790842117

2.2 Equations of Motion

While there are different formulations of classical mechanics, our goal is usually to find the correct equation of motion for each system. Solutions of the equation of motion describe how the objects in our system move as time passes.

For simplicity, let's restrict ourselves to one object moving in one dimension.[30]

[30] This means that we don't need vectors.

In mathematical terms, an equation of motion is a differential equation. This means that we have the first or second derivative of a function which describes the position of the object $x(t)$ on one side of the equation and something else on the other side. For example, the equation of motion for a free object is[31]

[31] We will discuss in detail below how we can derive the correct equation for any system.

$$\frac{d^2 x(t)}{dt^2} = 0. \quad (2.23)$$

But usually the equation of motion is more complicated. For example, if a force $F(x) = Cx(t)$ acts on the object, where C is some constant, the equation of motion becomes[32]

[32] This is the equation of motion for an object attached to a spring. This system is known as a harmonic oscillator and we will discuss it in more detail in Chapter 8.

$$\frac{d^2 x(t)}{dt^2} = Cx(t). \quad (2.24)$$

In either case, our goal is to find the correct function $x(t)$ which yields the same expression if we put it into the right-hand and left-hand side of the equation. There are sophisticated methods for solving differential equations and big cookbooks full of mathematical recipes which you can try. But sometimes, we can simply guess the right solution.

The equation of motion for the free object (Eq. 2.23) tells us that we need a function $x(t)$ which yields zero if we differentiate it twice. There are, in general, infinitely many such functions. In particular, $x(t) = a$ for any constant number a yields zero if we differentiate it twice. Similarly, $x(t) = bt$ for any constant number b yields zero. But higher order terms in t do not vanish

as we can check explicitly, e.g., for $x(t) = ct^2$:
$$\frac{d^2 x(t)}{dt^2} = \frac{d^2(ct^2)}{dt^2} = 2c \neq 0,$$
where c is some constant.

This means that the most general solution of the equation of motion for a free object (Eq. 2.23) reads[33]

[33] We will talk about the meaning of the constants a and b in a moment.

$$x(t) = a + bt \qquad (2.25)$$

since for any value of the two constants a and b we find
$$\underbrace{\frac{d^2 x(t)}{dt^2}}_{\text{Eq. 2.25}} = \frac{d^2(a+bt)}{dt^2} = 0 \quad \checkmark .$$

Finding solutions of the equation of motion for a non-free object (e.g., Eq. 2.24) is a bit more difficult. Specifically, for Eq. 2.24 we need a function $x(t)$ which yields itself times some constant if differentiated twice. Two functions with this special property are $\sin(\omega t)$ and $\cos(\omega t)$, where ω is an arbitrary number. We have

$$\frac{d}{dt} \sin(\omega t) = \omega \cos(\omega t)$$
$$\Rightarrow \quad \frac{d^2}{dt^2} \sin(\omega t) = \omega \frac{d}{dt} \cos(\omega t) = -\omega^2 \sin(\omega t) \qquad (2.26)$$
$$\frac{d}{dt} \cos(\omega t) = -\omega \sin(\omega t)$$
$$\Rightarrow \quad \frac{d^2}{dt^2} \cos(\omega t) = -\omega \frac{d}{dt} \sin(\omega t) = -\omega^2 \cos(\omega t) . \qquad (2.27)$$

Therefore, we try the ansatz
$$x(t) = A \cos(\omega t) \qquad (2.28)$$

in Eq. 2.24:
$$\frac{d^2 x(t)}{dt^2} = Cx(t)$$

$\circlearrowright\ x(t) = A\cos(\omega t)$, Eq. 2.28

$$\frac{d^2 A \cos(\omega t)}{dt^2} = CA \cos(\omega t)$$

$\circlearrowright\ \frac{d^2}{dt^2}\cos(\omega t) = -\omega^2 \cos(\omega t)$, Eq. 2.27

$$-A\omega^2 \cos(\omega t) = CA \cos(\omega t)$$

$\circlearrowright\ \cancel{A\cos(\omega t)}$

$$-\omega^2 = C . \qquad (2.29)$$

This tells us that our ansatz in Eq. 2.28 is indeed a solution of Eq. 2.24 if $\omega = \sqrt{-C}$. Analogously, we can check that $x(t) = B\sin(\omega t)$ is a solution too. And we can construct further solutions by using linear combinations of the form[34]

$$x(t) = A\cos(\omega t) + B\sin(\omega t). \tag{2.30}$$

[34] Take note that the constant ω is the same number in both terms because its value is fixed through the constant C appearing in the equation of motion (Eq. 2.24). (In Eq. 2.29 we found $\omega = \sqrt{-C}$.)

Take note that we haven't used any proper solution finding techniques here. Instead, we've guessed what the correct solutions could look like and then checked that they have the properties we are looking for. This is often possible and if not, we usually need to ask a friend in the math department anyway.

In other words, since solving a given equation of motion is a math problem, not a physics problem, we will often simply use the solutions some mathematician found and then investigate its properties.

Writing down the correct equation of motion and understanding its solutions are physics problems. But the technical task of finding these solutions is a pure math problem. And for most equations we are interested in, we can either guess the right solution directly, the solution is complicated but well known, or the equation can only be solved numerically.[35]

[35] You can recognize that a solution is complicated whenever it is named after a mathematician.

But as an example, let me show you one of the most famous solution strategies which works for many simple equations. For concreteness, let's consider the equation

$$\frac{d}{dt}x(t) = Cx(t), \tag{2.31}$$

where C is again some constant characterizing our system. We can solve this equation by formally multiplying it by dt and then integrating:[36]

[36] If you're unsure why we are allowed to do this, remember that dt simply means "a little bit of t". Moreover, $\ln(x)$ is the natural logarithm which is the inverse of the exponential function, i.e., $e^{\ln(x)} = x$.

$$\frac{d}{dt}x(t) = Cx(t)$$

$\quad\circlearrowright$ multiplying by dt

$$dx(t) = Cx(t)dt$$

$\quad\circlearrowright$ dividing by $x(t)$

$$\frac{dx(t)}{x(t)} = Cdt$$

$\quad\circlearrowright$ integrating

$$\int \frac{dx(t)}{x(t)} = \int Cdt$$

$\quad\circlearrowright\ \int \frac{dx(t)}{x(t)} = \ln(x(t))$

$$\ln(x(t)) = Ct + c$$

$\quad\circlearrowright$ exponentiating

$$x(t) = e^{Ct+c}$$

$\quad\circlearrowright\ e^{Ct+c} = e^c e^{Ct}$

$$x(t) = e^c e^{Ct}.$$

We've therefore learned that a function of the form $x(t) = e^c e^{Ct}$ solves Eq. 2.31. The method we used is known as **separation of variables** and it sometimes works for a particular type of differential equation. For more complicated equations, we need more sophisticated methods, but we will not discuss them here because, after all, our goal is solely to understand classical mechanics.

Now, what's the meaning of the constants a, b, A which appeared in the solutions?

These constants encode information about the **initial conditions**. We've seen above that for each equation of motion there are infinitely many solutions. For example, the function in Eq. 2.25 solves Eq. 2.23 for any choice of the constants a and b.[37] So an equation of motion alone does not tell us how a system will evolve. Instead, we need to supplement it with appropriate initial conditions. These conditions specify what the system looks like at one specific point in time. Only if this is known does the equation of motion tell us uniquely what happens afterwards.

[37] For your convenience: Eq. 2.25 reads
$$x(t) = a + bt$$
and Eq. 2.23 reads
$$\frac{d^2 x(t)}{dt^2} = 0.$$

For the free object described by Eq. 2.23, we need to specify its initial position and velocity. We can see this because we have two constants in the general solution (Eq. 2.25) that we need to determine. Mathematically, this follows because Eq. 2.23 is a second-order differential equation, which means that the second derivative of $x(t)$ appears in it.

For concreteness, let's say our object is at $x = 0$ m at $t = 0$ s and moves with $\frac{dx}{dt}(0) = 2\,\frac{m}{s}$.[38] This allows us to determine the constants a and b in Eq. 2.25:

[38] Take note that we do not necessarily need to choose $t = 0$ s as our initial moment in time. We could equally specify the location and velocity at some other moment in time like $t = 2$ s and then follow the same steps to determine the constants a and b. But $t = 0$ s is a convenient choice because it often makes the equations simpler since terms drop out if we put in a zero. Moreover, we can always choose $t = 0$ s as our initial moment in time because different choices simply correspond to a different choice of coordinate system. In other words, we always choose when the clock starts ticking. But choosing a different coordinate system cannot make any difference and hence, specifying the position and velocity at $t = 0$ is a completely general approach.

$$x(t) = a + bt$$

$$x(0) = a + b0 \stackrel{!}{=} 0 \quad \Rightarrow \quad a = 0. \qquad \text{initial condition } x(0) = 0 \qquad (2.32)$$

$$\frac{dx(t)}{dt} = \frac{d(a+bt)}{dt} = b$$

$$\frac{dx}{dt}(0) = b \stackrel{!}{=} 2\,\frac{m}{s} \quad \Rightarrow \quad b = 2\,\frac{m}{s}. \qquad \text{initial condition } \frac{dx}{dt}(0) = 2\,\frac{m}{s} \qquad (2.33)$$

Therefore, the correct function which describes our object reads

$$x(t) = a + bt$$

$$x(t) = 2\,\frac{m}{s}t. \qquad a = 0, b = 2\,\frac{m}{s}, \text{ (Eq. 2.32, Eq. 2.33)} \qquad (2.34)$$

We call the specific path described by a solution the **trajectory** of the object.

Of course, we can also imagine that our object is at $x = 3$ m at $t = 0$ s and moves with some other initial velocity like $\frac{dx}{dt}(0) = 0\,\frac{m}{s}$. Then, we can again determine the constants a and b in Eq. 2.25:

$$x(t) = a + bt$$

$$x(0) = a + b0 \stackrel{!}{=} 3 \text{ m} \quad \Rightarrow \quad a = 3 \text{ m}. \quad \circlearrowright \text{ initial condition } x(0) = 3 \text{ m} \quad (2.35)$$

$$\frac{dx(t)}{dt} = \frac{d(a+bt)}{dt} = b$$

$$\frac{dx}{dt}(0) = b \stackrel{!}{=} 0 \frac{\text{m}}{\text{s}} \quad \Rightarrow \quad b = 0 \frac{\text{m}}{\text{s}}. \quad \circlearrowright \text{ initial condition } \frac{dx}{dt}(0) = 0 \frac{\text{m}}{\text{s}} \quad (2.36)$$

Therefore, the correct function which describes our object for these initial conditions reads

$$x(t) = a + bt$$
$$x(t) = 3 \text{ m} \quad \circlearrowright \; a = 3 \text{ m}, b = 0 \frac{\text{m}}{\text{s}} \text{ (Eq. 2.35, Eq. 2.36)} \quad (2.37)$$

In words, this means that the object sits still at the location $x = 3$ m.

We can therefore see that there is a very direct connection between the two constants a, b and the initial position and velocity of the object. A better notation would therefore be $a \to x_0$ and $b \to v_0$:

$$x(t) = a + bt \quad \to \quad x(t) = x_0 + v_0 t. \quad (2.38)$$

This makes it immediately clear that x_0 describes the initial position of the object and v_0 describes its initial velocity.

We can now understand in physical terms why we get infinitely many solutions for our equation of motion in Eq. 2.25. For each possible initial position x_0 and each possible initial velocity v_0, we get a different trajectory.

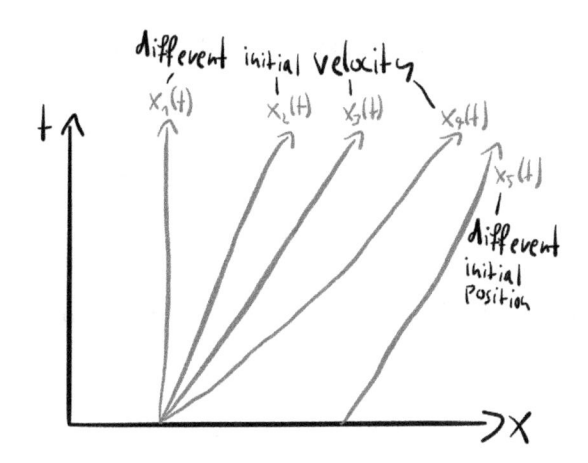

[Hand-drawn figure: trajectories $x_1(t)$, $x_2(t)$, $x_3(t)$, $x_4(t)$ labeled "different initial velocity" and $x_5(t)$ labeled "different initial position" on a t vs x plot.]

Completely analogously, we can determine the constants A and B which appear in our general solution (Eq. 2.30) of the second equation of motion that we considered (Eq. 2.24).[39]

For concreteness, let's say again that our object is at $x = 0$ m at $t = 0$ s and moves with $\frac{dx}{dt}(0) = 2\,\frac{m}{s}$. As soon as these initial conditions are fixed, we can calculate A and B explicitly:

[39] For your convenience: Eq. 2.24 reads
$$\frac{d^2x(t)}{dt^2} = Cx(t)$$
and Eq. 2.30 reads
$$x(t) = A\cos(\omega t) + B\sin(\omega t).$$
Moreover, recall that ω is not something that we need to determine using initial conditions because the value of this constant is determined completely by the constant C which describes a specific property of our system (e.g., the stiffness of a spring to which the object is attached).

$$x(t) = A\cos(\omega t) + B\sin(\omega t)$$
⤳ initial condition $x(0) = 0$
$$\therefore\ x(0) = A\cos(\omega 0) + B\sin(\omega 0) \stackrel{!}{=} 0$$
⤳ $\cos(0) = 1$ and $\sin(0) = 0$
$$\therefore\ x(0) = A \stackrel{!}{=} 0 \ \Rightarrow\ A = 0. \qquad (2.39)$$

$$\frac{d}{dt}x(t) = \frac{d}{dt}\left(A\cos(\omega t) + B\sin(\omega t)\right)$$
⤳ $\frac{d}{dt}\cos(\omega t) = -\omega\sin(\omega t)$ and $\frac{d}{dt}\sin(\omega t) = \omega\cos(\omega t)$
$$\therefore\ \frac{d}{dt}x(t) = -A\omega\sin(\omega t) + B\omega\cos(\omega t)$$
⤳ initial condition $\frac{dx}{dt}(0) = 2\,\frac{m}{s}$
$$\therefore\ \frac{dx}{dt}(0) = -A\omega\sin(\omega 0) + B\omega\cos(\omega 0) \stackrel{!}{=} 2\,\frac{m}{s}$$
⤳ $\cos(0) = 1$ and $\sin(0) = 0$
$$\therefore\ \frac{dx}{dt}(0) = B\omega \stackrel{!}{=} 2\,\frac{m}{s} \ \Rightarrow\ B = \frac{2\,\frac{m}{s}}{\omega} \qquad (2.40)$$

The trajectory of our object is therefore described by

$$x(t) = A\cos(\omega t) + B\sin(\omega t)$$

$$\curvearrowright \quad A = 0, B = \frac{2\frac{m}{s}}{\omega} \quad \text{(Eq. 2.39, Eq. 2.40)}$$

$$x(t) = \frac{2\frac{m}{s}}{\omega} \sin(\omega t). \tag{2.41}$$

In physical terms, this solution describes an oscillatory motion.[40]

[40] Eq. 2.24 is the equation of motion of an object attached to a spring. This solution therefore tells us that the object bounces back and forth.

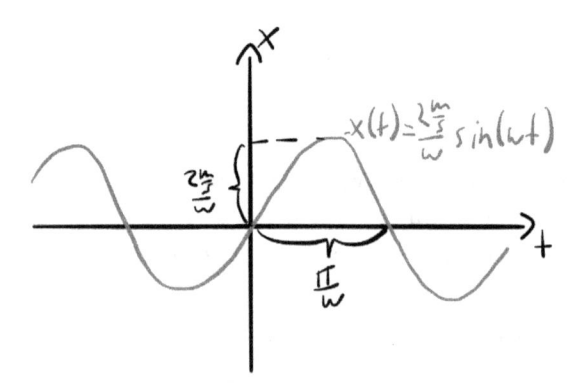

And once more, we can understand that we get a different trajectory for each possible initial position and velocity of the object.

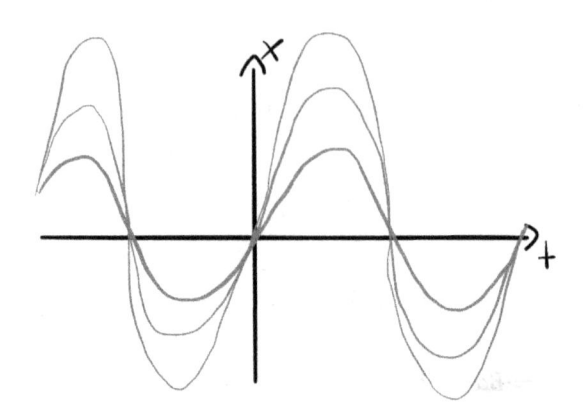

To summarize:

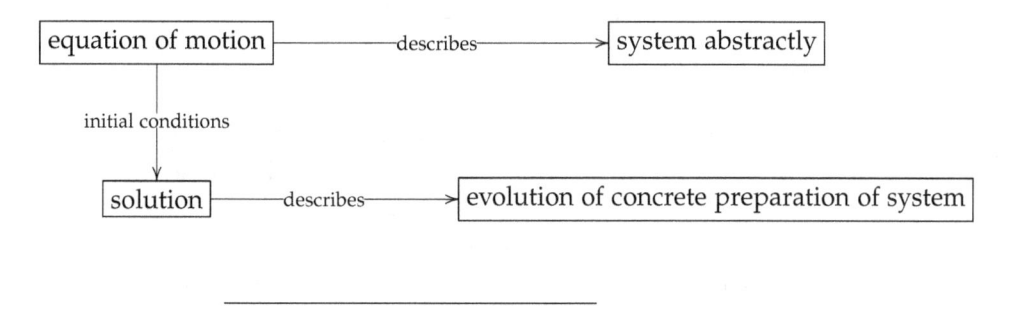

```
equation of motion ──describes──→ system abstractly
       │
initial conditions
       ↓
   solution ──describes──→ evolution of concrete preparation of system
```

The next thing we need to talk about is that it often makes sense to not consider the trajectories of each object individually. Instead, we can describe our system as a whole, no matter how many objects it consists of, using a *single* trajectory. This is what the next section is all about.

2.3 Mathematical Arenas

The simplest arena we can use to describe nature is, of course, our physical space.[41] We describe the location and the momentum of each object using an individual vector. These vectors all live in the same arena which we call physical space.[42]

For simplicity, let's consider an object which moves in just one dimension. Our mathematical arena is then simply a line ($\mathbb{R}$):

$$\xleftarrow{\qquad\vec{p}\quad\bullet\qquad\longrightarrow\quad}\vec{x}$$

Now, if we want to describe two objects which move in one dimension, the first method that comes to our mind is to use two vectors:

[41] By physical space, I mean the usual Euclidean three-dimensional space $\mathbb{R}^3$ or $\mathbb{R}$ (if for some reason our objects can only move in one dimension).

[42] All of this will make a lot more sense as soon as we talk about alternative arenas.

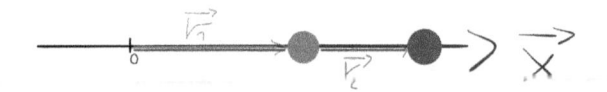

In addition to two vectors that keep track of the locations, we need two further vectors that keep track of the momenta.

This is what we do in the Newtonian formulation of classical mechanics. Such a description in physical space is handy since we can immediately understand everything that is going on in the system. Each vector is simply an arrow that points from one location to another. However, in practice, this approach is often laborious — especially when we are dealing with lots of objects.

So how else can we describe our system consisting of, for example, two objects that move along a line?

2.3.1 Configuration Space

What we need, mathematically, is a tool that allows us to keep track of the locations and momenta of the two objects. In the physical space description, we need four vectors to accomplish this: two for the locations and two for the momenta.

Using the following idea, we can describe the whole system with just two vectors.[43]

[43] We also need only two vectors if there are three or more objects in the system.

▷ First, we act as if there were a separate arena for each object:

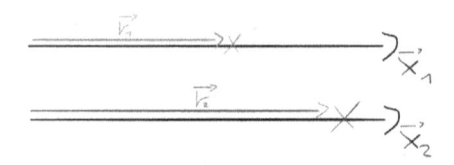

▷ Then we glue these separate spaces together:

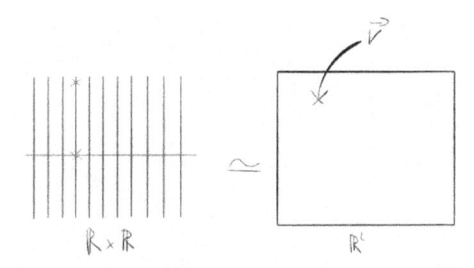

So for the example discussed above, this means that instead of just one line, we now use two. We say our first object moves along one line and the second object along another line. At each possible location of the first object, we need to take into account the possibility that the second object could be anywhere. Hence we need a complete copy of our line $\mathbb{R}$ that we use to keep track of the location of the second object at each location of the line $\mathbb{R}$ that we use to keep track of the location of the first object. Gluing a copy of $\mathbb{R}$ to each point of $\mathbb{R}$ yields a rectangle.[44].

So why is this a clever idea?

Well, instead of using two functions $(f(x), g(x))$, we can describe our whole system with just one vector $\vec{r} = (f(x), g(x))$. But this vector lives in a *higher-dimensional* space. So instead of pointing to a point on a line, this new vector $\vec{r}$ points to a point on a rectangle.

In the physical space description, we need N vectors to keep track of the locations of N objects. Using the idea of gluing the spaces together, we always only need one vector which lives in an $\mathbb{R}^N$-dimensional space. If the objects are allowed to move freely in three dimensions, our vector $\vec{r}$ lives in $\mathbb{R}^{3N}$ since we are gluing N times $\mathbb{R}^3$ together.

The resulting arena is known as **configuration space**. The basic idea is that instead of keeping track of the N individual objects in our system, we treat the system as a whole. We can imagine the whole system as just one point that moves through this higher-dimensional space called configuration space. Each point in configuration space corresponds to one specific configuration

[44] The mathematical name for this kind of construction is **product space** We will talk about another example of a product space in a moment.

the system can be in.

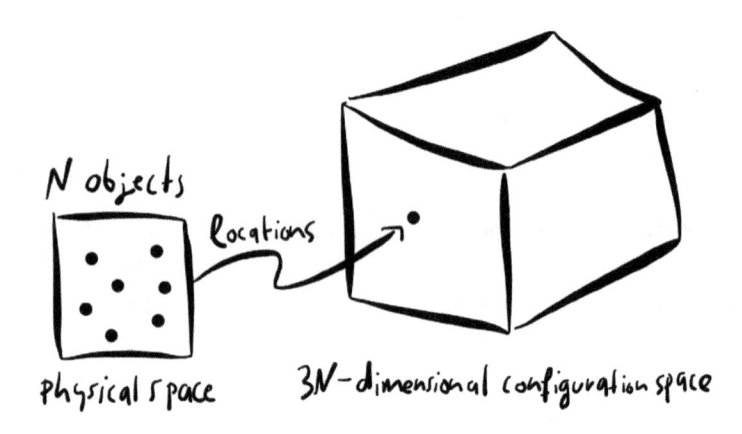

As time passes, the configuration of the system usually changes. This means that the point which describes the configuration of our system moves around. Therefore, the time evolution of a system is described in configuration space by a single path.

Let's have a look at two concrete examples.

The configuration space of a harmonic oscillator is simply a line:[45]

[45] We discuss the harmonic oscillator in Section 8.

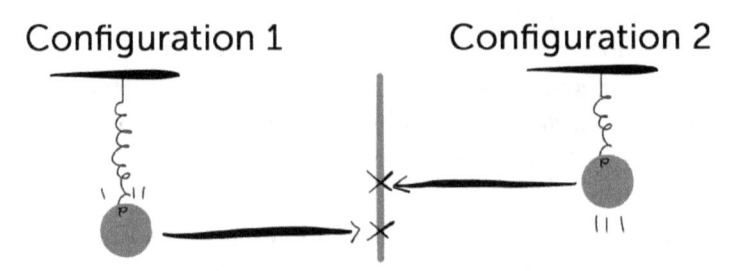

For a second harmonic oscillator, our configuration space is also a line which we rotate by 90° for reasons that will become clear in a moment:

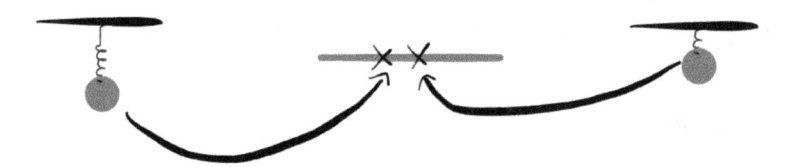

If we now consider the system that consists of the two harmonic oscillators, we need to attach the configuration space of the second object to each point of the configuration space of the first object. Again, what we end up with is a rectangle:

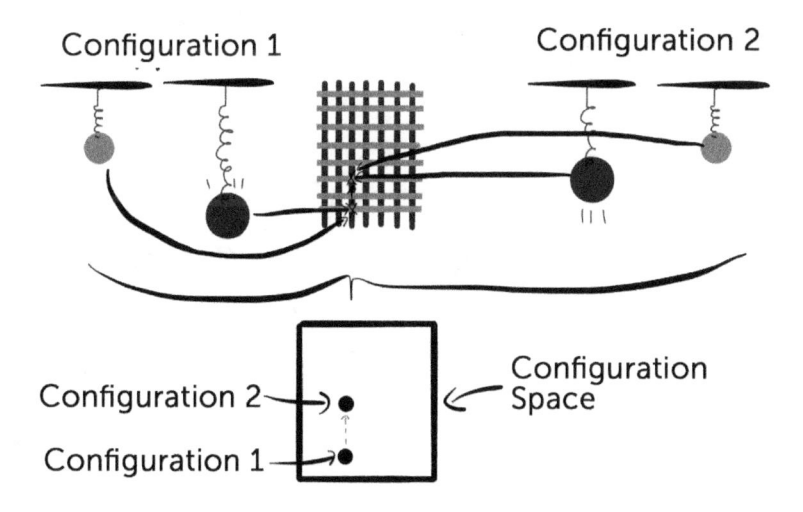

Our second example is a pendulum. The configuration space of a pendulum is a circle since it can rotate around its suspension:

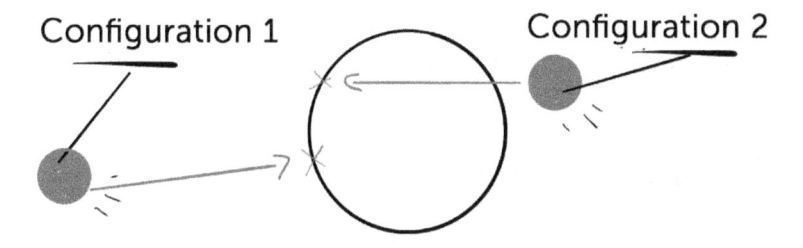

We can then construct the configuration space for a system that consists of two pendulums by attaching to each point of the configuration space of the first pendulum the configuration space of the second one. The result of this procedure is a torus:

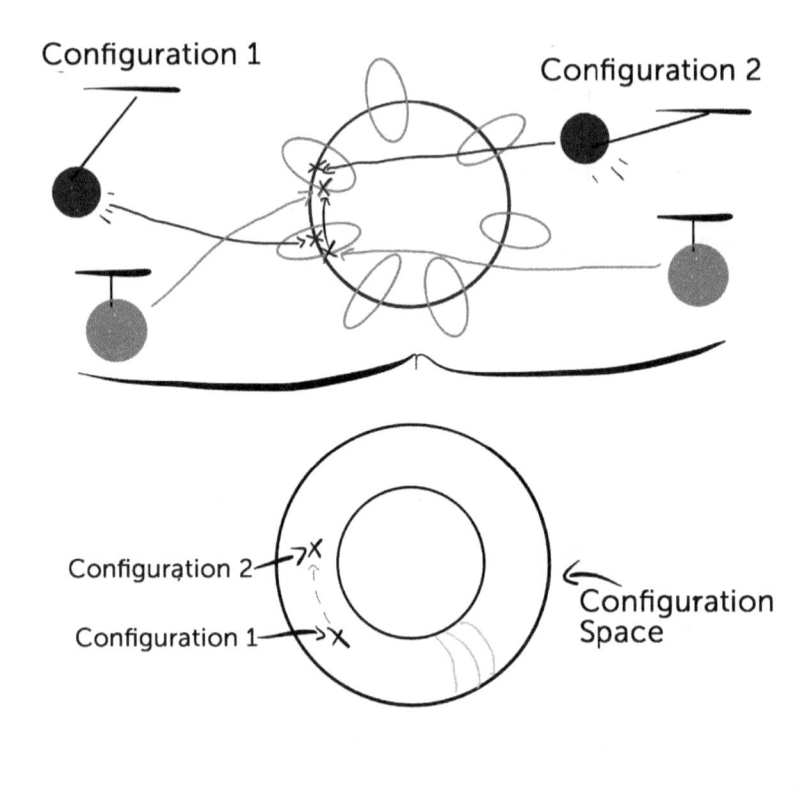

To summarize: while individual objects move in the three-dimensional physical space, the time evolution of a system as a whole takes place in a higher-dimensional configuration space. A single trajectory in configuration space describes the evolution of a system as a whole.

2.3.2 Phase Space

An important observation is that configuration space only keeps track of the locations of the various objects. But to describe the state of a system completely, we need additionally to keep track of *how* the objects move. Mathematically, this means that we need to keep track of their momenta. So in addition to a vector $\vec{r}$ that keeps track of the locations, we need a vector $\vec{p}$ that keeps track of the momenta.

This motivates the construction of the next mathematical arena which works completely analogously to how we constructed configuration space. However, this time we also act as if the momenta live in a different space and then glue the momentum spaces to our location spaces. As a result, we can describe the complete state (not just the configuration) of our system with a single vector.[46]

The resulting mathematical arena is known as **phase space**. Each point in phase space corresponds uniquely to one specific location and momentum of *each* object. So everything that is going on in the system is described by just one vector (or equally the point the vector points to) that moves through phase space.

[46] By the configuration of our system, we mean a complete description of all locations. In contrast, the state of a system corresponds to a complete description of all locations *and* all momenta.

Now, the price we have to pay for this is that the vector we use to describe the system lives in a $2 \times 3N$-dimensional space for N objects that move in three dimensions.

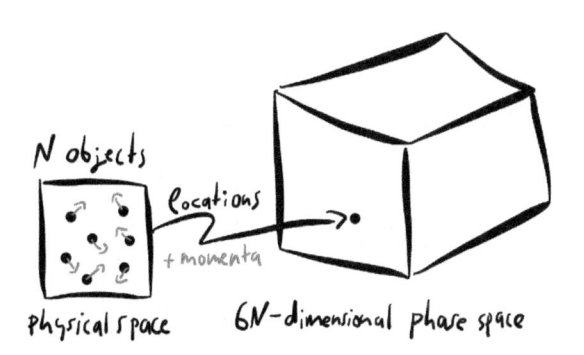

Phase space is notoriously difficult to visualize since even for just two objects moving in one dimension, phase space is already four-dimensional. However, for just one object in one dimension, it is possible.

Here's a concrete example: the phase space path of a swinging pendulum.

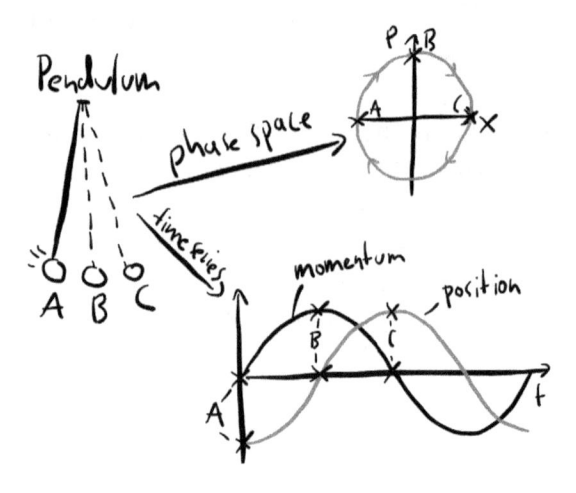

If we consider different paths of the pendulum in phase space (corresponding to different initial conditions), we end up with the following phase space portrait:

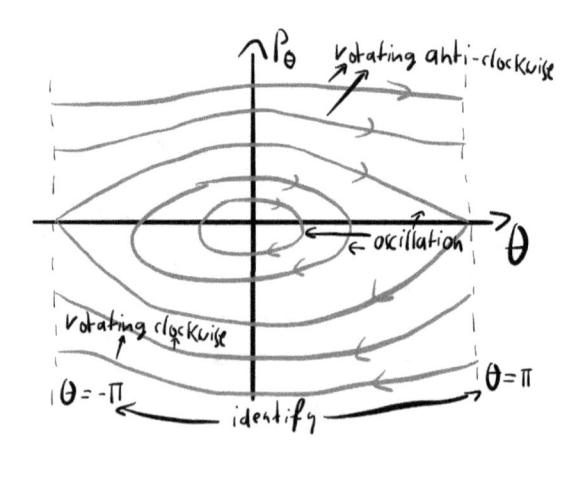

Now, before we move on and discuss how we can describe systems using the various mathematical arenas discussed in this section, let's summarize what we have learned so far.

▷ One possibility to describe nature is to keep track of everything using various vectors living in physical space.

▷ A bit more convenient is a description in configuration space. In configuration space, a single point is enough to keep track of the locations of all of the objects in our system.

▷ Even better is a description in phase space. Each phase space point corresponds to one specific state of the system (including all locations *and* all momenta).

Take note that these are really just different *mathematical* tools that allow us to describe a given system in different ways. It is up to you which one you like best.

However, the crucial point is the following:

> We can describe *any* given system using *any* of these mathematical arenas.

This is why there are different formulations of classical mechanics.

Specifically, we have the following well-known formulations:[47]

▷ Classical mechanics in physical space is what we call the **Newtonian formulation**.

▷ Classical mechanics in configuration space is what we call the **Lagrangian formulation**.

▷ Classical mechanics in phase space is what we call the **Hamiltonian formulation**.

Next, let's talk about these formulations.

[47] There is one additional important arena known as Hilbert space, and we can describe classical mechanics in Hilbert space too. The resulting formulation is known as the Koopman-von Neumann formulation and we will talk about it in Section 11.3.

3
Newtonian Mechanics

I'm pretty sure you're not reading this book to learn about Newtonian mechanics. The Newtonian formulation of classical mechanics is an extremely flexible and useful framework which allows us to describe almost anything. But at the same time, for anyone interested in fundamental questions, it's extremely unsatisfying. Newtonian mechanics is simply a collection of formulas ("laws") which allow us to describe things. But it offers no explanation for why these formulas are the right ones. In contrast, in the Lagrangian and Hamiltonian formalism, there are deeper principles from which not only the laws of mechanics, but really all modern theories of physics, can be derived. And to quote Steven Weinberg:[1] "After all, this is our aim in physics, not just to describe nature, but to explain nature." This is why physicists love the Lagrangian and Hamiltonian formalisms. Thus, I've tried to keep the following discussion to a minimum so that we can move on as quickly as possible to the good stuff.

[1] Steven Weinberg. What is quantum field theory, and what did we think it is? In *Conceptual Foundations of Quantum Field Theory. Proceedings, Symposium and Workshop, Boston, USA, March 1-3, 1996*, pages 241–251, 1996

The most important concept in the Newtonian formulation of mechanics is that of a force. A force is something which changes the momentum of an object. Mathematically, this means that

$$\boxed{\frac{d\vec{p}}{dt} = \vec{F}.} \tag{3.1}$$

This is **Newton's second law**.[2] Formulated differently:

[2] We'll talk about Newton's other laws in a moment.

> The rate at which the momentum of an object changes is exactly equal to the force acting on it.

Newton's little formula (Eq. 3.1) allows us to predict how any object will move starting from some specific position and velocity (initial conditions).

The definition of a force given here is an extremely general one. In Chapter 2, we've already discussed two simple examples.[3]

[3] Take note that these examples are especially simple because our object moves effectively only in one dimension, and therefore, we don't need vectors to describe the forces.

▷ The force a spring exerts on an object attached to it can be described by $F(x) = -kx$, where k is known as the spring constant and characterizes the spring in question.

▷ The (gravitational) force Earth exerts on an object we throw into the air can be described by $F(x) = -mg$, where m is the object's mass and $g \approx 9.81 \, \frac{m}{s^2}$ a constant

Historically, these formulas (and really most of the formulas we use in classical mechanics to describe forces) were discovered experimentally. We simply use them because we've discovered that by putting them into Newton's second law, we get the correct equations of motion.

In general, we need to find all forces $\vec{F}_1, \vec{F}_2, \ldots$ that act on the object in question. The total force $\vec{F}$ is then given by

$$\vec{F}(\vec{q}) = \sum_{i=1}^{n} \vec{F}_i(\vec{q}) = \vec{F}_1(\vec{q}) + \vec{F}_2(\vec{q}) + \ldots \tag{3.2}$$

For example, an object attached to a spring will realistically not only feel the force exerted by the spring but additionally will feel a gravitational force.

This is what we use on the right-hand side of Newton's second law (Eq. 3.1). On the left-hand side, we use the definition of the momentum $\vec{p} = m\dot{\vec{q}}$:

$$\frac{d}{dt}\vec{p} = \vec{F}$$

$$\circlearrowright \text{ Eq. 3.2 and } \vec{p} = m\dot{\vec{q}}$$

$$\frac{d}{dt}\left(m\dot{\vec{q}}\right) = \vec{F}_1(\vec{q}) + \vec{F}_2(\vec{q}) + \ldots . \tag{3.3}$$

This is the equation of motion for the object, and its solutions tell us explicitly how the object moves around.

While this procedure may sound simple, for most systems it isn't. There are two major difficulties. It's often cumbersome to write down all the forces acting on a given object in a consistent way. In particular, we need to take the various directions of the different forces into account. And secondly, solving the equation of motion is, for most realistic systems, an extremely difficult math problem.

We will see this explicitly in Part II when we discuss various systems in detail. But already here we take note that these difficulties are one of the main reasons people use alternative formulations of classical mechanics.[4]

Take note that if the mass of the object in question is constant, Newton's second law becomes

$$\frac{d}{dt}\vec{p} = \vec{F}$$

$$\circlearrowright \vec{p} = m\dot{\vec{q}}$$

$$\frac{d}{dt}\left(m\dot{\vec{q}}\right) = \vec{F}$$

$$\circlearrowright \text{ product rule}$$

$$\left(\frac{d}{dt}m\right)\dot{\vec{q}} + m\frac{d}{dt}\left(\dot{\vec{q}}\right) = \vec{F}$$

$$\circlearrowright m = \text{const.} \Rightarrow \frac{d}{dt}m = 0 \text{ and } \ddot{\vec{q}} \equiv \frac{d}{dt}\left(\dot{\vec{q}}\right)$$

$$m\ddot{\vec{q}} = \vec{F} . \tag{3.4}$$

In words, this means that whenever there is a non-zero acceleration of an object $\ddot{\vec{q}} \neq 0$, this change in the velocity $\ddot{\vec{q}} \equiv \frac{d}{dt}\left(\dot{\vec{q}}\right)$ is directly proportional to the total force $\vec{F}$ acting on it. The proportionality constant is what we call the mass of the object. This is why it's more difficult to push an object with large mass.

[4] There are additional reasons to consider these alternative formulations of classical mechanics. The Lagrangian formulation not only allows us to describe many systems in much simpler terms, but is at the same time conceptually beautiful and helps us to understand classical mechanics on a deeper level.

Now you probably wonder, since Eq. 3.1 is Newton's *second* law, what are the remaining laws?

Newton's first law states that, at least for one specific observer, an object which moves with constant velocity will always continue to move with this constant velocity unless a force acts on it. In particular, this implies that a particle at rest will stay at rest unless a force acts upon it.

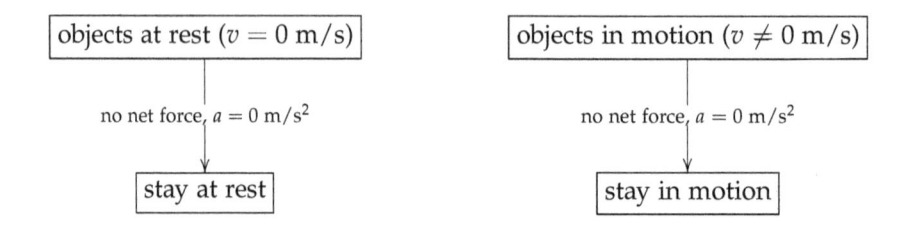

```
┌─────────────────────────────┐         ┌──────────────────────────────┐
│ objects at rest (v = 0 m/s) │         │ objects in motion (v ≠ 0 m/s)│
└─────────────────────────────┘         └──────────────────────────────┘
              │                                         │
    no net force, a = 0 m/s²                  no net force, a = 0 m/s²
              ▼                                         ▼
       ┌─────────────┐                          ┌───────────────┐
       │ stay at rest│                          │ stay in motion│
       └─────────────┘                          └───────────────┘
```

So in other words, no object likes to have its state of motion changed.

In this sense, Newton's first law is often called the law of inertia.

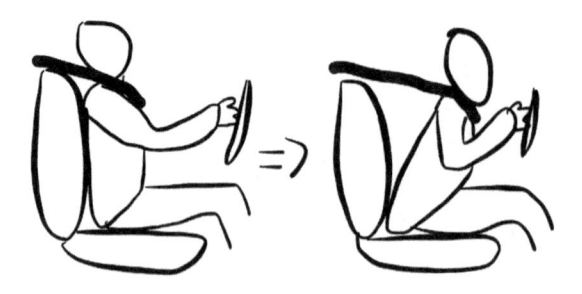

Although this law may seem really simple, it needs some explaining.

Newton's second law tells us that forces *can* change the momentum of objects. But Newton's first law tells us that forces are the *only* thing which change the momentum of objects. In other words, there is nothing else which causes a change in momentum.[5]

[5] One may argue that the force framework is so general that we can always define a force to account for any change in momentum that we observe.

Secondly, take note that if you observe an object while you are accelerating yourself, the object will appear to be accelerating as a result.

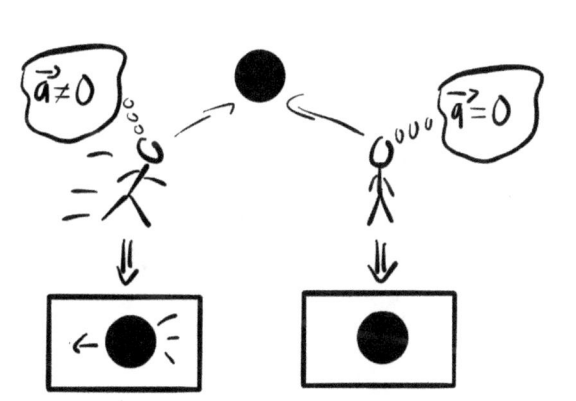

[6] All this may seem really pedantic or even trivial at the moment. But thoughts like these about how different observers perceive the same situation are at the heart of many of the most important insights in modern physics, e.g., special relativity. So for the moment you shouldn't worry about issues like this too much. But keep them in the back of your head as they will become a lot more important further down the road.

Therefore, objects will continue to move with uniform velocity unless acted upon by a force only for very specific observers. We call them **inertial observers**. In other words, an inertial observer is someone for whom an object with zero net force acting upon it is not accelerating. Newton's first law therefore says that for each object, at least one such observer exists. Only for inertial observers is Newton's second law (Eq. 3.1) true.

That's why Newton's first law, technically, needs to be established before we can use Newton's second law. In particular, Newton's first law establishes that for an inertial observer $\vec{F} = 0$ really implies $\dot{\vec{p}} = 0$.[6]

However, take note that we can describe how objects behave from the perspective of accelerating observers by introducing so-called fictitious forces.[7]

[7] We call them fictitious because they're a consequence of the reference frame and not of any interaction. However, for an observer within such a non-inertial system, they are as real as any other force. In fact, there is no possible way to determine the difference without looking outside of the system. The most famous fictitious force is probably the Coriolis force which can be used to explain many important meteorological effects. We need a fictitious force here because Earth is spinning and therefore any observer on Earth lives in an accelerating non-inertial frame. Moreover, in Einstein's theory of general relativity, gravity is, in some sense, a fictitious force. We will discuss this in more detail in Section 12.3.6.

In addition to the first and second law, there's Newton's third law. In Newton's own words: *"To any action there is always an opposite and equal reaction; in other words, the actions of two bodies upon each other are always equal and always opposite in direction."* Formulated differently, action = − reaction.

In more concrete terms, this third law tells us that whenever an object exerts a force $\vec{F}$ on another object, this second object always exerts the force $-\vec{F}$ on the first.

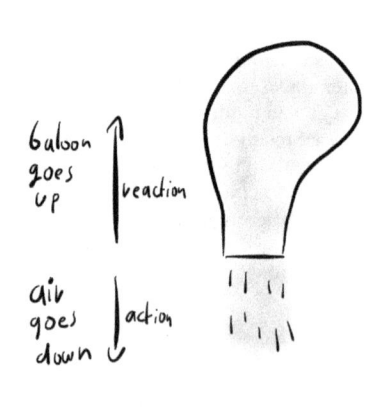

This phenomenon is sometimes hard to believe. For example, when we jump into the air there is a gravitational force $\vec{F}$ that pulls us back toward the ground. However, at the same time, we pull the Earth a little bit upward through a gravitational force of equal magnitude $-\vec{F}$. Since the Earth's mass is so huge, the effects of this force are basically unnoticeable and the resulting change in the Earth's velocity is tiny.

In practice, the equality of the forces two objects exert on each other are directly encoded in the formulas which describe the force in general. For example, the gravitational force an object with mass M located at $\vec{q}_1$ exerts on an object with mass m located at $\vec{q}_2$ is described by[8]

$$\vec{F}_{M \to m} = G \frac{mM}{|\vec{q}_1 - \vec{q}_2|^3} (\vec{q}_1 - \vec{q}_2), \qquad (3.5)$$

where G is a constant known as the gravitational constant.

---

[8] This formula may look frightening. Simply put, it says that the gravitational force between two objects is directly proportional to their masses and is inversely proportional to their distance $r \equiv |\vec{q}_1 - \vec{q}_2|$ squared. (The farther they are away from each other, the smaller is the resulting force). Thus, the absolute magnitude of the force can be written as

$$F = G \frac{mM}{r^2}.$$

The remaining stuff in the formula

$$\frac{\vec{q}_1 - \vec{q}_2}{|\vec{q}_1 - \vec{q}_2|}$$

yields a vector of length one which points from the location of the second object to the location of the first object.

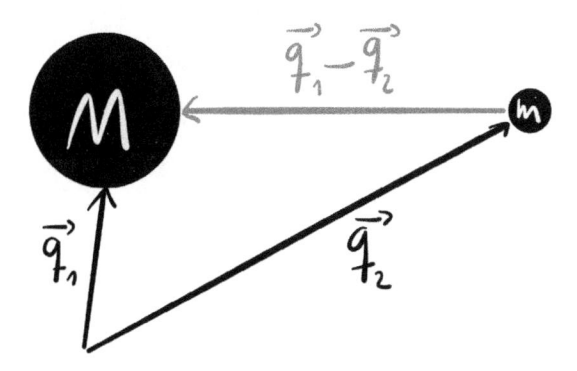

If we want to calculate the force that the second object exerts on the first, we only need to switch their roles in the formula $\vec{q}_1 \leftrightarrow \vec{q}_2$, and $m \leftrightarrow M$:[9]

$$\vec{F}_{m \to M} = G \frac{Mm}{|\vec{q}_2 - \vec{q}_1|^3} (\vec{q}_2 - \vec{q}_1) \tag{3.6}$$

[9] Take note that switching $m \leftrightarrow M$ makes no difference, but $\vec{q}_1 \leftrightarrow \vec{q}_2$ changes the direction of the resulting vector.

This is a force of equal magnitude which points in the opposite direction.

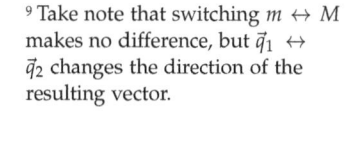

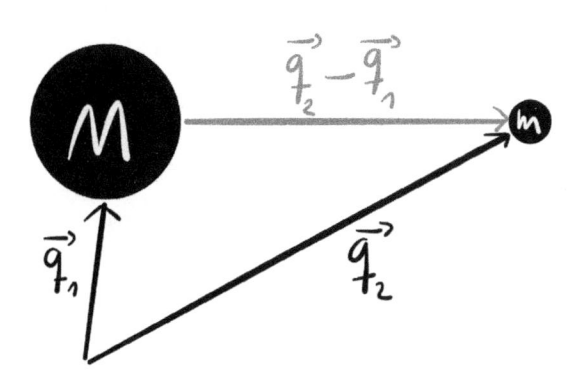

That's all we need to know for now about Newtonian mechanics and we are ready to discuss alternative formulations of classical mechanics.

4
Lagrangian Mechanics

The main idea of the Lagrangian formalism is that:

> Nature is lazy.

This little fact allows us to derive the correct equations of motion for *any* system.[1] We will discuss how we can do this in a moment. But first, let's talk about a simple example which nicely demonstrates the laziness of nature.

[1] The idea is so general that we can even use it to derive the correct equations of motion in quantum field theory.

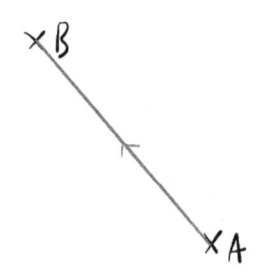

Let's say we ask ourselves: how does light travel between two points A and B?

Well, in a vacuum, light travels in a straight line:

This is certainly an economical choice. But the laziness of nature becomes even more obvious when A is in a vacuum and B in some medium, say, water.

In such a situation, light no longer travels in a simple straight line. Instead, its path looks roughly as follows:

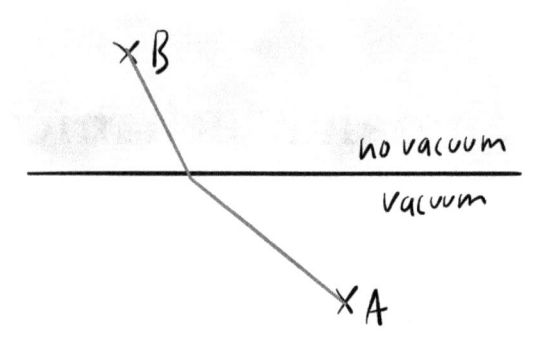

To understand this, imagine that you're a rescue swimmer who sees someone drowning in the water. Which path would you choose to get to the swimmer as fast as possible? An important factor which you need to take into account is that you're much slower in water. So a straight path is certainly not the fastest route. But minimizing the swimming distance isn't the optimal choice either because the total distance is quite long for this path. Instead, the optimal path is a trade-off between these two extremes.

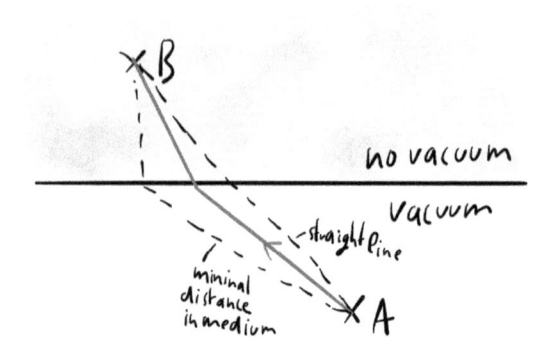

This is exactly the path light travels between two points. Light always travels the path between two points which requires the least travel time. For this exact reason, we say that nature is lazy.[2]

However, massive objects, in general, do not travel the path between two points which requires the least travel time.[3] Nevertheless, we can understand their behavior by using the idea that nature is lazy. This requires a different definition of what we mean by lazy. To that end, we introduce a new quantity called the action which describes how much diligence is necessary for each possible path between two points. This is what we will talk about next.

First of all, we can reformulate the main idea of the Lagrangian formulation in more technical terms by saying that:[4]

> Any system evolves in such a way that the action required is minimal.

The key observation which allows us to understand this is that there are always infinitely many possibilities how a given system can evolve from one initial state A to a specific final state B. For example, there are infinitely many possible paths a given object could take between two specific locations:

[2] This observation is known as Fermat's principle, named after Pierre de Fermat who first concluded that: *"Nature operates by means and ways that are easiest and fastest."*

[3] However, this is only true for slowly moving objects. Objects which move extremely fast are described by special relativity, and we can understand that massive objects follow the path which requires the least travel time too. But this fact gets hidden when we consider the approximation of slowly moving objects. We will talk about this in more detail in Chapter 6.

[4] Take note that sometimes the correct path is not the path with minimal action, but the path with maximal action or path for which the action is a saddle point. So in general, the correct path is always the path with extremal action (or stationary action). Please keep this in mind, but in the following we will always talk about minima of the action because it helps to reinforce the idea that nature is lazy.

And similarly, there are infinitely many paths that two or more objects could take between a fixed set of initial positions $(A_1, A_2, \ldots)$ and final positions $(B_1, B_2, \ldots)$:

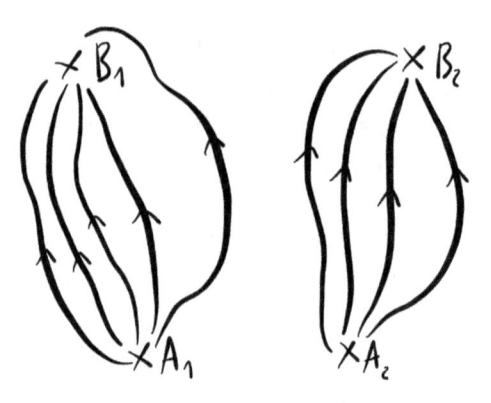

[5] We talked about configuration space in Section 2.3.

Since talking about lots of paths for several particles quickly becomes cumbersome, we will switch perspectives and talk about paths in configuration space.[5] A point in configuration space corresponds to a specific configuration of the system. Therefore, a path in configuration space between two fixed points X and Y corresponds to one specific possibility for how our system evolves between two fixed configurations.

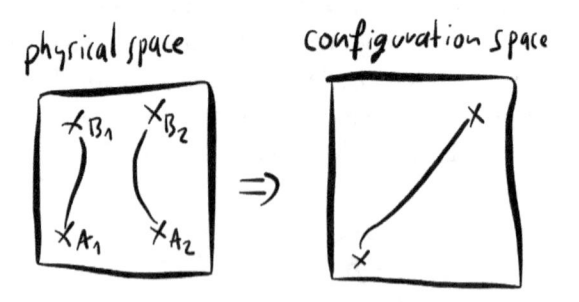

Using configuration space, we can therefore always talk about individual specific paths which describe how our system evolves as a whole.

Now, which path (in configuration space) is the right one?

The correct answer is that it's always the path with minimal action. This means that each of the infinitely many paths involve a specific amount of action. For example, we can imagine that one specific path $q_1(t)$ involves 8.73 $\frac{\text{kg} \cdot \text{m}^2}{\text{s}}$ of action, while a second path $q_2(t)$ involves 9.21 $\frac{\text{kg} \cdot \text{m}^2}{\text{s}}$ of action. We usually denote the action by S and thus write these statements in mathematical terms as:

$$S[q_1(t)] = 8.73 \frac{\text{kg} \cdot \text{m}^2}{\text{s}}$$

$$S[q_2(t)] = 9.21 \frac{\text{kg} \cdot \text{m}^2}{\text{s}}. \qquad (4.1)$$

Therefore, we can visualize the situation here as follows:

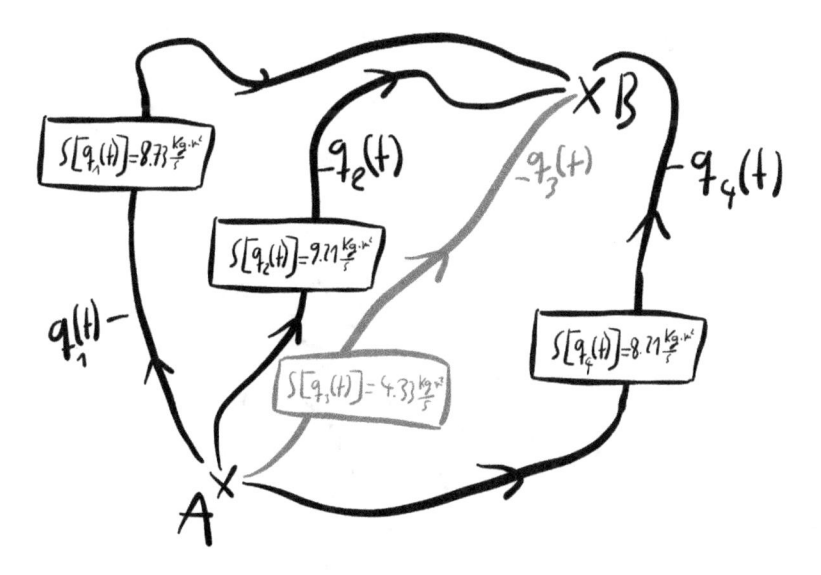

Nature prefers to minimize the action and hence, the path with minimal action (here $q_3(t)$) is "chosen."[6]

Before we can use this idea, we need to answer two crucial questions:

▷ How can we calculate the action?

▷ How can we find the path of least action?

[6] In this chapter, we will simply use this as an almost magical fundamental principle. But in Chapter 6, we will talk about exactly why the path with minimal action is the right one.

As soon as we've answered these questions, we are ready to tackle any classical mechanics system. In particular, using the formula for the action of a given system, we can calculate the path involving a minimum amount of action between a fixed initial and final configuration.[7] This path with minimal action describes how the system will evolve.

[7] Maybe you are wondering why we talk about fixed initial and final configurations. Especially if we need to specify the final configuration, how does all this help us to predict how a system will evolve? Isn't our goal to find the correct final configuration?
The logic here is that the least action principle allows us to understand how the system evolves between any fixed but *arbitrary* initial and final configuration. In this sense, the principle allows us to learn everything there is to learn about how the system evolves in general. And we can then use this knowledge to predict which final configuration is the right one for a given initial configuration (at least if we additionally specify the initial velocities of the various objects). In particular, we will see below that we can use the least action principle to derive the equation of motion for the system in question. By solving these equations, we can predict which final configuration is the right one for each possible initial condition.

[8] We will rewrite this more precisely below. Take note that only in classical mechanics does the action look like this. In quantum field theory, we can also describe systems by minimizing an action, but this requires a different definition of the action. We will discuss this in detail in Part III.

The short answer to the first question is that the action required for a specific path $q(t)$ between an initial configuration $q_i(t_i)$ and a final configuration, $q_f(t_f)$, is the integral over the Lagrangian $L = T - V$, where T denotes the kinetic energy and V the potential energy:[8]

$$\text{action}[q(t)] = S[q(t)] = \int_{t_i}^{t_f} dt L = \int_{t_i}^{t_f} dt (T - V). \quad (4.2)$$

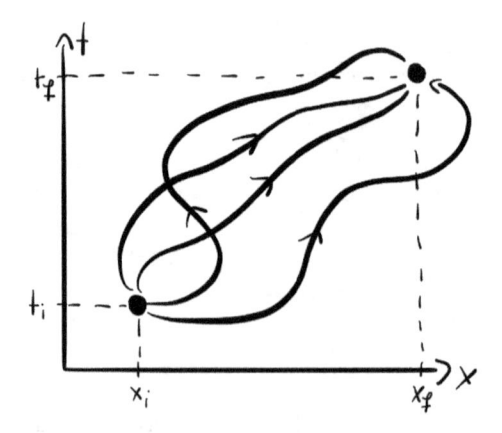

To answer the second question, we need a new mathematical toolbox known as variational calculus. We will talk about variational calculus in a moment, but first, let's talk about the action and Lagrangian in a bit more detail.

4.1 Action and the Lagrangian

First of all, why do we call the quantity

$$\int_{t_i}^{t_f} dt\, L = \int_{t_i}^{t_f} dt(T - V) \tag{4.3}$$

the action?[9]

There are two puzzle pieces that we need to understand here. First, we need to understand why we use an integral and secondly, why we integrate over the difference between the kinetic and potential energy. Let's talk about these puzzle pieces one after another.

[9] In this section, we want to understand the general idea behind the definition of the action and Lagrangian. We will talk about concrete examples in Part II.

To understand the first one, recall that an integral is something like a sum. We can imagine that the total time interval $T = t_f - t_i$ consists of N discrete time steps Δt.

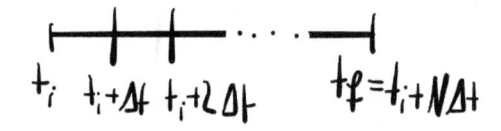

$$\vdash\!\!\!\!\vdash\!\!\!\!\vdash \cdots \dashv$$
$$t_i \quad t_i+\Delta t \quad t_i+2\Delta t \qquad t_f = t_i + N\Delta t$$

The action then reads

$$\sum_k^N L \Delta t = \sum_k^N (T - V) \Delta t. \tag{4.4}$$

In the limit $N \to \infty$ our discrete steps become smaller and smaller ($\Delta t \to dt$) and ultimately, we find the integral as given in Eq. 4.3. The integral formula is the right one because, as far as we know, time is continuous and time steps are infinitesimally small. But the discrete perspective is helpful because it allows us to discuss the definition of the action literally step by step.

We start at $t = t_i$ with one fixed initial configuration of the system $q_i(t_i)$.

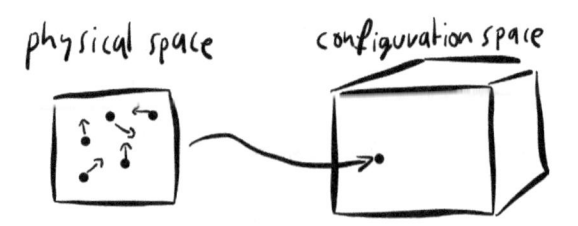

Then, during each time step there are infinitely many possibilities for how our system can evolve. Each possibility corresponds to a step in a specific direction in configuration space.[10]

[10] Only a few steps are shown in the picture, because, of course, we can't draw all of the infinitely many steps which are possible. Moreover, take note that if there is only one object in the system, each step corresponds to a real step in space. Only for many particles do we talk about the more abstract steps in configuration space.

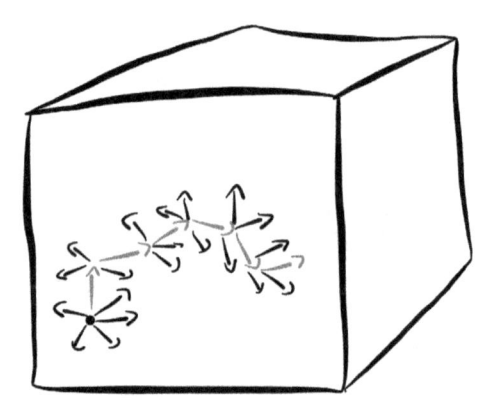

The main idea of the Lagrangian formalism is that each possible step involves a specific amount of action. We can calculate the action involved in each step by using the Lagrangian $L = T - V$. The step which accurately describes how our system evolves during the interval from t_i to $t_i + \Delta t$ is the step which involves a minimum amount of action. Analogously, for the second step there are infinitely many possible steps, and again each of them involves a specific amount of action. Once more our system moves in the configuration space direction which involves a minimum amount of action. This game goes on until, after N steps, we reach the final configuration q_f.

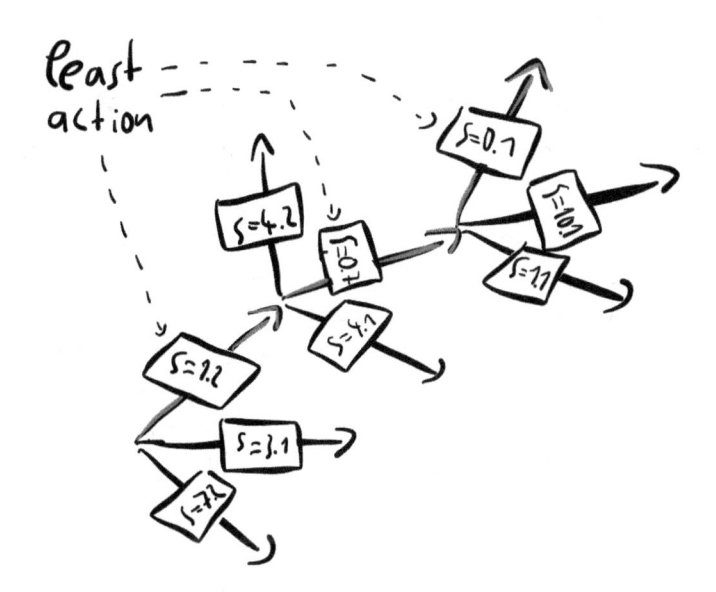

Therefore, the sum in Eq. 4.4 means that we calculate here the *total amount* of action involved in a specific path between q_i and q_f.[11] Since an integral is just a finely grained sum, we can conclude that in Eq. 4.3, we calculate the total amount of action necessary to move between two given configurations.[12]

Now, let's return to the question of why we integrate over $L = T - V$. What does the difference between kinetic and potential energy have to do with laziness?

To understand this, we need the fact that the total energy $E = T + V$ is always conserved.[13] But the two possible forms of energy T (kinetic) and V (potential) can change because only their sum has to remain unchanged. This means that as the system evolves from q_i to q_f, the total amount of energy can slosh back and forth between potential energy and kinetic energy.

Moreover, before we can understand the meaning of the Lagrangian, we need to recall that:[14]

▷ Kinetic energy is a measure for how much is going on in our system.

  – If all objects move around quickly, the kinetic energy is

[11] Reminder: Eq. 4.4 reads
$$\sum_k^N L \Delta t = \sum_k^N (T-V) \Delta t$$

[12] Reminder: Eq. 4.3 reads
$$\int_{t_i}^{t_f} dt L = \int_{t_i}^{t_f} dt (T-V).$$

Moreover, the path which correctly describes how our system evolves is the one which costs the minimum amount of total action.

[13] One of the most beautiful aspects of classical mechanics is that we can actually derive this. We will discuss how the conservation of energy follows from Noether's theorem in Chapter 10.

[14] This was already mentioned in Section 2.1.

high.
- In contrast, if all objects sit still, the kinetic energy is zero.
▷ Potential energy is a measure for how much could happen, but isn't.
- If all objects are placed on the top of a mountain they have a large (gravitational) potential energy and they could start moving quickly by falling down. But currently they don't. In this sense, their large potential energy is unrealized potential, which is why we use the word potential energy.
- In contrast, if all objects are placed on the floor, they have zero potential energy since they can't fall any farther down.

The Lagrangian $L = T - V$ depends crucially on the interplay between kinetic and potential energy and looks almost exactly like the total energy $E = T + V$. The only difference is the minus sign between kinetic and potential energy. This minus sign is really important because it means that the Lagrangian is not necessarily constant while E always remains unchanged.

The Lagrangian takes on large values whenever most of the energy is in kinetic form. And whenever most of the energy is in potential form, the Lagrangian is small because of the minus sign.

The Lagrangian is therefore a direct measure for the "liveliness" within a system at a specific moment in time. A high kinetic energy implies a large Lagrangian and that our system is extremely lively. A high potential energy implies a small Lagrangian and a less lively system.

The action is defined as the integral over the Lagrangian (Eq. 4.3) and is therefore a measure for the *total* "liveliness" within the system between two given configurations (q_i, q_f).

The statement that nature chooses the path which requires a minimum amount of action therefore tells us that nature prefers to minimize the "liveliness" within a system. In this sense, nature is really always as lazy as possible.

Let's discuss a concrete example.

How does a ball move when we throw it in the Earth's gravitational field?

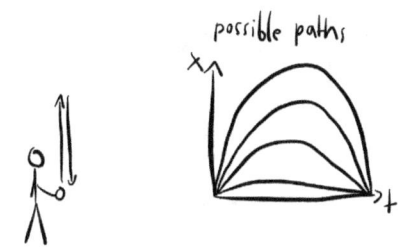

Near the bottom, the potential energy V is small and therefore the Lagrangian large (if we assume a fixed, given total energy). This means that the ball wants to move away from such an unwanted configuration quickly. But it doesn't want to move too quickly because quick movements imply a large kinetic energy and therefore a large value of the Lagrangian.

The ball prefers the configurations near the top of the trajectory because its potential energy is greatest there. But the ball will not spend too much time near the top because this would mean that it has to move extremely quickly downward, and this implies a large Lagrangian during this period. Moreover, take note that the ball cannot stay at the top indefinitely since we consider paths between a fixed initial position at a fixed initial time and a fixed final position at a fixed final time.

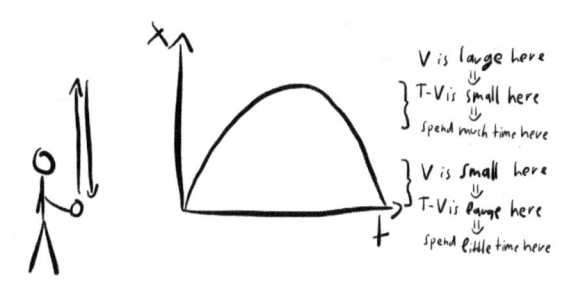

78 NO-NONSENSE CLASSICAL MECHANICS

Therefore, the correct path will be a parabola, and it will move quickly near the bottom and really slowly near the top. The parabola is a perfect compromise, analogous to what we discussed in the rescue swimmer example.[15]

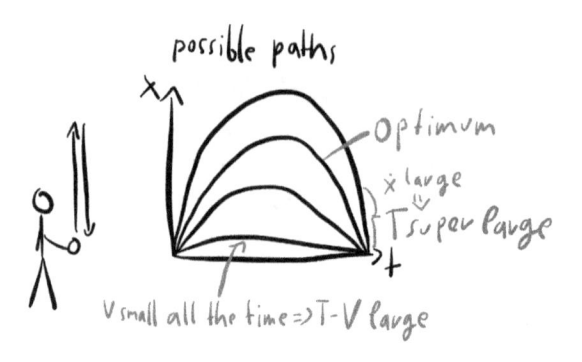

[15] As a reminder: a ball does not make any conscious choices. We are saying that the ball wants something because it helps us to understand the minimum action principle. We will discuss in Chapter 6 how objects "choose" the paths they are following.

One key observation before we move on is that the action $S[q(t)]$ is not a function but a **functional**. A function eats a *number x* and spits out a number $f(x)$:

In contrast, a functional eats a *function $f(x)$* and spits out a number $F[f(x)]$:

Specifically, the action functional $S[q(t)]$ yields a number for each path $q(t)$. We call this number the action of the path.

The Lagrangian is what we see if we peek inside the action functional:[16]

[16] To unclutter the notation, we neglect the units of all quantities.

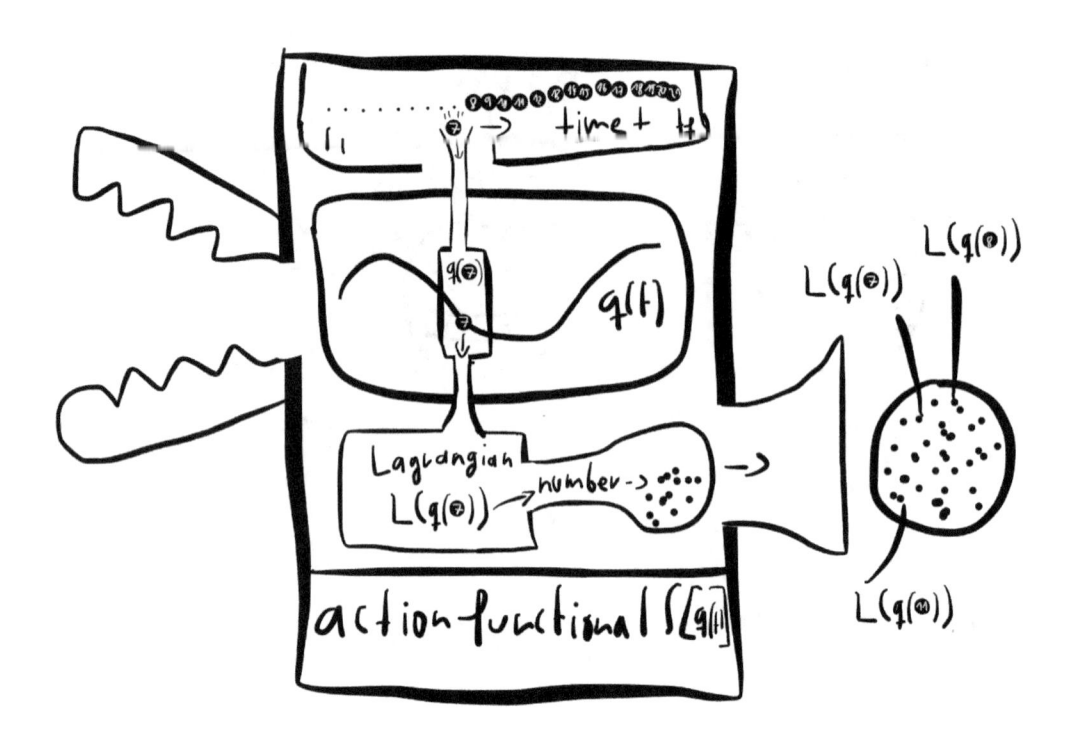

[17] If configuration space is one-dimensional, a location in configuration space is described by a single number. Otherwise we get multiple numbers $q^A(t)$.

[18] If we are dealing with multiple objects moving in three dimensions, the Lagrangian is a function of the various locations and velocities $L = L(q^A, \dot{q}^A)$, where A is an index which runs from 1 to $3N$ for N objects. For example q^1 can denote the x coordinate of the first object, q^4 can denote the x coordinate of the second object and q^2 can denote the y coordinate of the first object, etc. But in the following, to unclutter the notation, we will simply write $q(t)$ for the path in configuration space.

At each moment in time (e.g., $t = 7$) we can use the explicit formula for the specific path $q(t)$ to calculate where exactly in configuration space we are.[17] We then plug these numbers into the Lagrangian function $L(q(t), \dot{q}(t), t)$ and this yields a single number, say, $L(q(7), \dot{q}(7), 7) = 5$. The action functional S records these numbers (the values of the Lagrangian function) for each moment in time between t_i and t_f. The final number that the action functional spits out is a single number, which is the sum over all the individual values of the Lagrangian function.

A second key observation is that the Lagrangian is a function of the location q and the velocity $\frac{dq}{dt}$:[18] $L = L(q, \dot{q})$. In particular, this means that the Lagrangian does not depend on the acceleration $\frac{d^2q}{dt^2}$ or even higher derivatives.

To understand this, we need to recall that the usual formula for the kinetic energy of an object is

$$T = \frac{1}{2} m \left(\frac{dq}{dt} \right)^2, \tag{4.5}$$

where $\frac{dq}{dt}$ is the velocity of the object and m its mass. To simplify the notation, we again write derivatives with respect to time using a dot above the variable:

$$\dot{q} \equiv \frac{dq}{dt}. \tag{4.6}$$

The formula for the kinetic energy then reads

$$T = \frac{1}{2}m\dot{q}^2. \tag{4.7}$$

Moreover, the potential energy V usually only depends on the position of the object $V = V(q)$. For example, the potential energy in the Earth's gravitational field is

$$V = mgq, \tag{4.8}$$

where q denotes the height above the ground.

If we combine these two observations, we end up with the conclusion that the Lagrangian $L = T(\dot{q}) - V(q)$ is a function which only depends on the location q and the velocity $\dot{q}$:

$$L = L(q, \dot{q}). \tag{4.9}$$

In addition, take note that the potential sometimes varies in time $V = V(q, t)$. The Lagrangian then additionally depends on t:

$$L = L(q, \dot{q}, t). \tag{4.10}$$

Now, how can we find the path which involves a minimal amount of action?

4.2 Variational Calculus

For an ordinary function $f(x)$, we can find the minimum by calculating the zeroes of its derivative:

$$\frac{df(x)}{dx} \stackrel{!}{=} 0. \tag{4.11}$$

For example, for $f(x) = 3x^2 + x$, we calculate $\frac{df(x)}{dx} = 6x + 1$ and then find

$$6x + 1 \stackrel{!}{=} 0 \qquad \text{(condition in Eq. 4.11)}$$
$$\therefore \quad x = \frac{-1}{6}. \tag{4.12}$$

And indeed, a minimum of our function $f(x)$ is located at $x = \frac{-1}{6}$.

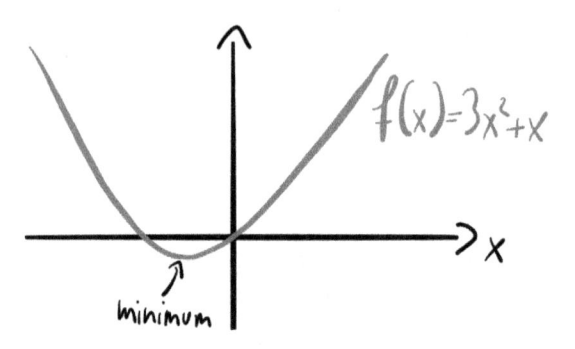

This method works because the derivative tells us something about the slope of $f(x)$ and the slope at a minimum is necessarily zero.[19]

[19] Take note that the slope is zero at a maximum or inflection point too. If you're unsure why the derivative tells us something about the slope, have a look at Appendix A.

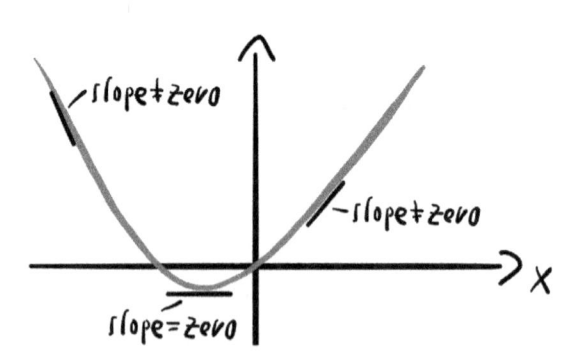

Now the bad news is that this simple method does not work for functionals like the action $S[q(t)]$. Instead, we need a new method to calculate the minimum of a functional.

To understand this alternative method, we need to take a step back and answer the question: what exactly characterizes a minimum?

Let's imagine that we have a function which describes the height of some terrain and want to find out where exactly the terrain height is a minimum.

The key observation is that if we stand at the minimum and look around, we will notice that it's going upward in all directions. This is necessarily the case because otherwise the point we are standing at wouldn't be a minimum.

This means that a minimum is characterized by its neighborhood. If all neighboring points lie higher, the point in question is a minimum.[20]

Let's use this idea to once more find the minimum of the function $f(x) = 3x^2 + x$ that we already considered above.[21]

We now pick one specific location $x = a$ and start investigating

[20] Take note that this criterion only tells us that we are dealing with a *local* minimum. There can be much deeper minima in some other region.

[21] We do this to demonstrate how the method works. In the following section, we will use it to derive the minimum of the action functional.

its neighborhood $a \to a + \epsilon$, where ϵ is an infinitesimally small (positive or negative) number. In general, we call ϵ a **variation**.

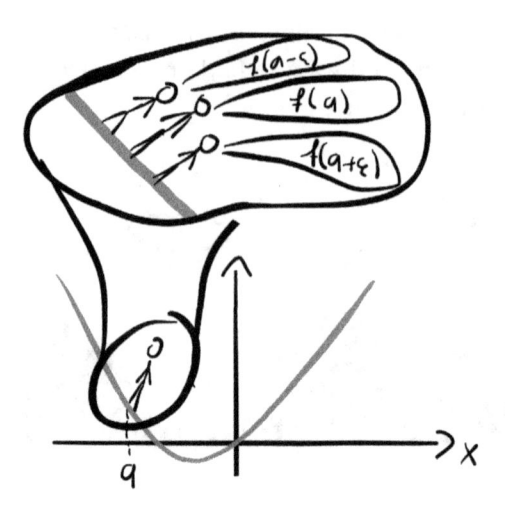

Putting this into the function yields

$$f(a + \epsilon) = 3(a + \epsilon)^2 + (a + \epsilon)$$
$$= 3(a^2 + 2a\epsilon + \epsilon^2) + a + \epsilon. \quad (4.13)$$

If the location a is a minimum, we can't get lower by going in any direction ϵ. Mathematically, this implies:[22]

[22] First order terms are all terms containing ϵ but not ϵ^2, ϵ^3, etc.

> All terms first order in ϵ must vanish.

Otherwise, for a negative ϵ the function value $f(a + \epsilon)$ would be smaller than $f(a)$ and therefore, a wouldn't be a minimum. To understand this, take note that if ϵ is an infinitesimally small number, we have $|\epsilon^2| \ll |\epsilon|$. This is true for any small number, e.g., $0.1^2 = 0.01 \ll 0.1$. Therefore, the negative shift due to a negative ϵ cannot be compensated by quadratic or even higher order terms in ϵ.[23]

[23] Don't worry if you're not completely satisfied with this explanation because in Section 12.2, we will discuss the real reason why we demand that first order terms vanish.

If we collect all terms linear in ϵ and demand that they vanish,

$$3 \cdot 2a\epsilon + \epsilon \stackrel{!}{=} 0$$

$$\text{cancel } \epsilon$$

$$\therefore \quad 6a + 1 \stackrel{!}{=} 0,$$

we find
$$a = \frac{-1}{6}. \qquad (4.14)$$
This specific location has exactly the property we are looking for (in its neighborhood all first order variations vanish, and therefore, it goes upward in all directions) and we can conclude that we've found a minimum.

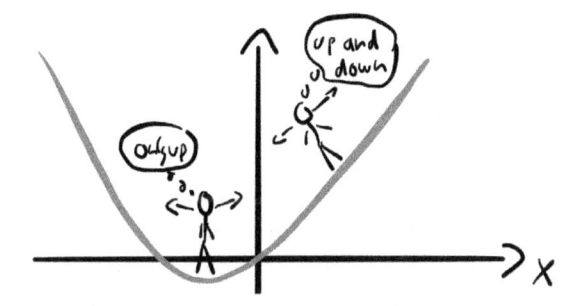

Of course, the result here is exactly equal to what we calculated using the standard method (Eq. 4.12). Thus, for ordinary functions what we've discovered is just another way of reaching the same conclusion. However, the variational method of finding minima can also be applied to functionals like the action $S[q(t)]$, not just functions. Take note that for functionals, our goal isn't to find a *location* like a which is the minimum of a function but instead, to find a *function* $q(t)$ which is the minimum of a functional. And this is what we will talk about in the next section.

But first, let's summarize the main lessons learned in this section.

▷ Minima are characterized by their neighborhood. If we are dealing with a minimum, it has to go upward everywhere in its neighborhood.

▷ Mathematically, this means that we can find minima by making a specific choice $x = a$ and then varying it $a \to a + \epsilon$. If a is a minimum, all first order variations ϵ must vanish.

▷ Through this condition, we can find locations a which are minima.

4.3 The Euler-Lagrange Equation

We learned above that the main idea of the Lagrangian formalism is that the path of least action in configuration space correctly describes how a given system evolves. Moreover, we have learned that the **action functional**[24] (Eq. 4.3)

$$S[q(t)] \equiv \int_{t_i}^{t_f} dt L\Big(q(t), \dot{q}(t)\Big) \qquad (4.15)$$

is a mathematical object which assigns a number to each possible path $q(t)$ between two fixed configurations $(q_i(t_i), q_f(t_f))$.[25]

Therefore, our task is to find a method which allows us to calculate the path $q_m(t)$ for which the action functional is a minimum. This path $q_m(t)$ correctly describes the evolution of our system. Luckily, we can derive a method which allows us to find $q_m(t)$ for any system by repeating everything we did in the previous section.

We start again with a concrete choice $q(t)$ and consider small variations around this specific path

$$q(t) \to q(t) + \epsilon(t), \qquad (4.16)$$

where ϵ is again an infinitesimally small variation.

[24] As discussed at the end of Section 4.1, the Lagrangian is in general a function of the path $q(t)$ and its velocity $\dot{q}$ but not of higher derivatives like $\ddot{q}$, i.e., does not depend on the acceleration.

[25] As before, we write $q(t)$ instead of $q^A(t)$, etc., to unclutter the notation. In other words, we use $q(t)$ and $\dot{q}(t)$ as a convenient notation for a path and the velocity in the possibly high-dimensional configuration space.

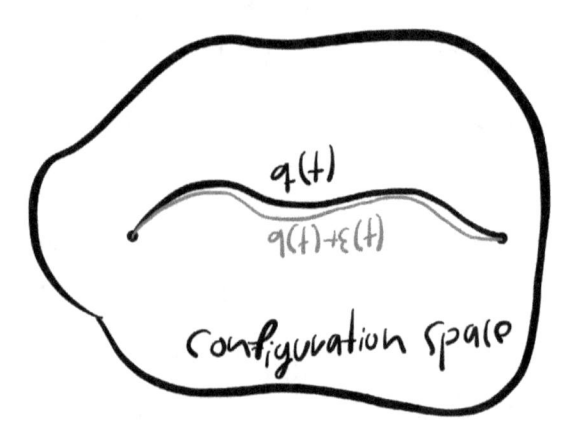

Moreover, since the Lagrangian not only depends on $q(t)$ but also on the velocity $\dot{q}(t)$, we need to consider velocity variations,

too:
$$\dot{q}(t) \to \dot{q}(t) + \dot{\epsilon}(t). \tag{4.17}$$

We consider variations between two *fixed* configurations ($q_i(t_i)$, $q_f(t_f)$). Therefore, the variation ϵ has to vanish at t_i and t_f:

$$0 = \epsilon(t_i) = \epsilon(t_f). \tag{4.18}$$

Analogously to what we did in the previous section, we use these variations explicitly

$$S = \int_{t_i}^{t_f} dt L\Big(q(t) + \epsilon(t), \dot{q}(t) + \dot{\epsilon}(t)\Big). \tag{4.19}$$

The key idea is again that our specific path $q(t)$ is indeed a minimum of the action if all terms that are first order in ϵ vanish. This yields a condition which allows us to identify the correct path $q(t)$ which is a minimum of the action, analogously to how in the previous section we were able to find the location at which a given function has a minimum.

We could do this for each possible Lagrangian L individually. But since this is quite cumbersome, it makes sense to try to move forward with a general Lagrangian. We can do this by using the Taylor expansion[26]

$$L(q + \epsilon, \dot{q} + \dot{\epsilon}) = L(q, \dot{q}) + \epsilon \frac{\partial L}{\partial q} + \dot{\epsilon} \frac{\partial L}{\partial \dot{q}} + \ldots, \tag{4.20}$$

where the dots indicate higher order terms in the expansion. Putting this Taylor expansion of the Lagrangian into the action (Eq. 4.19) yields

$$S = \int_{t_i}^{t_f} dt L\Big(q(t) + \epsilon(t), \dot{q}(t) + \dot{\epsilon}(t)\Big)$$

$$\overset{\text{Taylor expansion}}{=} \int_{t_i}^{t_f} dt \Big(L(q, \dot{q}) + \epsilon \frac{\partial L}{\partial q} + \dot{\epsilon} \frac{\partial L}{\partial \dot{q}} + \ldots\Big).$$

All additional terms in the Taylor expansion are proportional to ϵ^2, $\dot{\epsilon}^2$ or even higher powers. Therefore, we already have everything we need to use our idea that minima are characterized by vanishing *first order* variations.

[26] The Taylor expansion is explained in Appendix F. In words, it tells us that the value of a function at a neighboring point is given approximately by the value of the function at the original point plus the rate of change times the distance we are going. For a function which only depends on one variable $f = f(x)$, this means that its value at the point $x + \epsilon$ is approximately

$$f(x + \epsilon) \approx f(x) + \epsilon \frac{\partial f}{\partial x}.$$

If we are dealing with a function which depends on multiple variables $g = g(x, y)$, we need to take the rate of change in all directions into account as we move from our original point to the new point

$$g(x + \epsilon, y + \tilde{\epsilon}) \approx g(x, y) + \epsilon \frac{\partial g}{\partial x}$$
$$+ \tilde{\epsilon} \frac{\partial g}{\partial y}.$$

Moreover, take note that to unclutter the notation, we do not write the arguments of $q = q(t)$, $\epsilon = \epsilon(t)$, etc., explicitly.

So again, we collect all terms first order in the variations and demand that they vanish:

$$\int_{t_i}^{t_f} dt \left[\epsilon \frac{\partial L}{\partial q} + \dot{\epsilon} \frac{\partial L}{\partial \dot{q}} \right] \stackrel{!}{=} 0. \quad (4.21)$$

The path $q(t)$ for which this is true is a minimum of the action.

The key idea is that we can rewrite this condition by using a few mathematical tricks and derive a specific condition for the function $q(t)$ this way.[27] This condition is the equation of motion which allows us to predict how systems evolve in general.

[27] In particular, we can get rid of the nasty integral.

So, first of all, we integrate by parts the second term on the right-hand side[28]

$$\int_{t_i}^{t_f} dt \, \dot{\epsilon} \frac{\partial L}{\partial \dot{q}} = \int_{t_i}^{t_f} dt \left(\frac{d}{dt} \epsilon \right) \frac{\partial L}{\partial \dot{q}}$$

$$= \epsilon \frac{\partial L}{\partial \dot{q}} \bigg|_{t_i}^{t_f} - \int_{t_i}^{t_f} dt \, \epsilon \frac{d}{dt} \left(\frac{\partial L}{\partial \dot{q}} \right). \quad (4.22)$$

[28] We will see why this is a clever idea in a moment. Integration by parts is a direct consequence of the product rule and derived in Appendix A.2.

Since the variation $\epsilon(t)$ vanishes for $t = t_i$ and $t = t_f$ (Eq. 4.18), the first term on the right-hand side in Eq. 4.22 vanishes:

$$\epsilon \frac{\partial L}{\partial \dot{q}} \bigg|_{t_i}^{t_f} = 0. \quad (4.23)$$

Therefore, we can write Eq. 4.21 as

$$\int_{t_i}^{t_f} dt \left[\epsilon \frac{\partial L}{\partial q} + \dot{\epsilon} \frac{\partial L}{\partial \dot{q}} \right] \stackrel{!}{=} 0$$

↷ Eq. 4.22 and Eq. 4.23

$$\therefore \int_{t_i}^{t_f} dt \left[\epsilon \frac{\partial L}{\partial q} - \epsilon \frac{d}{dt} \left(\frac{\partial L}{\partial \dot{q}} \right) \right] \stackrel{!}{=} 0$$

↷ factoring out ϵ

$$\therefore \int_{t_i}^{t_f} dt \, \epsilon \left[\frac{\partial L}{\partial q} - \frac{d}{dt} \left(\frac{\partial L}{\partial \dot{q}} \right) \right] \stackrel{!}{=} 0. \quad (4.24)$$

Now we're almost finished. We only need to recall that if $q(t)$ is indeed the path of least action that we are looking for, the condition must be correct for *any* possible variation $\epsilon = \epsilon(t)$. But this can only be correct if, in the last line of Eq. 4.24, the expression between the two big square brackets vanishes:

$$\boxed{ \frac{\partial L}{\partial q} - \frac{d}{dt} \left(\frac{\partial L}{\partial \dot{q}} \right) \stackrel{!}{=} 0 } \quad (4.25)$$

This equation is the famous **Euler-Lagrange equation**. We can use it for any given Lagrangian L to derive the corresponding equation of motion.[29] Solutions of this equation of motion correctly describe how a system evolves.

[29] We will discuss below how this works concretely.

Before we discuss a few concrete examples, take note that, in general, our function $q(t)$ describes a path in a high-dimensional configuration space. This means that we get a condition like this for each coordinate $q_A(t)$:[30]

$$\frac{\partial L(q_A, \dot{q}_A)}{\partial q_A} - \frac{d}{dt}\left(\frac{\partial L(q_A, \dot{q}_A)}{\partial \dot{q}_A}\right) \stackrel{!}{=} 0 \qquad (4.26)$$

[30] Recall that $q(t)$ is only our shorthand notation. In general, a path in configuration space needs to be described by multiple coordinates. For example, for N freely-moving objects, we need $3N$ coordinates, i.e., the index A runs from 1 to $3N$.

Let's try to understand the meaning of the Euler-Lagrange equation by using a simple example.[31]

[31] Further examples will be discussed in detail in Part II.

4.3.1 Meaning of the Euler-Lagrange Equation

The easiest example is, of course, a system which consists of just one object with no external potential $V = 0$. For such a free object, the Lagrangian reads

$$L = T - V = T = \frac{1}{2}m\dot{q}^2. \tag{4.27}$$

The Euler-Lagrange equation (Eq. 4.25) then tells us

$$\frac{\partial L}{\partial q} - \frac{d}{dt}\left(\frac{\partial L}{\partial \dot{q}}\right) = 0$$

$$\circlearrowleft \quad L = \frac{1}{2}m\dot{q}^2$$

$$\therefore \quad \frac{\partial(\frac{1}{2}m\dot{q}^2)}{\partial q} - \frac{d}{dt}\left(\frac{\partial(\frac{1}{2}m\dot{q}^2)}{\partial \dot{q}}\right) = 0$$

$$\circlearrowleft \quad \frac{\partial(\frac{1}{2}m\dot{q}^2)}{\partial q} = 0$$

$$\therefore \quad -\frac{d}{dt}\left(\frac{\partial(\frac{1}{2}m\dot{q}^2)}{\partial \dot{q}}\right) = 0$$

$$\circlearrowleft \quad \left(\frac{\partial(\frac{1}{2}m\dot{q}^2)}{\partial \dot{q}}\right) = m\dot{q}$$

$$\therefore \quad -\frac{d}{dt}(m\dot{q}) = 0$$

$$\circlearrowleft \quad \frac{d}{dt}\dot{q} = \ddot{q} \text{ and assuming } m = \text{const.}$$

$$\therefore \quad m\ddot{q} = 0.$$

This is exactly the equation of motion for a free object that we also get by using Newton's second law

$$\frac{d}{dt}p = F$$

$$\circlearrowleft \quad F = 0 \text{ for a free object}$$

$$\therefore \quad \frac{d}{dt}p = 0$$

$$\circlearrowleft \quad p = m\dot{q} \text{ is the momentum for a single object}$$

$$\therefore \quad \frac{d}{dt}(m\dot{q}) = 0$$

$$\circlearrowleft \quad \frac{d}{dt}\dot{q} = \ddot{q} \text{ and assuming } m = \text{const.}$$

$$\therefore \quad m\ddot{q} = 0.$$

This little calculation is not only an important consistency check. It also allows us to understand the Euler-Lagrange equation a little better. In particular, we've seen that the first term $\frac{\partial L}{\partial q}$ yields zero if there is no potential because the kinetic energy only depends on $\dot{q}$ and not on q. Therefore, this term describes the forces F in the system because it is only non-zero if there are forces. In particular, for a general potential $V = V(q)$, this first term yields

$$\frac{\partial L}{\partial q} = \frac{\partial \left(T(\dot{q}) - V(q)\right)}{\partial q} = -\frac{\partial V(q)}{\partial q} \equiv F. \qquad (4.28)$$

Moreover, we've seen that the second term $\frac{d}{dt}\left(\frac{\partial L}{\partial \dot{q}}\right)$ yields the time derivative of the momentum $\frac{d}{dt}p \equiv \frac{d}{dt}m\dot{q}$ for a single object. This motivates us to propose that the term between the parentheses describes, in general, the **momentum**:[32]

$$p \equiv \frac{\partial L}{\partial \dot{q}}. \qquad (4.29)$$

[32] Take note that this quantity is not always the usual momentum. We will talk about this subtlety below.

With this in mind, we can rewrite the Euler-Lagrange equation as follows:

$$\frac{\partial L}{\partial q} - \frac{d}{dt}\left(\frac{\partial L}{\partial \dot{q}}\right) = 0$$

$$\quad \circlearrowright \text{ rearranging}$$

$$\therefore \frac{d}{dt}\left(\frac{\partial L}{\partial \dot{q}}\right) = \frac{\partial L}{\partial q}$$

$$\quad \circlearrowright \text{ Eq. 4.28 and Eq. 4.29}$$

$$\therefore \frac{d}{dt}p = F. \qquad (4.30)$$

This is exactly Newton's second law![33] While there are many subtleties which we are glossing over here, it is very helpful to keep in mind that $\frac{\partial L}{\partial \dot{q}}$ yields the momentum (Eq. 4.29) and that the Euler-Lagrange equation therefore says:

[33] Maybe it helps to recall that we can understand a force F as something which originates from an underlying potential V (Eq. 2.15):

$$F = -\frac{\partial V(q)}{\partial q}.$$

In a similar sense, we can argue that Newton's second law (or the equation of motion in general) originates from an underlying Lagrangian.

> The rate of change of the momentum equals the force.

Now, before we move on, we need to talk about one subtlety.

The quantity in Eq. 4.29 that we simply called the momentum, is more properly called the **conjugate momentum**, **canonical momentum** or **generalized momentum**. It is necessary to emphasize that there is a close connection to a specific variable q. For example, if we describe our system using angles instead of Cartesian coordinates, the corresponding conjugate momenta would be angular momenta. Moreover, for an object with electric charge e which moves in a magnetic potential $\vec{A}$, Eq. 4.29 yields[34]

$$\vec{p} = m\dot{\vec{q}} + e\vec{A},. \tag{4.31}$$

This is not just the usual momentum $\vec{p} = m\vec{v}$. What we therefore learn is that to include the effects of the magnetic potential, we need to replace our ordinary momentum with the more general canonical momentum.

In general, the canonical momentum is a measure of how responsive a Lagrangian is to changes in the velocity.

Similarly, the quantity in Eq. 4.28 that we simply called the force is more properly called the **generalized force**. Again, this is necessary because it clearly depends on which coordinates we use. If we use angles instead of Cartesian coordinates, Eq. 4.28 does not yield an ordinary force but a torque.[35]

[34] Don't worry if you don't know what a magnetic potential is. This formula is only shown here to emphasize that Eq. 4.29 does not always yield what we call momentum in the Newtonian formulation.

[35] A bit more abstractly, we can say that a generalized force describes the applied work done per unit displacement in the direction defined by the (generalized) coordinate q.

5
Hamiltonian Mechanics

In the Newtonian formulation of classical mechanics, we describe a given system by using for each object and each force $\vec{F}$ a vector in physical space $\vec{r}^A(t)$. In the Lagrangian formulation, we describe the system using one big vector $\vec{q}(t)$ which describes a path in configuration space and a Lagrangian function $L(\vec{q}, \dot{\vec{q}})$. This approach has the big advantage that everything which influences the dynamics within the system is described by a single function $L(\vec{q}, \dot{\vec{q}})$.

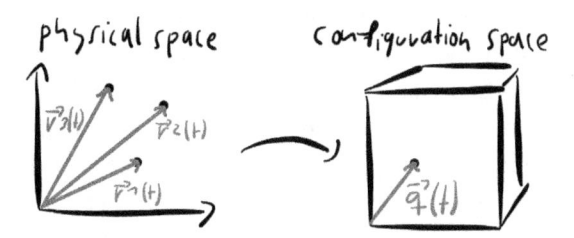

Now, as discussed in Section 2.3, there is another useful mathematical arena known as phase space which allows us to derive a third formulation of classical mechanics. The main observation which motivates us to introduce this new space is that a point in configuration space only encodes information about the locations of the various objects. But to describe everything

[1] We will see below why we use momenta when we formulate classical mechanics in phase space and not velocities. Moreover, take note that this is by no means a hard argument in favor of the Hamiltonian formalism. Of course, we *can* describe the state of systems using configuration space as demonstrated by the Lagrangian formalism. But this requires that we use the velocities $\vec{\dot{q}}(t)$ as additional input data. Since this data is not part of configuration space, it's much harder to develop a geometrical understanding. In other words, in configuration space the velocities appear as a somewhat awkward surplus structure and by switching to phase space we can put this additional input data on the same footing as the locations.

[2] In contrast, in configuration space, multiple states correspond to the same point. All states for which our objects are located at the same positions but possibly having different momenta, correspond to the same point in configuration space:

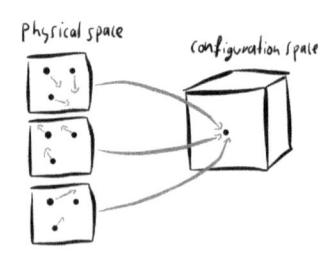

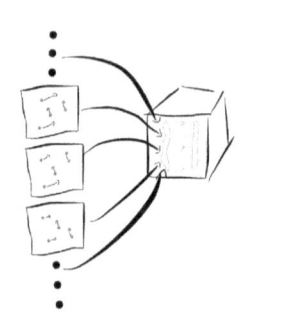

[3] The most important theorem in this context is known as Liouville's theorem, which is the topic of Section 11.2.4.

that is going on in a system at a specific moment in time, we not only need to keep track of the locations of objects, but also of their momenta (or alternatively their velocities). In other words, we need both, the position and momentum $(\vec{q}(t), \vec{p}(t))$ (or $(\vec{q}(t), \vec{\dot{q}}(t))$), in order to determine the future behavior of that system, not just $\vec{q}(t)$.[1]

In phase space, each point corresponds to one specific state that the system can be in $(\vec{q}(t), \vec{p}(t))$ and not just a configuration $q(t)$.[2]

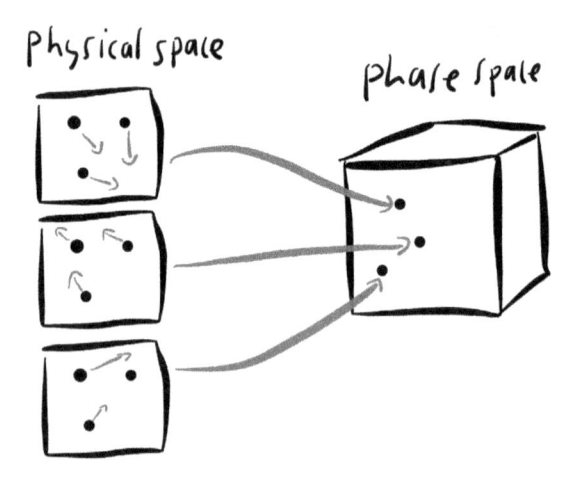

This allows us to get a geometrical understanding of how a specific initial state evolves as time passes. And this is especially useful whenever we are dealing with uncertainty. In a real system, we are never 100% sure in which initial state a given system is since there is always experimental uncertainty. In phase space, this means that our initial state is not exactly a point. We need to take a region of phase space into account.

Then, using that probabilities must always add up to 100%, we can derive powerful and beautiful theorems which allow us to understand how uncertainty evolves as time passes.[3]

All this leads to the beautiful interpretation that we can understand the evolution of a system in classical mechanics as the flow of some kind of (incompressible) probability fluid in phase space.

Now, how can we describe classical mechanics in phase space?

5.1 Hamilton's Equations

While there are various ways of deriving the fundamental equations which describe classical mechanics in phase space, we will start with arguably the simplest one. The starting point for this derivation are our results from the previous section.

In the Hamiltonian formulation of classical mechanics, we describe the evolution of our system as a path in phase space. The new thing about phase space is that we not only use the various locations q_i to specify a point, but at the same time use the momenta p_i of the objects.

The first key idea is that we act as if q_i and p_i are completely independent variables. Of course, for a single object we have $p_i = m\dot{q}_i$, and we will see in a moment that there is always a close connection between q_i and p_i. But for the moment, we act as if they are truly independent. Take note that this is not possible for the velocity $\dot{q}_i$ because the velocity is always simply the rate of change of the location $\frac{d}{dt}q_i(t) = \dot{q}_i(t)$. In contrast, the relationship between the location and the (generalized) momentum is not always so direct. For example, as mentioned at the end of the previous chapter, for a charged object moving in a magnetic potential $\vec{A}$, the (generalized) momentum is $\vec{p} = m\dot{\vec{q}} + e\vec{A}$ (Eq. 4.31). This quantity certainly can be completely independent of $q(t)$ because $\vec{A}$ can change independently.[4]

This is, in some sense, what we mean when we say that in phase space, we not only have an axis for each q_i but also for

[4] Again, the detailed form of this formula and what exactly a magnetic potential is, is not important for us here. The point to take away is that when we speak of momentum, we mean the generalized momentum that is defined in Eq. 4.29:
$$p \equiv \frac{\partial L}{\partial \dot{q}}.$$
If we evaluate this formula explicitly for certain Lagrangians, we find a quantity which can be completely independent of $q(t)$. This is why we can treat $q(t)$ and the generalized momentum $p(t)$ as independent variables. Moreover, take note that the difference between a partial derivative like $\frac{\partial L}{\partial q}$ and a total derivative like $\frac{dp}{dt}$ is discussed in detail in Appendix A.3.

[5] For example, for a free object the x coordinate of an object is independent of the y and z coordinate. That's why we have three axes to describe the location of a free object.

each p_i.[5] The momenta are on a completely equal footing with the locations.

Secondly, we recall the definition of the momentum (Eq. 4.29)

$$p \equiv \frac{\partial L}{\partial \dot{q}}. \tag{5.1}$$

Using this definition, we can write the Euler-Lagrange equation as

$$\frac{dp}{dt} = \frac{\partial L}{\partial q}. \tag{5.2}$$

Moreover, the second defining equation in the Lagrangian formalism is the definition of the velocity as the rate of change of the location[6]

$$\frac{dq}{dt} = \dot{q}. \tag{5.3}$$

[6] It may seem strange why this point is emphasized here. We will see below, however, that we really need this equation.

Our goal is to rewrite these two equations (Eq. 5.2, Eq. 5.3) in such a way that they only depend on p and no longer on $\dot{q}$. We can do this by using the explicit definition of the momentum.

First of all, we can invert Eq. 5.1 to get a formula for the velocity in terms of the momentum: $\dot{q} = \dot{q}(q, p)$.[7] We can then use this result to derive an equation from which $\dot{q}$ has been eliminated.

[7] For example, when we have $p = \dot{q}m$, we can calculate that $\dot{q} = \frac{p}{m}$. This is a function $\dot{q} = \dot{q}(p)$.

But we need to be careful. The Lagrangian is a function which depends on q and $\dot{q}$: $L = L(q, \dot{q})$. Therefore, in general, we don't get the same result when we take the derivative of $L(q, \dot{q})$ and when we take the derivative of the new function $\tilde{L}(q, p)$:

$$\frac{\partial \tilde{L}(q, p)}{\partial q} \neq \frac{\partial L(q, \dot{q})}{\partial q}, \tag{5.4}$$

where

$$\tilde{L}(q, p) \equiv L\left(q, \dot{q}(q, p)\right). \tag{5.5}$$

In words, $\tilde{L}(q, p)$ is the function that we get if we use the formula $\dot{q} = \dot{q}(q, p)$ to eliminate $\dot{q}$ from $L(q, \dot{q})$. In particular, take note that, in general,

$$\tilde{L}(q, p) \neq L(q, p). \tag{5.6}$$

For example, for the free Lagrangian (Eq. 4.27)

$$L(q, \dot{q}) = \frac{m\dot{q}^2}{2} \tag{5.7}$$

and the explicit formula for the momentum ($p = m\dot{q}$), we find:

$$\tilde{L}(q,p) \stackrel{(5.5)}{=} L\Big(q, \dot{q}(q,p)\Big)$$

$\curvearrowright$ $L(q,\dot{q}) = \frac{m\dot{q}^2}{2}$

$$= \frac{m\Big(\dot{q}(q,p)\Big)^2}{2}$$

$\curvearrowright$ $\dot{q} = \frac{p}{m}$

$$= \frac{m\left(\frac{p}{m}\right)^2}{2}$$

$\curvearrowright$

$$= \frac{p^2}{2m}. \tag{5.8}$$

Therefore[8]

$$L(q,p) = \frac{mp^2}{2} \neq \frac{p^2}{2m} = \tilde{L}(q,p). \tag{5.9}$$

[8] If this is unclear, please have a look at Appendix D.

So when we calculate the derivative, we find

$$\frac{\partial \tilde{L}(q,p)}{\partial q} \stackrel{(5.5)}{=} \frac{\partial L\Big(q, \dot{q}(q,p)\Big)}{\partial q}$$

$\curvearrowright$ chain rule

$$\therefore \frac{\partial \tilde{L}(q,p)}{\partial q} = \frac{\partial L(q,\dot{q})}{\partial q} + \frac{\partial L(q,\dot{q})}{\partial \dot{q}} \frac{\partial \dot{q}(q,p)}{\partial q}$$

$\curvearrowright$ $p \equiv \frac{\partial L}{\partial \dot{q}}$, Eq. 5.1

$$\therefore \frac{\partial \tilde{L}(q,p)}{\partial q} = \frac{\partial L(q,\dot{q})}{\partial q} + p \frac{\partial \dot{q}(q,p)}{\partial q}$$

$\curvearrowright$ rearranging terms

$$\therefore \frac{\partial L(q,\dot{q})}{\partial q} = \frac{\partial \tilde{L}(q,p)}{\partial q} - p \frac{\partial \dot{q}(q,p)}{\partial q}$$

$\curvearrowright$ $\frac{\partial}{\partial q} p\dot{q} = p \frac{\partial}{\partial q} \dot{q}$ because $\frac{\partial}{\partial q} p = 0$

$$\therefore \frac{\partial L(q,\dot{q})}{\partial q} = \frac{\partial}{\partial q} \Big(\tilde{L}(q,p) - p\dot{q}(q,p) \Big). \tag{5.10}$$

This is the equation that we need to use to eliminate $\dot{q}$ from the right-hand side in Eq. 5.2

$$\frac{dp}{dt} = \frac{\partial L}{\partial q}$$

$\curvearrowright$ Eq. 5.10

$$= \frac{\partial}{\partial q} \Big(\tilde{L}(q,p) - p\dot{q}(q,p) \Big)$$

$\curvearrowright$ definition $H \equiv p\dot{q}(q,p) - \tilde{L}(q,p)$

$$= -\frac{\partial H}{\partial q}. \tag{5.11}$$

We can see that the new function

$$H \equiv p\dot{q}(q,p) - \tilde{L}(q,p) \tag{5.12}$$

which we defined here, determines the time-evolution of the momentum p and is known as the **Hamiltonian function** or simply the **Hamiltonian**.[9] While in the Lagrangian formulation, the Lagrangian L is the most central object, in the Hamiltonian formulation, the Hamiltonian H is central.[10]

Following similar steps, we can rewrite the second defining equation of Lagrangian mechanics (Eq. 5.3).[11] To do this, we calculate the derivative of $\tilde{L}(q,p)$ with respect to p:[12]

$$\frac{\partial \tilde{L}(q,p)}{\partial p} \stackrel{(5.5)}{=} \frac{\partial L(q, \dot{q}(q,p))}{\partial p}$$

⟳ chain rule

$$\therefore \frac{\partial \tilde{L}(q,p)}{\partial p} = \frac{\partial L(q,\dot{q})}{\partial \dot{q}} \frac{\partial \dot{q}}{\partial p}$$

⟳ $p \equiv \frac{\partial L}{\partial \dot{q}}$, Eq. 5.1

$$\therefore \frac{\partial \tilde{L}(q,p)}{\partial p} = p \frac{\partial \dot{q}}{\partial p}$$

⟳ $\frac{\partial}{\partial p}(p\dot{q}) = p\frac{\partial \dot{q}}{\partial p} + \dot{q}$

$$\therefore \frac{\partial \tilde{L}(q,p)}{\partial p} = \frac{\partial}{\partial p}(p\dot{q}) - \dot{q}$$

⟳ rearranging terms

$$\therefore \frac{\partial \tilde{L}(q,p)}{\partial p} - \frac{\partial}{\partial p}(p\dot{q}) = -\dot{q}$$

⟳ factoring out $\frac{\partial}{\partial p}$

$$\therefore \frac{\partial}{\partial p}\left(\tilde{L}(q,p) - p\dot{q}\right) = -\dot{q}$$

⟳ $H \equiv p\dot{q}(q,p) - \tilde{L}(q,p)$

$$\therefore \frac{\partial H}{\partial p} = \dot{q}. \tag{5.13}$$

This result allows us to eliminate $\dot{q}$ from Eq. 5.3, which then becomes

$$\frac{dq}{dt} = \dot{q}$$

⟳ Eq. 5.13

$$\therefore \frac{dq}{dt} = \frac{\partial H}{\partial p}. \tag{5.14}$$

---

[9] We will see in Section 5.1.1 why we define the Hamiltonian function like this. Moreover, take note that mathematically, the Hamiltonian and the Lagrangian are related by a Legendre transform. The Legendre transform is discussed in Appendix B.

[10] We will discuss the meaning of H in more detail below.

[11] For your convenience: Eq. 5.3 reads
$$\frac{dq}{dt} = \dot{q}.$$

[12] We will see in a moment why this is useful.

The two equations (Eq. 5.11, Eq. 5.14) we derived by eliminating $\dot{q}$ in favor of p from the two defining equations of Lagrangian mechanics (Eq. 5.2, Eq. 5.3) are known as **Hamilton's equations**. Since they are so important, we recite them here for further convenience:[13]

$$\boxed{\begin{aligned}\frac{dp}{dt} &= -\frac{\partial H}{\partial q} \\ \frac{dq}{dt} &= \frac{\partial H}{\partial p}\end{aligned}} \quad (5.15)$$

[13] Don't worry if you didn't like the derivation of Hamilton's equations that we discussed above because we will next discuss a second way to derive them.

Take note how symmetrical these two equations are in how they treat q and p. As mentioned at the beginning of this chapter, putting the locations and momenta on an equal footing is one of the main motivations behind the Hamiltonian formalism.

While in the Lagrangian formalism, we also have a simple equation for the rate of change of the location (Eq. 5.3): $\frac{dq}{dt} = \dot{q}$, there is no similarly simple equation for $\frac{d\dot{q}}{dt}$. In other words, there is no general way to dig $\frac{d\dot{q}}{dt}$ out of the Euler-Lagrange equations (Eq. 5.2). But in contrast, by switching variables, we get a simple equation for the time evolution of q and a simple equation for the time evolution of p.[14]

As before, if there are multiple objects in the system moving in three dimensions, we need to take all their locations and momenta into account. Hamilton's equations then read

$$\boxed{\begin{aligned}\frac{dp_i}{dt} &= -\frac{\partial H}{\partial q_i} \\ \frac{dq_i}{dt} &= \frac{\partial H}{\partial p_i}\end{aligned}} \quad (5.16)$$

[14] Mathematically, we have first-order evolution equations in the Hamiltonian formalism while in the Newtonian and Langrangian formulations of classical mechanics, we have second-order differential equations. The price we have to pay is that we have twice as many equations in the Hamiltonian formalism since we've doubled the size (dimensions) of the space we are describing our system in:

configuration space $\to$ phase space.

For example, for N free particles configuration space is $3N$-dimensional but phase space is $6N$-dimensional.

The index i is used to distinguish all the different momentum coordinates $p_1, p_2, p_3, p_4, \ldots$ and all the different position coordinates $x_1, x_2, x_3, x_4, \ldots$.

Before we move on and discuss the meaning of Hamilton's equations in a bit more detail, let's talk about an alternative way of deriving them.

In the Lagrangian formalism, our central object is the action which is defined as the integral over the Lagrangian (Eq. 4.15)[15]

$$S \equiv \int_{t_i}^{t_f} L \, dt. \tag{5.17}$$

[15] As before, we restrict ourselves to one object moving in one dimension. The derivation for a general set of positions q_i, velocities $\dot{q}_i$ and momenta p_i follows analogously.

Above, we've already learned that the fundamental function in the Hamiltonian formalism is the Hamiltonian, which is defined as (Eq. 5.12)[16]

$$H \equiv p\dot{q} - L. \tag{5.18}$$

[16] Mathematically, the Hamiltonian function is the Legendre transform of the Lagrangian function. The Legendre transform is discussed in Appendix B.

We can use this explicit relationship between the Hamiltonian function H and Lagrangian L to derive Hamilton's equations completely analogously to how we derived the Euler-Lagrange equation in Section 4.3.

When we rearrange the terms in Eq. 5.18, we find

$$L = p\dot{q} - H. \tag{5.19}$$

Using this, we can rewrite the action (Eq. 5.17) in terms of the Hamiltonian:

$$S = \int_{t_i}^{t_f} dt\, L$$

$$= \int_{t_i}^{t_f} dt \left(p\dot{q} - H \right). \tag{5.20}$$

↷ Eq. 5.19

This is useful because now we can once more use the least action principle to derive the correct equations of motion.[17]

[17] In Section 4.3, we used the least action principle to derive the Euler-Lagrange equation.

Again, our goal is to find the path which minimizes the action. But take note that now the functional S assigns a value to each path $(Q(t) = (q(t), p(t)))$ in *phase space*. In other words, we are now searching for a path in phase space and not for a path in configuration space. To find this path, we again consider small variations around some arbitrary but fixed path $(q(t), p(t)) \to (q(t) + \epsilon(t), p(t) + \tilde{\epsilon}(t))$. The main idea is then once more that we can find the correct path by demanding that all terms that are first-order in the variations must vanish.[18]

[18] This is the key idea of variational calculus which we discussed in Section 4.2.

HAMILTONIAN MECHANICS

[figure: 3D plot with axes p, q, t showing path $(q(t), p(t))$ and varied path $(q(t)+\epsilon(t), p(t)+\tilde\epsilon(t))$ between t_i and t_f, with q_i and q_f marked]

Putting the variations into Eq. 5.20 and using the Taylor expansion yields[19]

$$S = \int_{t_i}^{t_f} dt \left(p\dot q - H(q,p)\right) = \int_{t_i}^{t_f} dt \left(p\frac{d}{dt}q - H(q,p)\right)$$

↓ variations

$$\to \quad S = \int_{t_i}^{t_f} dt \left((p+\tilde\epsilon)\frac{d}{dt}(q+\epsilon) - H(q+\epsilon, p+\tilde\epsilon)\right)$$

↓ Taylor expansion

$$= \int_{t_i}^{t_f} dt \Bigg((p+\tilde\epsilon)\frac{d}{dt}(q+\epsilon)$$

$$- H(q,p) - \epsilon\frac{\partial H(q,p)}{\partial q} - \tilde\epsilon\frac{\partial H(q,p)}{\partial p} - \ldots\Bigg)$$

↓ rearranging terms

$$= \int_{t_i}^{t_f} dt \Bigg(p\frac{dq}{dt} - H + \tilde\epsilon\left(\frac{d}{dt}q - \frac{\partial H}{\partial p}\right) + \ldots$$

$$- \epsilon\frac{\partial H}{\partial q} + p\frac{d\epsilon}{dt} + \tilde\epsilon\frac{d\epsilon}{dt}\Bigg). \tag{5.21}$$

This looks like a huge mess, but luckily all we are interested in are the terms first order in ϵ and $\tilde\epsilon$.[20]

In the last line of Eq. 5.21, we have a term proportional to $\frac{d\epsilon}{dt}$. But we can turn it into a term proportional to ϵ by integrating

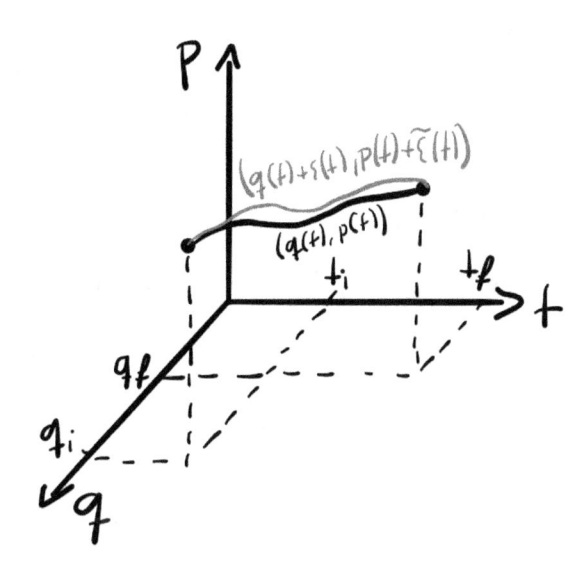

[19] The Taylor expansion that we use for H is completely analogous to the expansion in Eq. 4.20.

[20] Recall that we are looking for the path for which the action is a minimum. To get a condition which allows us to calculate this path, we use the fact that minima are characterized by vanishing first order variations.

[21] Integration by parts is explained in Appendix A.2. Moreover, we will see in a moment why it makes sense to use it here. The boundary term

$$\epsilon p\Big|_{t_i}^{t_f} \equiv \epsilon(t_f)p(t_f) - \epsilon(t_i)p(t_i)$$

vanishes because we are considering variations between *fixed* initial and final configurations. Mathematically this means that $\epsilon(t_i) = 0$ and $\epsilon(t_f) = 0$ (Eq. 4.18).

[22] We want to factor out ϵ because this tells us which terms have to vanish.

by parts[21]

$$\int_{t_i}^{t_f} dt\, p \frac{d\epsilon}{dt} = \epsilon p\Big|_{t_i}^{t_f} - \int_{t_i}^{t_f} dt \frac{dp}{dt}\epsilon$$

$$\qquad \epsilon(t_i) = \epsilon(t_f) = 0, \text{ (Eq. 4.18)}$$

$$= -\int_{t_i}^{t_f} dt \frac{dp}{dt}\epsilon. \qquad (5.22)$$

By using this, we can factor out ϵ in the last line of Eq. 5.21:[22]

$$S = \int_{t_i}^{t_f} dt \left(p\frac{dq}{dt} - H + \tilde{\epsilon}\left(\frac{d}{dt}q - \frac{\partial H}{\partial p}\right) + \ldots \right.$$

$$\left. - \epsilon\frac{\partial H}{\partial q} + p\frac{d\epsilon}{dt} + \tilde{\epsilon}\frac{d\epsilon}{dt} \right)$$

this is Eq. 5.21

$$\qquad \text{Eq. 5.22}$$

$$= \int_{t_i}^{t_f} dt \left(p\frac{dq}{dt} - H + \tilde{\epsilon}\left(\frac{d}{dt}q - \frac{\partial H}{\partial p}\right) + \ldots \right.$$

$$\left. - \epsilon\frac{\partial H}{\partial q} - \frac{dp}{dt}\epsilon + \tilde{\epsilon}\frac{d\epsilon}{dt} \right)$$

$$\qquad \text{factoring out } -\epsilon$$

$$= \int_{t_i}^{t_f} dt \left(p\frac{dq}{dt} - H + \tilde{\epsilon}\left(\frac{d}{dt}q - \frac{\partial H}{\partial p}\right) + \ldots \right.$$

$$\left. - \epsilon\left(\frac{\partial H}{\partial q} + \frac{dp}{dt}\right) + \tilde{\epsilon}\frac{d\epsilon}{dt} \right). \qquad (5.23)$$

Here, the dots indicate higher order terms in the Taylor expansion. We can ignore them because we are only interested in terms linear in ϵ and $\tilde{\epsilon}$.[23]

[23] All higher order terms are proportional to ϵ^2, $\tilde{\epsilon}^2$, ϵ^3, $\tilde{\epsilon}^3$, etc.

The correct path describing our system is the one for which the terms linear in ϵ and $\tilde{\epsilon}$ vanish. However, we need to be careful because ϵ and $\tilde{\epsilon}$ are completely independent variations of the path. While in Section 4.3 we also considered variations of the velocity $\dot{q} \to \dot{q} + \dot{\epsilon}$, this is not an independent variation because $\dot{\epsilon} \equiv \frac{d\epsilon}{dt}$. In other words, the variation of the velocity function $\dot{q}$ is automatically fixed once we specify the variation of the path q. But in phase space, we treat the position and momentum as independent variables and therefore assume that we can vary ϵ and $\tilde{\epsilon}$ independently.

This means that the terms linear in ϵ and $\tilde{\epsilon}$ only vanish, in general, if the following two conditions are fulfilled:[24]

$$\frac{\partial H}{\partial q} + \frac{d}{dt}p \stackrel{!}{=} 0$$
$$\frac{\partial H}{\partial p} - \frac{d}{dt}q \stackrel{!}{=} 0. \qquad (5.24)$$

[24] These are the terms proportional to ϵ and $\tilde{\epsilon}$ in Eq. 5.23.

The path $(q(t), p(t))$ which fulfills these two conditions is the correct path which minimizes the action and therefore describes the evolution of our system. These equations are exactly Hamilton's equations (Eq. 5.15).

To summarize:

▷ In the Hamiltonian formulation of classical mechanics, we describe the evolution of a system as a path in phase space.

▷ This requires that we eliminate the velocity $\dot{q}$ in favor of the momentum p. When we do this in the defining equations of the Lagrangian formulation (Eq. 5.2, Eq. 5.3) we get Hamilton's equations (Eq. 5.15).

▷ Alternatively, we can start again from the action and eliminate the Lagrangian in favor of the Hamiltonian (Eq. 5.19). If we then calculate (using variational calculus) which condition a path in phase space $(q(t), p(t))$ minimizing this new action must fulfill, we find Hamilton's equations once more.

The following diagram illustrates the relationship between the derivations discussed above.

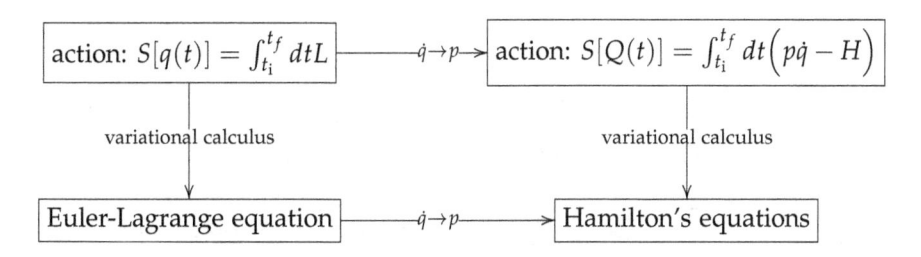

```
┌──────────────────────────────┐        ┌──────────────────────────────────────┐
│ action: S[q(t)] = ∫_{t_i}^{t_f} dt L │──q̇→p──▶│ action: S[Q(t)] = ∫_{t_i}^{t_f} dt(pq̇ − H) │
└──────────────┬───────────────┘        └──────────────────┬───────────────────┘
               │ variational calculus                      │ variational calculus
               ▼                                           ▼
┌──────────────────────────────┐        ┌──────────────────────────────────────┐
│   Euler-Lagrange equation    │──q̇→p──▶│         Hamilton's equations         │
└──────────────────────────────┘        └──────────────────────────────────────┘
```

Next, let's talk about the meaning of Hamilton's equations.

5.1.1 Meaning of Hamilton's Equations

Again, the best way to get a feeling for the meaning of the Hamiltonian function H and Hamilton's equations is to consider a simple example. Let's consider a system which consists of only one object and some general potential $V = V(q)$. The Lagrangian for this object reads $L = T - V = \frac{1}{2}m\dot{q}^2 - V(q)$.

We can calculate the corresponding Hamiltonian by using the definition in Eq. 5.12 and the definition of the momentum in Eq. 4.29. First of all, we need to calculate the momentum explicitly:

$$p \equiv \frac{\partial L}{\partial \dot{q}} \quad \text{definition of the momentum, (Eq. 4.29)}$$

$$\circlearrowleft \quad L = \frac{1}{2}m\dot{q}^2 - V(q)$$

$$= \frac{\partial \left(\frac{1}{2}m\dot{q}^2 - V(q)\right)}{\partial \dot{q}}$$

$$\circlearrowleft \quad \frac{\partial V(q)}{\partial \dot{q}} = 0 \text{ and } \frac{\partial \dot{q}^2}{\partial \dot{q}} = 2\dot{q}$$

$$= m\dot{q} \tag{5.25}$$

This implies that

$$\dot{q} = \frac{p}{m}. \tag{5.26}$$

Using this result, we can derive the Hamiltonian:

$$H = p\dot{q} - L \quad \text{this is Eq. 5.12}$$

$$\circlearrowleft \quad L = \frac{1}{2}m\dot{q}^2 - V(q)$$

$$= p\dot{q} - \left(\frac{1}{2}m\dot{q}^2 - V(q)\right)$$

$$\circlearrowleft \quad \dot{q} = \frac{p}{m}, \text{ (Eq. 5.26)}$$

$$= p\frac{p}{m} - \left(\frac{1}{2}m\left(\frac{p}{m}\right)^2 - V(q)\right)$$

$$\circlearrowleft \quad \text{rearranging terms}$$

$$= \frac{p^2}{2m} + V(q). \tag{5.27}$$

This is exactly the total energy of the object:[25]

$$H = \frac{p^2}{2m} + V(q) = T + V = \text{kinetic energy} + \text{potential energy!}$$

Similar results can be obtained for many systems and we can therefore summarize that often:[26]

---

[25] The formula
$$T = \frac{p^2}{2m} = \frac{1}{2}m\dot{q}^2$$
describes exactly the kinetic energy, where we used once more that $\dot{q} = \frac{p}{m}$ (Eq. 5.26).

[26] Take note that this is not correct for open systems, i.e., systems in which energy is not conserved. (A closed system is one for which no energy is leaking out.) This is the case, for example, whenever the potential is time-dependent $V = V(q, t)$. For such systems, the Hamiltonian (like the Lagrangian) is merely an abstract, but useful, function which allows us to describe the system.

> The Hamiltonian represents the total energy.

Next, let's try to understand the physical meaning of Hamilton's equations.

Once more, it's instructive to use the simple example that we discussed above. Putting the explicit form of the Hamiltonian that we calculated in Eq. 5.27 into Hamilton's first equation (Eq. 5.15) yields

$$\frac{dp}{dt} = -\frac{\partial H}{\partial q} \qquad \text{this is Hamilton's first equation (Eq. 5.15)}$$

$$= -\frac{\partial \left(\frac{p^2}{2m} + V(q)\right)}{\partial q} \qquad \circlearrowright H = \frac{p^2}{2m} + V(q),\ \text{Eq. 5.27}$$

$$= -\frac{\partial V(q)}{\partial q}. \qquad \circlearrowright \frac{\partial p}{\partial q} = 0 \qquad (5.28)$$

This is exactly Newton's second law! Therefore, since the derivative of the potential with respect to the location q yields the force ($F = -\frac{\partial V}{\partial q}$), we can conclude that Hamilton's first equation effectively tells us that:

> The rate of change of momentum equals the force.

Hamilton's second equation (Eq. 5.15) reads

$$\frac{dq}{dt} = \frac{\partial H}{\partial p}. \qquad (5.29)$$

Using the explicit Hamiltonian we derived above, we can evaluate it explicitly:

$$\frac{dq}{dt} = \frac{\partial H}{\partial p}$$

↪ $H = \frac{1}{2}\frac{p^2}{m} + V(q)$, Eq. 5.27

$$= \frac{\partial\left(\frac{1}{2}\frac{p^2}{m} + V(q)\right)}{\partial p}$$

↪ $\frac{\partial V(q)}{\partial p} = 0$ and $\frac{\partial p^2}{\partial p} = 2p$

$$= \frac{p}{m}. \tag{5.30}$$

Therefore, we can now understand that the purpose of Hamilton's second law is to establish a relationship between the momentum and the rate of change of the position. In other words, Hamilton's second equation tells us what the momentum really is for our system.[27]

[27] This is analogous to how the second fundamental equation in the Lagrangian formalism (Eq. 5.3) $\frac{dq}{dt} = \dot{q}$ tells us how the velocity is defined in terms of the location.

All this can be summarized perfectly as follows:

> *Hamilton's Equations show how the q_i's and p_i's undergo a 'dance to the music of time', a dance in which, as some q_i's or p_i's increase in value, others decrease in value, but always such as to keep the energy constant (in conservative systems), and always such as to keep the total action minimized, both instant by instant, and over the whole path between 'surfaces-of-common-action'. This 'dance' is governed by one function, H, - that is to say, while H is different for different systems (orbiting planets, a statistical ensemble, an electrical circuit, positrons orbiting an atomic antinucleus, a spinning top, juggling pins, a flowing river and so on), yet within any one system there is just one overarching function (there is no need for individual functions, H_1, H_2,...,H_n).*
>
> Jennifer Coopersmith, 2017, The Lazy Universe

Before we summarize what we've learned in this and all previous chapters, let's discuss one alternative way for how we can understand why the Hamiltonian is so important. What we will learn in the following section is a first glance at the deeper structure underlying classical mechanics.

5.2 Hamilton's General Equation

First of all, we can imagine that sometimes we are not only interested in the locations and momenta of the various objects in the system but other quantities too. For instance, the temperature or how the kinetic energy evolves as time passes can be interesting things to investigate.

As discussed at the beginning of this chapter, in the Hamiltonian formulation of classical mechanics, we describe our system using phase space. This implies that quantities like the temperature or kinetic energy are functions of the locations q_i and momenta p_i.[28]

[28] Recall that the defining feature of phase space is that we use the locations and momenta of all objects as coordinates.

But how can we calculate the time evolution of such functions depending on the locations $q_i(t)$ and momenta $p_i(t)$?

For simplicity, let's restrict ourselves to one object moving in one dimension. Then the total rate of change of a function $F = F(q(t), p(t))$ along a single object's trajectory reads[29]

[29] This is the total derivative of the function F. If you're unfamiliar with the distinction between the total and partial derivative, see Appendix A.

$$\frac{d}{dt}F(q,p) = \frac{\partial F(q,p)}{\partial q}\frac{dq}{dt} + \frac{\partial F(q,p)}{\partial p}\frac{dp}{dt}. \quad (5.31)$$

Using Hamilton's equations (Eq. 5.15), we can rewrite this result as follows:

$$\frac{d}{dt}F(q,p) = \frac{\partial F(q,p)}{\partial q}\frac{dq}{dt} + \frac{\partial F(q,p)}{\partial p}\frac{dp}{dt}$$

$$= \frac{\partial F(q,p)}{\partial q}\frac{\partial H(q,p)}{\partial p} - \frac{\partial F(q,p)}{\partial p}\frac{\partial H(q,p)}{\partial q}. \quad (5.32)$$

$\frac{dp}{dt} = -\frac{\partial H}{\partial q}, \quad \frac{dq}{dt} = \frac{\partial H}{\partial p}$ (Eq. 5.15)

In words, this means that the time evolution of a general function in phase space is completely determined by the Hamiltonian function H![30]

[30] We will make this more concrete in a moment.

Since the structure that appears on the right-hand side here is so important, it is conventional to introduce a more compact notation. We therefore introduce the **Poisson bracket** $\{\,,\,\}$ of two phase space functions $A(q,p)$, $B(q,p)$ by defining:

$$\{A,B\} \equiv \frac{\partial A}{\partial q}\frac{\partial B}{\partial p} - \frac{\partial A}{\partial p}\frac{\partial B}{\partial q}. \tag{5.33}$$

A helpful way of thinking about the Poisson bracket is that it describes a natural product of two phase space functions:[31]

$$A \circ B \equiv \{A,B\} \equiv \frac{\partial A}{\partial q}\frac{\partial B}{\partial p} - \frac{\partial A}{\partial p}\frac{\partial B}{\partial q}. \tag{5.34}$$

[31] If you don't find this helpful, simply ignore this remark. Technically, the Poisson bracket is not the product but actually the (Lie) derivative in phase space. So $\{A,B\}$ yields the phase space derivative of A with respect to B, i.e., $\partial_B A \equiv \{A,B\}$. We will discuss in Section 11.2 that a phase space function, like the Hamiltonian H, defines a flow (a vector field) in phase space if we put it into the Poisson bracket. Therefore, by calculating the Poisson bracket $\{A,B\}$, we find the derivative of A in the direction defined by the flow of B.

This means that you can't combine two functions in phase space which describe properties of our system arbitrarily and expect to get something that describes another useful property of the system. But if you calculate the Poisson bracket of the two functions, you'll get something sensible. In this sense, the Poisson bracket is the natural product (the correct rule to combine things) in the phase space of a specific system.[32]

[32] We will see in Chapter 10 that any function F for which

$$\{F,H\} = 0,$$

where H is the Hamiltonian function, represents a conserved quantity. Moreover, let's assume we find another function G for which

$$\{G,H\} = 0$$

also holds (i.e., another conserved quantity). The quantity we then get by putting F and G into the Poisson bracket

$$\{F,G\} = I$$

will be a conserved quantity, too:

$$\{I,H\} = 0.$$

This result is known as the Poisson theorem. In this sense, the natural product that allows us to combine conserved quantities in such a way that we get new conserved quantities is indeed given by the Poisson bracket.

Using the definition of the Poisson bracket, we can write Eq. 5.32 more compactly

$$\frac{d}{dt}F = \{F,H\}$$

$$= \frac{\partial F}{\partial q}\frac{\partial H}{\partial p} - \frac{\partial F}{\partial p}\frac{\partial H}{\partial q} \checkmark. \quad \text{check using Eq. 5.33} \tag{5.35}$$

This equation

$$\boxed{\frac{d}{dt}F = \{F,H\}} \tag{5.36}$$

describes the time evolution of a general phase space function and we call it **Hamilton's equation of motion**.

We don't introduce a new name for this equation because the equations which we called Hamilton's equations previously are simply special cases of this general formula. For instance, one of the simplest examples of a phase space function is certainly $q(t)$. For $F = q$, we find

$$\frac{d}{dt}q = \{q, H\} \qquad \text{this is Eq. 5.36 with } F \to q$$

$$\phantom{\frac{d}{dt}q} = \frac{\partial q}{\partial q}\frac{\partial H}{\partial p} - \frac{\partial q}{\partial p}\frac{\partial H}{\partial q} \qquad \text{definition of Poisson bracket (Eq. 5.33)}$$

$$\phantom{\frac{d}{dt}q} \qquad\qquad\qquad \frac{\partial q}{\partial q} = 1 \text{ and } \frac{\partial q}{\partial p} = 0$$

$$\phantom{\frac{d}{dt}q} = \frac{\partial H}{\partial p}. \tag{5.37}$$

This is exactly Hamilton's second equation (Eq. 5.15). Analogously, for $F = p$ we find

$$\frac{d}{dt}p = \{p, H\} \qquad \text{this is Eq. 5.36}$$

$$\phantom{\frac{d}{dt}p} = \frac{\partial p}{\partial q}\frac{\partial H}{\partial p} - \frac{\partial p}{\partial p}\frac{\partial H}{\partial q} \qquad \text{definition of Poisson bracket (Eq. 5.33)}$$

$$\phantom{\frac{d}{dt}p} \qquad\qquad\qquad \frac{\partial p}{\partial q} = 0 \text{ and } \frac{\partial p}{\partial p} = 1$$

$$\phantom{\frac{d}{dt}p} = -\frac{\partial H}{\partial q}. \tag{5.38}$$

This is exactly Hamilton's first equation (Eq. 5.15).

So, to summarize:

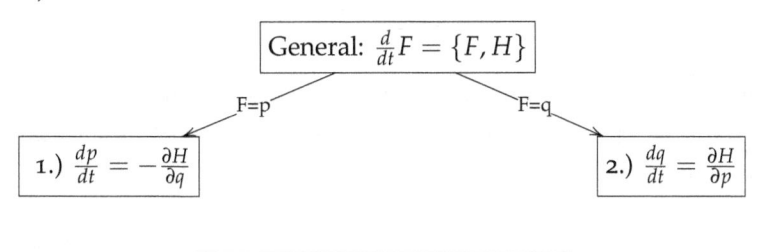

General: $\frac{d}{dt}F = \{F, H\}$

1.) $\frac{dp}{dt} = -\frac{\partial H}{\partial q}$ (F=p)   2.) $\frac{dq}{dt} = \frac{\partial H}{\partial p}$ (F=q)

To understand our new equation (Eq. 5.36) a little better, we rewrite it as follows:

$$\frac{d}{dt}F = \{F, H\}$$

$$\qquad\qquad \text{multiplying by } dt$$

$$dF = \{F, H\}dt$$

$$\qquad\qquad \text{switching to finite intervals}$$

$$\Delta F = \{F, H\}\Delta t. \tag{5.39}$$

The object on the left-hand side describes how much F changes during the interval Δt. In other words, as time passes from t to $t + \Delta t$, our function F becomes $F + \Delta F$. Therefore, $\{F, H\}$ yields the rate of change of F.

So one way to understand Hamilton's general equation of motion is by imagining that we have a new kind of object $\{\,, H\}$ (an operator) which eats any function F on phase space ($\{F, H\}$) and spits out the correct time evolution of F.

For example, as we've just discovered, for $F = p$ we get Hamilton's first equation and for $F = q$ we get Hamilton's second equation.

So the Hamiltonian and the Poisson bracket together yield something which allows us to calculate how functions evolve as time passes.[33] In this sense, it is conventional to say that Hamilton's equation of motion tells us that:

[33] From a more abstract perspective, the Hamiltonian together with the Poisson bracket yields a vector field in phase space. This vector field tells us how functions get pushed along as time passes.

> The Hamiltonian *generates* time evolution in phase space.

You'll probably still find this formulation somewhat strange and it takes some time getting used to it. But be assured that this is an extremely powerful perspective.[34]

[34] To spoil the surprise: we can understand other important quantities like the momentum or angular momentum analogously. In particular, momentum generates spatial translations and angular momentum generates rotations. And this perspective leads us directly to quantum mechanics. We will explore the connection between the physical (conserved) quantities and how they generate changes in more detail in Chapter 10.

In particular, in quantum mechanics this way of thinking is incredibly important. For example, the equation which describes the time-evolution of observables in quantum mechanics reads

$$\frac{d\hat{F}}{dt} = -\frac{i}{\hbar}[\hat{F}, \hat{H}], \qquad (5.40)$$

where $[\hat{F}, \hat{H}] = \hat{F}\hat{H} - \hat{H}\hat{F}$ is known as the **commutator bracket**. This equation is known as the **Heisenberg equation**. In some sense, the Heisenberg equation is completely analogous to Hamilton's general equation that we derived above. The main difference is that we use a different bracket.[35] In fact, many textbooks introduce quantum mechanics by proposing the replacement rule

[35] As an aside: mathematically, the Poisson bracket and commutator bracket are examples of a larger category of brackets known as Lie brackets.

$$\text{Poisson bracket} \quad \{F, H\} \quad \rightarrow \quad \text{Commutator} \quad [\hat{F}, \hat{H}].$$

Here's one final comment before we summarize everything we've learned so far.

Sometimes we are dealing with a function in phase space which not only depends on q and p, but also explicitly on t. For example, this is necessarily the case if there is a time-dependent potential $V = V(q,t)$. The total rate of change then reads[36]

$$\frac{d}{dt}F(q,p,t) = \frac{dq}{dt}\frac{\partial F}{\partial q} + \frac{dp}{dt}\frac{\partial F}{\partial p} + \frac{\partial F}{\partial t}. \tag{5.41}$$

[36] This is the total derivative of a general function. The difference between a total and partial derivative is discussed in Appendix A.3.

In words, this means that we get an additional term which takes this additional explicit dependence on t into account. Then, following exactly the same steps we can derive

$$\boxed{\frac{d}{dt}F(q,p,t) = \{F,H\} + \frac{\partial F}{\partial t}.} \tag{5.42}$$

This is Hamilton's equation of motion for a phase space general function $F = F(q,p,t)$.

6

Summary

Let's summarize what we've learned so far.

In the previous three chapters, we discussed three different formulations of classical mechanics. In the Newtonian formulation, we keep track of objects using vectors in physical space. In the Lagrangian formulation, we describe a system using a path in configuration space. And, in the Hamiltonian formulation, we use a path in phase space. While the goal in classical mechanics is usually to derive (and solve) the equations of motion for a given system, the way we derive them in the various formulations is quite different.

In the **Newtonian formulation**, the equation of motion for any system can be calculated by adding all forces acting on some object to the right-hand side of Newton's second law (Eq. 3.1)

$$\frac{d}{dt}\vec{p} = \vec{F}.$$

(6.1)

In words it tells us:[1]

> The rate at which the momentum of an object changes is exactly equal to the force acting on it.

[1] Reminder: this law is supplemented by Newton's first and third law. The first law establishes that there is always an observer for whom Newton's second law holds. (We call observers for whom Newton's second law holds inertial observers. Speaking colloquially, these inertial observers are observers which are not accelerated themselves.) Newton's third law tells us that for every force $\vec{F}$ exerted by some object A on another object B, there is necessarily a force $-\vec{F}$ of equal magnitude but pointing in the opposite direction exerted by B on A.

In the **Lagrangian formulation**, we can calculate the equation of motion for a given system by using the Euler-Lagrange equation (Eq. 4.25)

$$\boxed{\frac{\partial L}{\partial q} - \frac{d}{dt}\left(\frac{\partial L}{\partial \dot{q}}\right) = 0,} \quad (6.2)$$

where L is the Lagrangian. The Lagrangian is the difference between the kinetic energy T and the potential energy V:[2]

[2] Take note that this formula for the Lagrangian only works for non-relativistic classical mechanics. In special relativity or other theories like quantum field theory, the Lagrangian looks quite different. We will discuss this in more detail in Chapter 12.

$$L(q, \dot{q}, t) = T(\dot{q}) - V(q, t). \quad (6.3)$$

In words, the Euler-Lagrange equation is a condition which the path $q(t)$ that minimizes the action functional (Eq. 4.15)

$$S[q(t)] \equiv \int_{t_i}^{t_f} dt L\Big(q(t), \dot{q}(t), t\Big) \quad (6.4)$$

has to fulfill. In other words, a solution of the Euler-Lagrange equation yields a path which is a minimum of the action. The key idea at the heart of the Lagrangian formulation is that nature is lazy and hence the path of least action is the correct path that describes how our system behaves.

Here $q(t)$ describes a path in configuration space and therefore if we want to use it to describe concrete systems, we need to recall that we've really lots of Euler-Lagrange equations (Eq. 4.26)[3]

[3] For example, we get three Euler-Lagrange equations for each object moving freely in three dimensions. For N objects moving freely, configuration space is $3N$-dimensional and therefore we have $3N$ Euler-Lagrange equations, one for each component (i.e., for each direction in configuration space we can vary our path in $q^A \to q^A + \epsilon^A(t)$).

$$\boxed{\frac{\partial L(q^A, \dot{q}^A, t)}{\partial q^A} - \frac{d}{dt}\left(\frac{\partial L(q^A, \dot{q}^A, t)}{\partial \dot{q}^A}\right) \stackrel{!}{=} 0,} \quad (6.5)$$

where, for example,

$$(q^1, q^2, q^3, q^4, q^5, q^6, q^7, \ldots) \equiv (x^1, y^1, z^1, x^2, y^2, z^2, x^3, \ldots)$$

and x^1 denotes the x-coordinate of the first object, x^2 the x-coordinate of the second object, etc.

In the **Hamiltonian formalism**, we can calculate the equation of motion for any given system by using **Hamilton's equations**

(Eq. 5.15)

$$\boxed{\begin{aligned}\frac{dp}{dt} &= -\frac{\partial H}{\partial q} \\ \frac{dq}{dt} &= \frac{\partial H}{\partial p}\end{aligned}}\,, \qquad (6.6)$$

where H is the Hamiltonian. The Hamiltonian is the sum of the kinetic energy T and the potential energy V:[4]

$$H(q,p) = T(p) + V(q,t). \qquad (6.7)$$

Again, whenever we want to use Hamilton's equations to describe concrete systems, we need to remember that we need to take all coordinates of phase space into account

$$\boxed{\begin{aligned}\frac{dp_i}{dt} &= -\frac{\partial H}{\partial q_i} \\ \frac{dq_i}{dt} &= \frac{\partial H}{\partial p_i}\end{aligned}}. \qquad (6.8)$$

[4] More generally, the Hamiltonian is defined as the Legendre transform (see Appenidx B) of the Lagrangian. For some systems, the Hamiltonian function is not equal to the total energy. This is the case because the Hamiltonian always represents a conserved quantity, but the total energy isn't necessarily conserved within a given system, e.g., when energy is pumped from the outside into the system or energy is leaking out. We will discuss this in more detail in Chapter 10.

We discussed two derivations of Hamilton's equations. First, we derived it by switching from $\dot{q}$ to p in the Euler-Lagrange equation by using the explicit formulas of the form $\dot{q} = \dot{q}(q,p)$. And secondly, we derived it by switching variables in the action functional (Eq. 5.20):

$$S[q(t), p(t)] = \int_{t_i}^{t_f} dt \left(p\dot{q} - H \right). \qquad (6.9)$$

If we use variational calculus to find the minimum of this functional, we find Hamilton's equations.

Moreover, we derived the more general **Hamilton equation** (Eq. 5.36):

$$\boxed{\frac{d}{dt}F = \{F, H\}} \qquad (6.10)$$

where $\{\,,\,\}$ denotes the Poisson bracket (Eq. 5.34):

$$\{A, B\} \equiv \frac{\partial A}{\partial q}\frac{\partial B}{\partial p} - \frac{\partial A}{\partial p}\frac{\partial B}{\partial q}. \qquad (6.11)$$

If we use $F = q$ and $F = p$ in Hamilton's general equation, we find Hamilton's equations (Eq. 5.15) once more. But Eq. 5.36 is more general because it allows us to calculate the time evolution of *any* phase space function.

Now it's time to see how all this works in practice. In the following chapters, we will talk about the most important systems in classical mechanics and how we can describe them using the frameworks that we have just talked about.

Part II
Essential Systems and Tools

"The only way to learn it is to do it."

Archimedes

PS: You can discuss the content of Part II with other readers and give feedback at www.nononsensebooks.com/cm/bonus.

7
Algorithms and Tools

In this second part of the book, we discuss how we can use the various formulations of classical mechanics to describe concrete systems.

Let me shortly outline our plan for this chapter.

We will start by talking about the Newtonian, Lagrangian, and Hamiltonian algorithms. This means that we will discuss in quite general terms how we can use Newton's second law, the Euler-Lagrange equation and Hamilton's equations to derive the equations of motion.

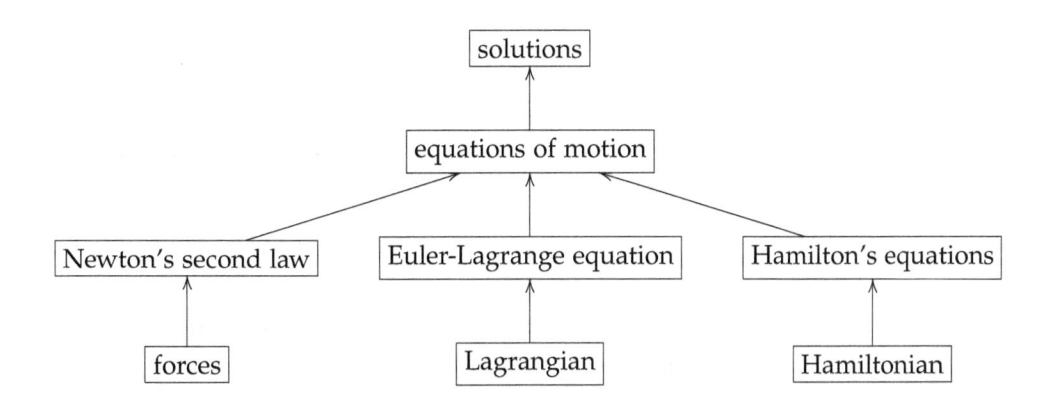

In addition, we will discuss several tools which allow us to simplify problems. For example, one of the most powerful tools that we can use to simplify problems is our freedom to choose a suitable coordinate system.

However, especially in the Newtonian formalism, switching coordinates can sometimes be more difficult than solving the problem in an inconvenient one.

In contrast, in the Hamiltonian and Lagrangian formalism, switching coordinates is easy because the fundamental equations (Hamilton's equations, Euler-Lagrange equation) have the same *form* no matter which coordinates we choose.[1]

[1] We will discuss this in detail below.

In this context, it is important to recall that the main difference between the various formalisms is that we use different mathematical arenas.

While in the Newtonian formalism, we use coordinates to describe vectors in physical space, in the Lagrangian formalism, we use coordinates in configuration space, and in the Hamiltonian formalism, we use coordinates in phase space. Since the structure of these spaces is quite different, the transformations which allow us to switch coordinates work quite differently.

For this reason, we introduce special names for these different kinds of coordinate transformations:

▷ A switch of coordinates in physical space is an ordinary **coordinate transformation.**

▷ A switch of coordinates in configuration space is a **point transformation.**

▷ A switch of coordinates in phase space is a **canonical transformation.**[2]

[2] Canonical transformation is really the umbrella term for all transformations that leave Hamilton's equations unchanged.

Diagrammatically this means:

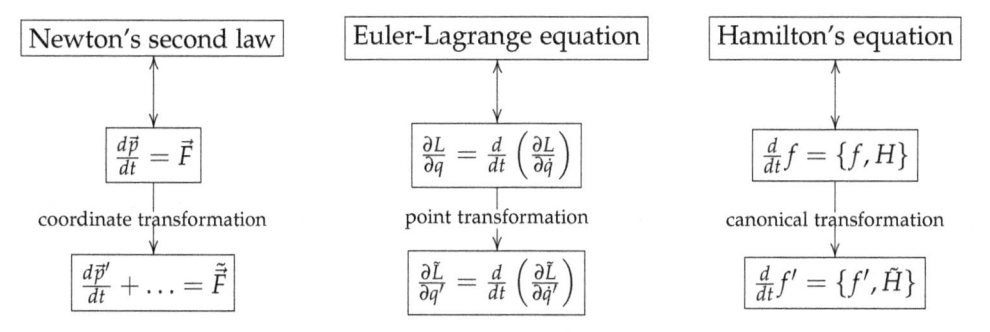

Newton's second law	Euler-Lagrange equation	Hamilton's equation
$\frac{d\vec{p}}{dt} = \vec{F}$	$\frac{\partial L}{\partial q} = \frac{d}{dt}\left(\frac{\partial L}{\partial \dot{q}}\right)$	$\frac{d}{dt}f = \{f, H\}$
coordinate transformation	point transformation	canonical transformation
$\frac{d\vec{p}'}{dt} + \ldots = \tilde{\vec{F}}$	$\frac{\partial \tilde{L}}{\partial q'} = \frac{d}{dt}\left(\frac{\partial \tilde{L}}{\partial \dot{q}'}\right)$	$\frac{d}{dt}f' = \{f', \tilde{H}\}$

Here $\tilde{\vec{F}}$, $\tilde{L}$ and $\tilde{H}$ denote the forces, Lagrangian, and Hamiltonian, in new coordinates $(p', q', \dot{q}')$, respectively. Moreover, f and f' are phase space functions.

The main point is that in the Newtonian formalism, depending on the coordinate system, new terms appear on the left-hand side and therefore Newton's second law takes on a completely different form. Alternatively, these new terms can be understood as new (fictitious) forces.[3]

[3] In non-inertial coordinate systems, we need to take so-called fictitious forces into account.

But we cannot only modify the coordinate systems we use to describe the locations and momenta of objects. We can also modify the coordinate systems we use to describe the somewhat more abstract spaces that the action and Lagrangian live in. To understand this, recall that the action functional assigns a specific number to each possible path. Mathematically, these numbers are points on the real line $\mathbb{R}$. However, it doesn't matter where we put the origin of our real line coordinate system. In other words, it doesn't matter which absolute number we assign to each path. Only the relative differences between the numbers for different paths are important because the only reason we care about the action is to find the path of least action. If we shift all numbers assigned to individual paths equally by a constant factor, it doesn't make any difference because the path of least action stays the path of least action.

Moreover, since the Lagrangian and Hamiltonian are directly connected to the action functional, we have similar freedom in how we choose the absolute numbers the Lagrangian and

[4] We will talk about gauge transformation in detail in Section 7.2.3 and Section 7.3.2.

Hamiltonian function spit out. It is conventional to call these kinds of transformations **gauge transformations**.[4]

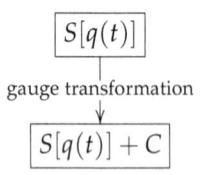

Action	Lagrangian	Hamiltonian
$S[q(t)]$	$L(q,\dot{q},t)$	$H(q,p,t)$
gauge transformation ↓	gauge transformation ↓	gauge transformation ↓
$S[q(t)] + C$	$L(q,\dot{q},t) + \frac{d}{dt}F(q,t)$	$H(q,p,t) - \frac{\partial}{\partial t}F(q,t)$

[5] We restrict ourselves to just two examples but discuss them carefully step by step. This should give you a solid first idea of how the various formalisms are used in practice. We will not discuss additional or more advanced examples for two reasons. First, you'll probably feel obliged to go through all of the examples presented here before you move on to Part III. But the content of Part III is extremely important, so I tried to keep the hurdle to get there as small as possible. And second, there are usually no deep truths to be learned from complicated applications. It's the same story but told in more complicated terms. The only thing you really learn by studying many complicated problems is how to solve complicated problems. There are no further deep insights regarding the structure and interpretation of classical mechanics. Moreover, the skills you get by studying lots of problems usually don't help when you engage in research projects. Exactly solvable problems are usually only found in textbooks. So for real research projects you need a very different skillset. A great way to learn about these real-world problem solving skills is by reading *Street-Fighting Mathematics* by Sanjoy Mahajan. But, of course, if your goal is to pass some university exam, you'll need to study a lot more examples. You can find many worked examples in the books recommended in Chapter 13.

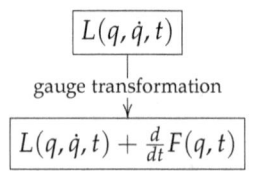

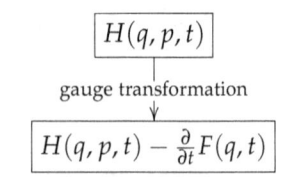

In addition, we will discuss two concrete examples: the harmonic oscillator and the pendulum.[5] We will derive the equation of motion for both systems step by step using all three formulations. Moreover, we will discuss how we can solve the resulting equations.

Let's now talk about all this in more detail.

7.1 The Newtonian Algorithm

In the Newtonian formalism, our main task is to identify all of the forces which act on a given object. Afterwards, we need to write down formulas which describe the strength and direction of the various forces. Then, given these concrete formulas, we can use Newton's second law (Eq. 3.1)

$$\frac{d\vec{p}}{dt} = \vec{F},\qquad(7.1)$$

where $\vec{p} = m\vec{v}$, to find the acceleration $\vec{a} = \dot{\vec{v}}$. Finally, using the acceleration, we can calculate how exactly the object in question moves around. Formulated more technically, given the acceleration $\vec{a}$, we can derive its velocity $\vec{v}$ and location $\vec{r}$ since $\vec{a} = \dot{\vec{v}} = \ddot{\vec{r}}$.

While this procedure may sound straightforward, there are two main difficulties:

▷ Finding all of the forces acting on all of the objects is not an easy task. Usually, the forces point in various directions and it's quite easy to lose track of things.

▷ Solving the equation of motion can be challenging or even impossible. Depending on the forces, the differential equation we end up with by using Newton's second law can be extremely difficult to solve exactly.[6]

Let's consider a simple example.

We stand at the top of the Leaning Tower of Pisa and let a ball fall to the ground. How is the ball moving?[7]

The only force acting on our ball (neglecting air resistance) is gravity. Let's choose our coordinate system such that the z-axis points upward toward the sky and its origin is exactly at the point from which we release the ball. The gravitational force can

[6] The problems usually handed to students are exactly those for which the equation of motion can be solved quite easily. But for most real-world applications, the equations need to be evaluated numerically.

[7] In other words, our task is to describe the movement of a freely falling object.

then be described by the formula

$$\vec{F} = \begin{pmatrix} 0 \\ 0 \\ -mg \end{pmatrix} \quad (7.2)$$

Newton's second law (Eq. 3.1) tells us

$$\frac{d\vec{p}}{dt} = \vec{F}$$

⟩ Eq. 7.2 and $\vec{p} = m\vec{v}$

$$\frac{d(m\vec{v})}{dt} = \begin{pmatrix} 0 \\ 0 \\ -mg \end{pmatrix}$$

⟩ $m = $ const.

$$m \frac{d}{dt} \begin{pmatrix} v_x \\ v_y \\ v_z \end{pmatrix} = \begin{pmatrix} 0 \\ 0 \\ -mg \end{pmatrix}$$

⟩ $\not{m}$

$$\frac{d}{dt} \begin{pmatrix} v_x \\ v_y \\ v_z \end{pmatrix} = \begin{pmatrix} 0 \\ 0 \\ -g \end{pmatrix}. \quad (7.3)$$

Our task is to solve these three equations of motion. Luckily, we can simply integrate the equations twice since gravity is constant:

$$\frac{d}{dt} \begin{pmatrix} v_x \\ v_y \\ v_z \end{pmatrix} = \begin{pmatrix} 0 \\ 0 \\ -g \end{pmatrix}$$

⟩ $\int_0^t dt'$ and $t \to t'$

$$\int_0^t dt' \frac{d}{dt'} \begin{pmatrix} v_x \\ v_y \\ v_z \end{pmatrix} = \int_0^t dt' \begin{pmatrix} 0 \\ 0 \\ -g \end{pmatrix}$$

⟩

$$\begin{pmatrix} v_x(t) \\ v_y(t) \\ v_z(t) \end{pmatrix} - \begin{pmatrix} v_x(0) \\ v_y(0) \\ v_z(0) \end{pmatrix} = \begin{pmatrix} 0 \\ 0 \\ -gt \end{pmatrix}$$

⟩ $\int_0^t dt'$

$$\int_0^t dt' \begin{pmatrix} v_x(t') \\ v_y(t') \\ v_z(t') \end{pmatrix} - \int_0^t dt' \begin{pmatrix} v_x(0) \\ v_y(0) \\ v_z(0) \end{pmatrix} = \int_0^t dt' \begin{pmatrix} 0 \\ 0 \\ -gt \end{pmatrix}$$

⟩

$$\begin{pmatrix} x(t) \\ y(t) \\ z(t) \end{pmatrix} - \begin{pmatrix} x(0) \\ y(0) \\ z(0) \end{pmatrix} - \begin{pmatrix} v_x(0)t \\ v_y(0)t \\ v_z(0)t \end{pmatrix} = \begin{pmatrix} 0 \\ 0 \\ -\frac{1}{2}gt^2 \end{pmatrix}. \quad (7.4)$$

Next, we need to determine the integration constants

$$v_x(0), v_y(0), v_z(0), x(0), y(0), z(0)$$

using concrete initial conditions.

Since we don't throw the ball, its initial velocity is zero

$$\begin{pmatrix} v_x(0) \\ v_y(0) \\ v_z(0) \end{pmatrix} = \begin{pmatrix} 0 \\ 0 \\ 0 \end{pmatrix}. \qquad (7.5)$$

Moreover, since we put the origin of our coordinate system exactly at the spot from which we release the ball, we also have

$$\begin{pmatrix} x(0) \\ y(0) \\ z(0) \end{pmatrix} = \begin{pmatrix} 0 \\ 0 \\ 0 \end{pmatrix}. \qquad (7.6)$$

Therefore, for our concrete situation, Eq. 7.4 becomes

$$\begin{pmatrix} x(t) \\ y(t) \\ z(t) \end{pmatrix} = \begin{pmatrix} 0 \\ 0 \\ -\tfrac{1}{2}gt^2 \end{pmatrix}. \qquad (7.7)$$

These three functions correctly describe how the ball falls down to the ground. Since there is no force acting in the x and y direction, the location on these axes remains constant. In addition, gravity pulls the ball in the negative z-direction and that's why the ball moves downward.

7.2 The Lagrangian Algorithm

If we want to use the Lagrangian formalism to describe a given system, our main task is to write down the correct Lagrangian L. As discussed already in Part I, in classical mechanics the Lagrangian is simply the difference between the kinetic and potential energy:

$$L(q, \dot{q}) = T(\dot{q}) - V(q). \qquad (7.8)$$

For the kinetic energy of an object we use $T = \tfrac{1}{2}m\dot{q}^2$. Therefore, the main difficulty is to specify the potential energy $V(q)$ of the object in question which possibly consists of various contributions $V(q) = V_1(q) + V_2(q) + \ldots$ analogous to how there can be multiple forces.

[8] We use here and in the following the notation $v_q = \dot{q}$.

Afterwards, we can use the Euler-Lagrange equation (Eq. 4.25)[8]

$$\frac{\partial L}{\partial q} = \frac{d}{dt}\left(\frac{\partial L}{\partial v_q}\right), \qquad (7.9)$$

to derive the equation of motion. Our final task is then once more to solve the equation of motion.

Again, as a concrete example, let's consider the ball that we drop from the Leaning Tower of Pisa.

The Lagrangian for this system reads

$$L = T - V = \frac{1}{2}m\vec{v}^2 - mgz$$

$$= \frac{1}{2}m(v_x^2 + v_y^2 + v_z^2) - mgz \qquad (7.10)$$

where we used Eq. 2.21 for the potential energy of the ball in the Earth's gravitational field.

The Euler-Lagrange equation (Eq. 4.25) for the x-coordinate tells us

$$\frac{\partial L}{\partial x} = \frac{d}{dt}\left(\frac{\partial L}{\partial v_x}\right)$$

↱ Eq. 7.10

$$\frac{\partial\left(\frac{1}{2}m(v_x^2 + v_y^2 + v_z^2) - mgz\right)}{\partial x} = \frac{d}{dt}\left(\frac{\partial\left(\frac{1}{2}m(v_x^2 + v_y^2 + v_z^2) - mgz\right)}{\partial v_x}\right)$$

$$0 = m\frac{d}{dt}v_x. \qquad (7.11)$$

Following exactly the same steps, we find for the y-coordinate

$$0 = m\frac{d}{dt}v_y. \qquad (7.12)$$

And using the Euler-Lagrange equation (Eq. 4.25) for the z-coordinate yields

$$\frac{\partial L}{\partial z} = \frac{d}{dt}\left(\frac{\partial L}{\partial v_z}\right)$$

Eq. 7.10

$$\frac{\partial\left(\frac{1}{2}m(v_x^2+v_y^2+v_z^2)-mgz\right)}{\partial z} = \frac{d}{dt}\left(\frac{\partial\left(\frac{1}{2}m(v_x^2+v_y^2+v_z^2)-mgz\right)}{\partial v_z}\right)$$

$$-mg = m\frac{d}{dt}v_z. \tag{7.13}$$

These three equations (Eq. 7.11, Eq. 7.12, Eq. 7.13) are exactly the equations of motion that we already derived using the Newtonian formalism (Eq. 7.3). So it may seem as if the Lagrangian formalism were merely another way of doing the same thing. However, take note that in the derivations above, we didn't have to think about vectors at all. This is one advantage of the Lagrangian formalism.

In addition, the Lagrangian formulation of classical mechanics is always a good choice whenever we are dealing with a system which is subject to constraints. This may sound abstract, but constraints are really common. For example, if we want to describe a pendulum we need to take into account that the object attached at the end of the string cannot move around freely. Instead, the object always remains attached to the string and this is what we call a constraint.[9] Moreover, for any object which slides along some surface we need to take the constraint that it will not fall through the surface into account.

[9] We will discuss the pendulum in detail in Chapter 9.

[10] We will see this explicitly in Section 9.1.

While we can incorporate constraints in the Newtonian formalism by introducing so-called constraint forces, this is quite cumbersome.[10] In contrast, in the Lagrangian framework, we only need to add one or several terms to the Lagrangian which follow automatically once we've identified all constraints for the given system. Afterwards, we can calculate the equations of motion, as usual, by using the Euler-Lagrange equation. Why and how this works is what we will talk about in the next section.

7.2.1 Constraints

We are dealing with constraints whenever an object is attached to other objects or to its surroundings. As a result, the objects influence each other and can't move around freely. A typical example is when an object is forced to slide along a wire or along a specific surface.

In mathematical terms, a constraint is a relationship between coordinates. For example, for a mass attached to a circular loop with radius l, we have the constraint

$$x^2 + y^2 = l^2. \tag{7.14}$$

In words this constraint encodes that our mass is only allowed to move in a circle.

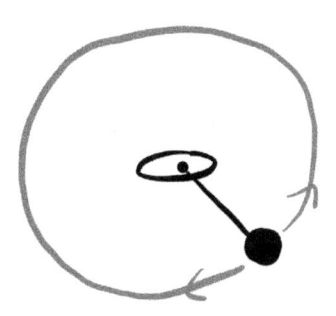

More generally, a constraint is a formula of the form[11]

$$f(q_1, q_2, \ldots, t) = \text{const.} \tag{7.15}$$

The trick which allows us to incorporate constraints in the Lagrangian formalism is known as the method of Lagrange multipliers and works as follows.[12]

First of all, we rewrite our constraint equation such that we have zero on the right-hand side:

$$f(q_1, q_2, \ldots, t) = \text{const.}$$
$$\therefore \quad f(q_1, q_2, \ldots, t) - \text{const.} = 0$$
$$\therefore \quad g(q_1, q_2, \ldots, t) = 0, \tag{7.16}$$

where we defined a new function

$$g(q_1, q_2, \ldots, t) \equiv f(q_1, q_2, \ldots, t) - \text{const.}.$$

We then take the Lagrangian L_{free} that we would use if the object could move around freely without constraints and add a new term L_{con} which encodes the constraint

$$\begin{aligned} L_{\text{full}} &= L_{\text{free}} + L_{\text{con}} \\ &= L_{\text{free}} + \lambda g(q, t). \end{aligned} \tag{7.17}$$

This additional term has the form $L_{\text{con}} = \lambda g(q, t)$, where $g(q, t)$ is the function that describes the constraint and λ is something new which we call a **Lagrange multiplier**.[13]

This is a clever trick because if we treat λ as a new coordinate, the Euler-Lagrange equation (Eq. 4.25) for λ tells us

---

[11] It is conventional to call this type of constraint a **holonomic constraint**. Take note that there are other types of constraints which cannot be written like this. Such constraints are called **non-holonomic constraints**. One example is when a constraint can only be formulated in terms of an inequality

$$f(q_1, q_2, \ldots) \geq \text{const.}.$$

We encounter this kind of constraint, for example, when we want to describe an object moving under the influence of gravity in the region outside of some sphere of radius R. Mathematically, this implies

$$x^2 + y^2 + z^2 \geq R^2.$$

Another instance when we encounter non-holonomic constraints is when they depend on the rates of change

$$f(q, \dot{q}, t) = \text{const.}$$

in such a way that we can't integrate the equation to get something of the form $f(q, t) = \text{const.}$ However, since there is no general method to solve systems with non-holonomic constraints, we need to evaluate them for each system individually, and we will not discuss them any further. Additionally, take note that it is conventional to call a constraint which does not explicitly depend on t **scleronomic** (Greek for "rigid") and a constraint with explicit dependence on t **rheonomic** (Greek for "moving").

[12] Lagrange multipliers are discussed in more general terms in Appendix C.

[13] Here, as usual, q is a shorthand notation for all coordinates $q = (q_1, q_2, \ldots)$.

$$\frac{\partial L}{\partial \lambda} = \frac{d}{dt}\left(\frac{\partial L}{\partial \dot\lambda}\right)$$

⟩ Eq. 7.17

$$\frac{\partial\left(L_{\text{free}} + \lambda g(q,t)\right)}{\partial \lambda} = \frac{d}{dt}\left(\frac{\partial\left(L_{\text{free}} + \lambda g(q,t)\right)}{\partial \dot\lambda}\right)$$

⟩ λ and $\dot\lambda$ do not appear in L_{free}.

$$\frac{\partial\left(\lambda g(q,t)\right)}{\partial \lambda} = \frac{d}{dt}\left(\frac{\partial\left(\lambda g(q,t)\right)}{\partial \dot\lambda}\right)$$

⟩ $\dot\lambda$ appears nowhere.

$$\frac{\partial\left(\lambda g(q,t)\right)}{\partial \lambda} = 0$$

⟩

$$g(q,t) = 0. \tag{7.18}$$

This is exactly our constraint formula (Eq. 7.16). Therefore, we can see that by adding the constraint term $\lambda g(q,t)$ to the Lagrangian, the constraint becomes a part of our Lagrangian description.

In addition, by using the Euler-Lagrange equation (Eq. 4.25) for our ordinary coordinates q, we find

$$\frac{\partial L}{\partial q} = \frac{d}{dt}\left(\frac{\partial L}{\partial \dot q}\right)$$

⟩ Eq. 7.17

$$\frac{\partial\left(L_{\text{free}} + \lambda g(q,t)\right)}{\partial q} = \frac{d}{dt}\left(\frac{\partial\left(L_{\text{free}} + \lambda g(q,t)\right)}{\partial \dot q}\right)$$

⟩ $g = g(q,t)$

$$\frac{\partial\left(L_{\text{free}} + \lambda g(q,t)\right)}{\partial q} = \frac{d}{dt}\left(\frac{\partial L_{\text{free}}}{\partial \dot q}\right)$$

⟩ rearranging

$$\frac{\partial L_{\text{free}}}{\partial q} + \lambda \frac{\partial g(q,t)}{\partial q} = \frac{d}{dt}\left(\frac{\partial L_{\text{free}}}{\partial \dot q}\right). \tag{7.19}$$

In Section 4.3.1, we've learned that the term on the right-hand side $\frac{d}{dt}\left(\frac{\partial L_{\text{free}}}{\partial \dot q}\right)$ is analogous to $\frac{dp}{dt}$ in the Newtonian formalism. Moreover, the first term on the left-hand side $\frac{\partial L_{\text{free}}}{\partial q}$ describes the forces.[14] Therefore, the last line in Eq. 7.19 tells us that the effect of the constraint (Eq. 7.16) is to add new forces to the equation of motion. In other words, $\lambda \frac{\partial g(q,t)}{\partial q}$ yields exactly the constraint

[14] Recall that Newton's second law reads $F = \frac{dp}{dt}$.

forces, which are cumbersome to find in the Newtonian formulation of classical mechanics.[15]

This demonstrates nicely why the Lagrangian formalism is extremely powerful. But in the following section we will see that it gets even better. By using a specific feature of the Lagrangian formalism, the description of a system involving constraints becomes even simpler.

To summarize: by using appropriate Lagrange multiplier terms, we can transform the variational problem with constraints to a variational problem without constraints.

As a final comment before we move on, take note that if there is more than one constraint

$$g_1(q,t) = 0$$
$$g_2(q,t) = 0$$
$$\vdots \qquad (7.20)$$

we need to add a Lagrange multiplier term for each constraint:

$$L_{\text{full}} = L_{\text{free}} + \lambda_1 g_1(q,t) + \lambda_2 g_2(q,t) + \ldots \qquad (7.21)$$

Then using the Euler-Lagrange equation for each Lagrange multiplier $\lambda_1, \lambda_2, \ldots$ yields the corresponding constraint equation. Moreover, using the Euler-Lagrange equation for the regular coordinate q yields the equation of motion including all constraint forces:

$$\frac{\partial L_{\text{free}}}{\partial q} + \lambda_1 \frac{\partial g_1(q,t)}{\partial q} + \lambda_2 \frac{\partial g_2(q,t)}{\partial q} + \ldots = \frac{d}{dt}\left(\frac{\partial L_{\text{free}}}{\partial \dot{q}}\right). \qquad (7.22)$$

[15] Once we've derived the equation of motion including the constraint forces, we can proceed as usual, i.e., search for solutions, etc.

7.2.2 Point Transformations and Generalized Coordinates

In the previous section, we've learned that whenever we want to describe a system which is subject to constraints, we simply need to add Lagrange multiplier terms to the Lagrangian. This is already a lot easier than figuring out the constraint forces in the Newtonian formalism. But there is one additional trick we can use in the Lagrangian formalism which makes the whole issue even simpler.

First of all, it's a truism that:

> Nature doesn't care about how we describe her.

This means that we can describe *any* system in *any* formulation of classical mechanics using *any* coordinate system.[16]

But while we have the freedom to choose any coordinate system, not all choices are equally good. In some coordinate systems, the description of a given system can appear incredibly complicated, while in a better suited coordinate system, the description becomes almost trivial.

For example, when we want to describe a pendulum we can, of course, use the usual Cartesian coordinates $(x(t), y(t))$ to describe the location of the bob at any given moment t. But describing the pendulum is much easier using the angle $\phi(t)$ which parameterizes the distance from the equilibrium position. We will see this explicitly in Chapter 9.

In general, we call a transformation from one set of coordinates in configuration space $q = (q_1, q_2, \ldots)$ to a new set of coordinates $q' = (q'_1, q'_2, \ldots)$ a **point transformation**.[17]

We can use the freedom to choose a suitable coordinate system in any formulation of classical mechanics. However, switching coordinate systems is not always easy. In particular, in the Newtonian formalism, we need to transform Newton's second law

---

[16] However, take note that the number of independent coordinates must be at least equal to the number of degrees of freedom within the system. For example, to describe a free particle moving freely in three dimensions, you'll always need three coordinates, say, (x, y, z) or (r, ϕ, θ). Moreover, the formulas which tell us how the new and old coordinates are related must be sufficiently smooth and invertible. This means that it must be possible to switch back to the old coordinates. For example, you can't use $x \to 0$, $y \to 0$, and $z \to 0$. In other words, you can't use completely wild new coordinates. Any reasonable choice of new coordinates will do. In technical terms, the condition on the map from the old to new coordinates is that it has to be a diffeomorphism.

[17] Take note that point transformations must be reversible, i.e., it must be possible to switch back to the original coordinates without problems. In other words, nothing is allowed to get lost through the transformation. Moreover, the total number of generalized coordinates is fixed and determines the degrees of freedom in the system. We need a specific name because later we will talk about transformations in phase space, and like everything else, coordinate transformations work a little differently in phase space. Thus, coordinate transformations in configuration space $q \to q'$ are called **point transformations** and coordinate transformations in phase space $(q, p) \to (q', p')$ are called **canonical transformations** (or alternatively contact transformations). We will talk about canonical transformations in Section 7.3.1.

(Eq. 3.1) by using the transformation rules explicitly.[18] This is often extremely cumbersome.

In contrast, in the Lagrangian formalism switching coordinates is always much simpler no matter how complicated your coordinate choice is. This is where the Lagrangian formalism shows its real strength. In particular:[19]

> The Euler-Lagrange equation (Eq. 4.25) is valid for *any* choice of coordinates.

For example, if we want to describe a given system using Cartesian coordinates (x, y), the equations

$$\frac{\partial L(x, \dot{x}, y, \dot{y})}{\partial x} = \frac{d}{dt}\left(\frac{\partial L(x, \dot{x}, y, \dot{y})}{\partial \dot{x}}\right)$$
$$\frac{\partial L(x, \dot{x}, y, \dot{y})}{\partial y} = \frac{d}{dt}\left(\frac{\partial L(x, \dot{x}, y, \dot{y})}{\partial \dot{y}}\right) \quad (7.23)$$

yield the correct equations of motion in terms of $x(t)$ and $y(t)$.

And if we want to use polar coordinates (r, ϕ), we can use[20]

$$\frac{\partial \tilde{L}(r, \dot{r}, \phi, \dot{\phi})}{\partial r} = \frac{d}{dt}\left(\frac{\partial \tilde{L}(r, \dot{r}, \phi, \dot{\phi})}{\partial \dot{r}}\right)$$
$$\frac{\partial \tilde{L}(r, \dot{r}, \phi, \dot{\phi})}{\partial \phi} = \frac{d}{dt}\left(\frac{\partial \tilde{L}(r, \dot{r}, \phi, \dot{\phi})}{\partial \dot{\phi}}\right)$$
$$(7.24)$$

to calculate the correct equations of motion in terms of $r(t)$ and $\phi(t)$. In other words, no matter which coordinates we choose, the Euler-Lagrange equations always look exactly the same.[21]

It is conventional to rephrase this by saying that in the Lagrangian formalism, we are free to use **generalized coordinates**. In particular, these generalized coordinates are not necessarily quantities with dimension "length". For example, angles ϕ do not have dimension length. Moreover, they are coordinates which we calculate through suitable combinations of the traditional Cartesian coordinates in *configuration space*. As a result, the generalized coordinates that we construct this way

[18] For example, the acceleration vector in Cartesian coordinates $\vec{a}$ which appears in Newton's second law $m\vec{a} = \vec{F}$ reads in spherical coordinates:
$$\vec{a} = \left(\ddot{r} - r\dot{\theta}^2 - r\dot{\phi}^2 \sin^2\theta\right)\vec{e}_r$$
$$+ \left(r\ddot{\theta} + 2\dot{r}\dot{\theta} - r\dot{\phi}^2 \sin\theta\cos\theta\right)\vec{e}_\theta$$
$$+ \left(r\ddot{\phi}\sin\theta + 2\dot{r}\dot{\phi}\sin\theta + 2r\dot{\theta}\dot{\phi}\cos\theta\right)\vec{e}_\phi.$$

Moreover, switching to a new coordinate system which accelerates relative to the old coordinate system usually requires page-long calculations.

[19] We will discuss why this is true below.

[20] In general, we have
$$\tilde{L}(q', \dot{q}') = L\left(q(q'), \dot{q}(q', \dot{q}')\right),$$
where q' and $\dot{q}'$ represent our new coordinates and $q, \dot{q}$ represent the old ones. If you're unsure why and what this means, have a look at Appendix D.

[21] Formulated differently, the Euler-Lagrange equation maintains its general form. In technical terms, we say the Euler-Lagrange equation is **covariant** under arbitrary point transformations $q \to q'$. Covariant means that the *form* of the equation remains unchanged, i.e., no new terms appear after the transformation. But take note that covariant does not mean invariant. The Euler-Lagrange equation is only invariant under invariance transformations for which
$$\tilde{L}(q', \dot{q}') = L(q, \dot{q})$$
holds. Only then, the form of the Euler-Lagrange equation is not only preserved, but it's actually completely equivalent. We discuss this in more detail in Chapter 10 and Appendix D.

are sometimes difficult to interpret because they are simply a convenient way to characterize the *whole* system and there is no longer a clear connection to *individual* objects within the system.[22]

Before we check that this is true, let's discuss why this is an extremely powerful observation.

In the previous section, we've learned that in the Lagrangian formalism, we describe constraints using Lagrange multiplier terms. We use the Lagrangian that we would use if the object in question could move around freely L_{free} and then add a new term L_{con} which encodes the constraint

$$L_{\text{full}} = L_{\text{free}} + L_{\text{con}}. \qquad (7.25)$$

This additional term has the form $L_{\text{con}} = \lambda f(q)$, where λ is our Lagrange multiplier and $f(q)$ a function that describes the constraint.[23]

An extremely clever idea is that we can use our freedom to choose suitable coordinates to pick a specific coordinate system for which the Lagrangian becomes especially simple.

In particular, we can choose new coordinates q' such that $\tilde{f}(q') = f(q(q'))$ vanishes.[24] In words, this means that in terms of the new coordinates q' our object can move around freely. There are no longer any restrictions when it comes to how our object can move along the coordinate axis corresponding to the new q'. Formulated differently, if we find such coordinates for which the constraint term vanishes, we've found the most natural coordinates to describe our system.[25] And rather unsurprisingly, the description of the system in terms of these new coordinates is as simple as it gets.

For example, when we want to describe an object which slides along the surface of a sphere, mathematically, we have the constraint

$$R^2 = x^2 + y^2 + z^2. \qquad (7.26)$$

In words, this means that our object remains attached to the sur-

---

[22] We will discuss in a moment why we sometimes switch to such strange coordinates. The main idea is that we often use whatever coordinates simplify the Lagrangian as much as possible. Since the Euler-Lagrange equations remain unchanged, this is all we care about.

[23] Here, as usual, q is a shorthand notation for all coordinates $q = (q_1, q_2, \ldots)$. Moreover, for example, for a pendulum the function f reads $f(x,y) = x^2 + y^2 - l^2$ where l denotes the length of the pendulum.

[24] In words, this means that we use the explicit formulas which express q in terms of the new coordinates q' to eliminate them from f. Since we can choose our new coordinates q' cleverly such that $\tilde{f}(q') = 0$, we certainly have, in general, $f(q') \neq \tilde{f}(q')$. The expression $f(q')$ means that we use the old formula for f in terms of the old coordinates q and simply replace them directly by q' *without* using the transformation formulas. This is not possible in general, but is possible for some choices of q'. We will see this explicitly in Section 9.2.

[25] Take note that many authors call those coordinates for which the constraint terms vanish *the* generalized coordinates.

face of the sphere. If we now switch to spherical coordinates[26]

$$x = R\sin\phi\cos\theta$$
$$y = R\sin\phi\sin\theta$$
$$z = R\cos\phi \qquad (7.27)$$

the constraint equation becomes trivially true, as we can check explicitly:

$$R^2 = x^2 + y^2 + z^2$$

$$\circlearrowright \text{ Eq. 7.27}$$

$$R^2 = (R\sin\phi\cos\theta)^2 + (R\sin\phi\sin\theta)^2 + (R\cos\phi)^2$$

$$\circlearrowright$$

$$R^2 = R^2\sin^2\phi(\cos^2\theta + \sin^2\theta) + (R\cos\phi)^2$$

$$\circlearrowright \cos^2\theta + \sin^2\theta = 1$$

$$R^2 = R^2\sin^2\phi + R^2\cos^2\phi$$

$$\circlearrowright \sin^2\phi + \cos^2\phi = 1$$

$$R^2 = R^2 \quad \checkmark.$$

[26] Take note that we do not switch to arbitrary spherical coordinates but keep $r = R$ fixed, where R is the radius of our sphere. These are the most natural coordinates for our problem because only for $r = R$ does the constraint become implicitly encoded in the coordinates.

This means that if we switch to these coordinates, the otherwise necessary constraint term $L_{\text{con}} = \lambda(x^2 + y^2 + z^2 - R^2)$ simply vanishes. Therefore, in terms of our new coordinates (θ and ϕ), our object can move around freely and we can describe it without any constraint. This is possible because by choosing the coordinates in Eq. 7.27, we've hard-coded the constraint into our description of the system.

A second reason why the fact that the Euler-Lagrange equations hold in any coordinate system is useful is that it allows us to calculate quite easily what a given system looks like for a non-inertial observer.[27] All we have to do is switch coordinates and then use, as usual, the Euler-Lagrange equation. In contrast, in the Newtonian formalism, we need to introduce so-called fictitious forces to describe what observers in non-inertial frames see.

[27] For an inertial observer an object with zero net force acting upon it is not accelerating. This is Newton's first law, which we discussed in Chapter 3. Non-inertial observers accelerate relative to inertial observers.

To summarize:

▷ The Lagrangian formalism makes it particularly easy to switch coordinates because the Euler-Lagrange equations have exactly the same form in all coordinate systems. Our only task when we want to use new coordinates is to rewrite the Lagrangian.

▷ This allows us to simplify problems involving constraints because we can choose coordinates for which the constraint terms vanish and the Lagrangian therefore becomes especially simple.

Now, why does the Euler-Lagrange equation look exactly the same no matter which coordinates we choose?

The Euler-Lagrange equation reads (Eq. 4.25)

$$\frac{\partial L}{\partial q} - \frac{d}{dt}\left(\frac{\partial L}{\partial \dot{q}}\right) = 0. \tag{7.28}$$

We want to show that no matter which transformation formulas we use,[28]

$$q' = q'(q, t)$$
$$\therefore \quad \dot{q}' = \dot{q}'(q, \dot{q}, t) \tag{7.29}$$

the Euler-Lagrange equation in terms of the new coordinates

$$\frac{\partial \tilde{L}}{\partial q'} - \frac{d}{dt}\left(\frac{\partial \tilde{L}}{\partial \dot{q}'}\right) = 0, \tag{7.30}$$

yields the correct equations of motion.[29] Here $\tilde{L}$ is the Lagrangian that we get by using the replacement rules (Eq. 7.29) in the original Lagrangian:[30]

$$\tilde{L}(q', \dot{q}', t) = L\Big(q(q', t), \dot{q}(q', \dot{q}', t), t\Big). \tag{7.31}$$

---

[28] As usual, q is our shorthand notation for all coordinates. This means that our transformation formulas read

$$q'_i = q'_i(q_1, q_2, q_3, \ldots),$$

i.e., each new coordinate q'_i is possibly a function of all old coordinates $q_1, q_2, \ldots$

[29] This is not trivial because there could be additional terms through the transformation $q \to q'$.

[30] We discussed this already in Section 5.1. (See Eq. 5.5.) Moreover, take note that in order to eliminate the original coordinates $q, \dot{q}$ from the Lagrangian, we need to invert the transformation formulas (Eq. 7.29). In other words, to substitute for $q, \dot{q}$ in L we need to know how exactly they are related to the new coordinates

$$q = q(q', t)$$
$$\therefore \quad \dot{q} = \dot{q}(q', \dot{q}', t).$$

Now, to show that the Euler-Lagrange equation holds for any choice of coordinates, we rewrite the left-hand side of the "new" Euler-Lagrange equation (Eq. 7.30) in such a way that we can see that this expression indeed vanishes provided that the Euler-Lagrange equation in terms of the original coordinates (Eq. 7.28) holds.

For this demonstration, we need a few puzzle pieces:[31]

[31] We need these puzzle pieces to translate the "new" Euler-Lagrange equation (Eq. 7.30) into the language of the old coordinates because only then can we use Eq. 7.28 to show that it is indeed correct. We will see below that each of the puzzle pieces is essential.

First:

$$\dot{q} = \frac{d}{dt} q(q', t)$$

$$= \frac{\partial q}{\partial q'} \frac{dq'}{dt} + \frac{\partial q}{\partial t} \qquad \curvearrowright \text{ chain rule}$$

$$= \frac{\partial q}{\partial q'} \dot{q}' + \frac{\partial q}{\partial t} \qquad \curvearrowright \frac{dq'}{dt} \equiv \dot{q}' \qquad (7.32)$$

which implies

$$\frac{\partial}{\partial \dot{q}'} \dot{q} \stackrel{(7.32)}{=} \frac{\partial}{\partial \dot{q}'} \left(\frac{\partial q}{\partial q'} \dot{q}' + \frac{\partial q}{\partial t} \right) \qquad \curvearrowright$$

$$= \frac{\partial q}{\partial q'} \,. \qquad (7.33)$$

Secondly:

$$\frac{\partial \tilde{L}}{\partial q'} \stackrel{(7.31)}{=} \frac{\partial L \left(q(q', t), \dot{q}(q', \dot{q}', t), t \right)}{\partial q'}$$

$$\curvearrowright \text{ chain rule}$$

$$= \frac{\partial L}{\partial q} \frac{\partial q}{\partial q'} + \frac{\partial L}{\partial \dot{q}} \frac{\partial \dot{q}}{\partial q'} \,. \qquad (7.34)$$

And thirdly:

$$\frac{\partial \tilde{L}}{\partial \dot{q}'} \stackrel{(7.31)}{=} \frac{\partial L \left(q(q', t), \dot{q}(q', \dot{q}', t), t \right)}{\partial \dot{q}'}$$

$$\curvearrowright \text{ chain rule}$$

$$= \frac{\partial L}{\partial \dot{q}} \frac{\partial \dot{q}}{\partial \dot{q}'}$$

$$\curvearrowright \text{ Eq. 7.33}$$

$$= \frac{\partial L}{\partial \dot{q}} \frac{\partial q}{\partial q'} \qquad (7.35)$$

[32] The product rule is derived in Appendix A.1.

which implies:[32]

$$\frac{d}{dt}\frac{\partial \tilde{L}}{\partial \dot{q}'} \stackrel{(7.35)}{=} \frac{d}{dt}\left(\frac{\partial L}{\partial \dot{q}}\frac{\partial q}{\partial q'}\right)$$

↳ product rule

$$= \left(\frac{d}{dt}\frac{\partial L}{\partial \dot{q}}\right)\frac{\partial q}{\partial q'} + \frac{\partial L}{\partial \dot{q}}\left(\frac{d}{dt}\frac{\partial q}{\partial q'}\right)$$

↳

$$= \left(\frac{d}{dt}\frac{\partial L}{\partial \dot{q}}\right)\frac{\partial q}{\partial q'} + \frac{\partial L}{\partial \dot{q}}\frac{\partial \dot{q}}{\partial q'}. \quad (7.36)$$

With these formulas at hand, let's rewrite the "new" Euler-Lagrange equation (Eq. 7.30)

$$0 \stackrel{!}{=} \frac{\partial \tilde{L}}{\partial q'} - \frac{d}{dt}\left(\frac{\partial \tilde{L}}{\partial \dot{q}'}\right)$$

↳ Eq. 7.34, Eq. 7.36

$$= \left(\frac{\partial L}{\partial q}\frac{\partial q}{\partial q'} + \frac{\partial L}{\partial \dot{q}}\frac{\partial \dot{q}}{\partial q'}\right) - \left(\left(\frac{d}{dt}\frac{\partial L}{\partial \dot{q}}\right)\frac{\partial q}{\partial q'} + \frac{\partial L}{\partial \dot{q}}\frac{\partial \dot{q}}{\partial q'}\right)$$

↳ $\frac{\partial L}{\partial \dot{q}}\frac{\partial \dot{q}}{\partial q'}$ (crossed out)

$$= \frac{\partial L}{\partial q}\frac{\partial q}{\partial q'} - \left(\frac{d}{dt}\frac{\partial L}{\partial \dot{q}}\right)\frac{\partial q}{\partial q'}$$

↳ rearranging terms

$$= \left(\frac{\partial L}{\partial q} - \frac{d}{dt}\frac{\partial L}{\partial \dot{q}}\right)\frac{\partial q}{\partial q'}$$

↳ Eq. 7.28

$$= (0)\frac{\partial q}{\partial q'}$$

↳

$$= 0 \checkmark \quad (7.37)$$

We've therefore demonstrated that the "new" Euler-Lagrange equation is indeed fulfilled provided that the Euler-Lagrange equations for the original coordinates (Eq. 7.28) holds. Therefore, Eq. 7.30 correctly describes the system in terms of the new coordinates.

However, take note that we've only shown that the Euler-Lagrange equation has the same *form* no matter which coordinates we use. In technical terms, we say the Euler-Lagrange equation is only *covariant* but not *invariant* under general coordinate transformations. Only for very special coordinate transformations is the Euler-Lagrange equation actually invariant. We will talk a bit more about the distinction between covariance and invariance in Chapter 10.[33]

[33] Covariance and invariance are also discussed in Appendix D.

There is one more thing that we need to talk about before we move on and discuss how we can solve problems using the Hamiltonian formalism.

7.2.3 Gauge Transformations

In addition to the point transformations we discussed in the previous section, we can discover even more freedom in how we describe things. In particular, not only can we change our configuration space coordinates q, but we can also change the Lagrangian *itself* and consequently also the action *itself* without changing the Euler-Lagrange equations. This is possible because the number which the action functional $S[q(t)]$ assigns to each path $q(t)$ is not really important. Instead, the only thing we really care about is the relative difference between the numbers assigned to different paths.

For example, if we add some number C to the action functional

$$S \to S + C, \qquad (7.38)$$

the action of all paths is shifted by the same constant amount. Therefore, the path of least action will still be the path of least action:

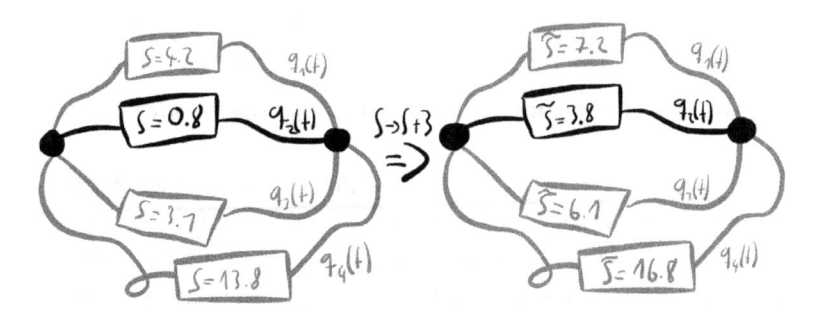

[34] The symbol $\mathbb{R}$ denotes the real numbers.

For a slightly different perspective, recall that the action functional assigns an ordinary number to each possible path. Mathematically, these numbers live simply on a number line denoted by $\mathbb{R}$.[34] However, it doesn't matter how we choose our coordinate system for this space (in which the values the action assigns to paths live). In particular, we can freely choose where we put the origin of this space.

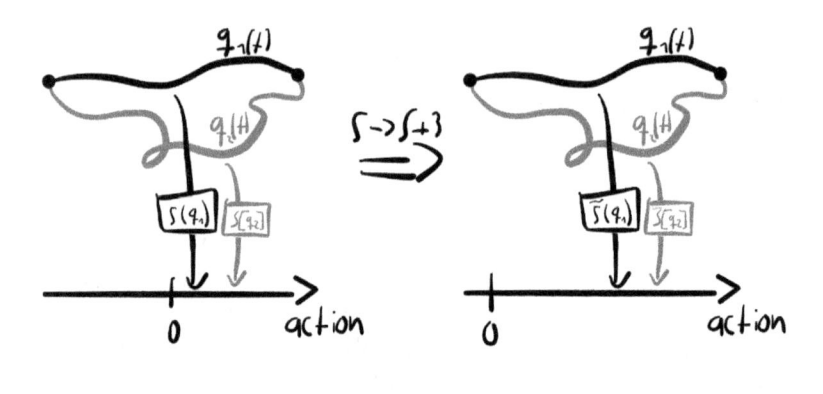

For the Lagrangian, this freedom implies that we can add new terms to the Lagrangian $L \to L + L_G$ as long as these terms L_G fulfill a few special conditions.

In particular, the Euler-Lagrange equation does not change when we add the total time derivative of *any* (differentiable) function F which only depends on the location q and possibly the time coordinate t to the Lagrangian[35]

[35] We will show this explicitly below.

$$L \to L' = L + \frac{dF(q,t)}{dt}.$$
(7.39)

We call this kind of transformation a **gauge transformation**.

To understand why we have this freedom, recall that we derived the Euler-Lagrange equation by invoking the principle of least action.[36] Solutions of the Euler-Lagrange equation yield the path of least action which correctly describes the time evolution

[36] We did this explicitly in Section 4.3.

of the system in question. Moreover, L and L' differ by $\dot{F} \equiv \frac{dF(q,t)}{dt}$. This allows us to calculate:

$$S[q(t)] = \int_{t_i}^{t_f} dt\, L\big(q(t), \dot{q}(t), t\big)$$

$$S'[q(t)] = \int_{t_i}^{t_f} dt\, L'\big(q(t), \dot{q}(t), t\big)$$

$$= \int_{t_i}^{t_f} dt\, \left(L\big(q(t), \dot{q}(t), t\big) + \dot{F}\big(q(t), t\big) \right)$$

$$\curvearrowright \int_a^b dx \frac{df(x)}{dx} = f(b) - f(a)$$

$$= \left(\int_{t_i}^{t_f} dt\, L \right) + F\big(q(t_f), t_f\big) - F\big(q(t_i), t_i\big)$$

$$= S + F\big(q(t_f), t_f\big) - F\big(q(t_i), t_i\big). \qquad (7.40)$$

But $F\big(q(t_f), t_f\big) - F\big(q(t_i), t_i\big)$ is a constant which only depends on $q(t_f), q(t_i), t_f$ and t_i.[37] The initial and final points are fixed. Therefore, this additional term in S' only provides an equal shift of the action for all paths. So the path of least action of S will also be the path of least action of S'.

Therefore, we can conclude that the Euler-Lagrange equation indeed does not change under gauge transformations (Eq. 7.39).

In addition, we can also check *explicitly* that the Euler-Lagrange equation (Eq. 4.25) remains unaffected if we add the total time-derivative of a function $F(q,t)$ to the Lagrangian:[38]

[37] Take note that we are only able to reach this conclusion because F does not depend on $\dot{q}$. If we use a function G which additionally depends on the velocities $G = G(q, \dot{q}, t)$, the additional term we get can be different for different paths because the initial and final velocities $\dot{q}(t_i), \dot{q}(t_f)$ are not fixed. Only the initial and final position in configuration space are fixed when we consider all possible paths between them. We discussed this in Section 4.3.

[38] As usual, feel free to skip the following somewhat lengthy demonstration.

$$\frac{\partial L'}{\partial q} = \frac{d}{dt}\left(\frac{\partial L'}{\partial \dot{q}} \right)$$

$$\curvearrowright L' = L + \frac{dF}{dt}$$

$$\frac{\partial \left(L + \frac{dF}{dt}\right)}{\partial q} = \frac{d}{dt}\left(\frac{\partial \left(L + \frac{dF}{dt}\right)}{\partial \dot{q}} \right)$$

$$\curvearrowright \text{rearranging terms}$$

$$\frac{\partial L}{\partial q} + \frac{\partial}{\partial q}\frac{dF}{dt} = \frac{d}{dt}\left(\frac{\partial L}{\partial \dot{q}} \right) + \frac{d}{dt}\frac{\partial}{\partial \dot{q}}\frac{dF}{dt}. \qquad (7.41)$$

We want to show that this is equal to

$$\frac{\partial L}{\partial q} = \frac{d}{dt}\left(\frac{\partial L}{\partial \dot{q}} \right), \qquad (7.42)$$

144 NO-NONSENSE CLASSICAL MECHANICS

which is the Euler-Lagrange equation for the original Lagrangian. In other words, if we can show that[39]

$$\frac{\partial}{\partial q}\frac{dF}{dt} = \frac{d}{dt}\frac{\partial}{\partial \dot{q}}\frac{dF}{dt}, \quad (7.43)$$

[39] These are the additional terms in Eq. 7.41 which need to vanish such that we get Eq. 7.42.

we've proven that the Euler-Lagrange equation remains unaffected by gauge transformations ($L \to L + \frac{dF(q,t)}{dt}$, Eq. 7.39).

To show this, we use the general formula for the total derivative of a function $F(q,t)$:[40]

[40] The total derivative is explained in Appendix A.3.

$$\frac{dF}{dt} = \frac{\partial F}{\partial q}\frac{dq}{dt} + \frac{\partial F}{\partial t}. \quad (7.44)$$

This yields

$$\frac{\partial}{\partial q}\frac{dF}{dt} \stackrel{!}{=} \frac{d}{dt}\frac{\partial}{\partial \dot{q}}\frac{dF}{dt} \quad \text{Eq. 7.43}$$

$$\frac{\partial}{\partial q}\left(\frac{\partial F}{\partial q}\frac{dq}{dt} + \frac{\partial F}{\partial t}\right) \stackrel{!}{=} \frac{d}{dt}\frac{\partial}{\partial \dot{q}}\left(\frac{\partial F}{\partial q}\frac{dq}{dt} + \frac{\partial F}{\partial t}\right) \quad \circlearrowleft \text{ Eq. 7.44}$$

$$(7.45)$$

On the right-hand side, we can use the fact that since F only depends on q and t but not on $\dot{q}$, we have $\frac{\partial}{\partial \dot{q}}\frac{\partial F}{\partial t} = 0$ and $\frac{\partial}{\partial \dot{q}}\frac{\partial F}{\partial q} = 0$. Moreover, $\frac{\partial}{\partial \dot{q}}\frac{dq}{dt} = 1$ since $\frac{dq}{dt} \equiv \dot{q}$. Therefore[41], $\frac{\partial}{\partial \dot{q}}\frac{\partial F}{\partial q}\frac{dq}{dt} = \frac{\partial F}{\partial q}$ and using this our equation reads

[41] We use the product rule

$$\frac{\partial}{\partial \dot{q}}\frac{\partial F}{\partial q}\frac{dq}{dt}$$
$$= \left(\frac{\partial}{\partial \dot{q}}\frac{\partial F}{\partial q}\right)\frac{dq}{dt} + \frac{\partial F}{\partial q}\left(\frac{\partial}{\partial \dot{q}}\frac{dq}{dt}\right)$$
$$= (0)\frac{dq}{dt} + \frac{\partial F}{\partial q}(1)$$
$$= \frac{\partial F}{\partial q}.$$

The product rule is derived in Appendix A.1.

$$\frac{\partial}{\partial q}\left(\frac{\partial F}{\partial q}\frac{dq}{dt} + \frac{\partial F}{\partial t}\right) \stackrel{!}{=} \frac{d}{dt}\frac{\partial F}{\partial q}. \quad (7.46)$$

To show that the left-hand and right-hand sides are indeed equal, we use our general formula for the total time derivative once more. This time, we use it for $\frac{\partial F}{\partial q} \equiv G(q,t)$ which appears on the right-hand side:[42]

[42] G is simply the name we give to the expression on the right-hand side.

$$\frac{d}{dt}G = \frac{\partial G}{\partial q}\frac{dq}{dt} + \frac{\partial G}{\partial t} \quad \text{general formula}$$

$$\circlearrowleft G \equiv \frac{\partial F}{\partial q}$$

$$\frac{d}{dt}\frac{\partial F}{\partial q} = \frac{\partial}{\partial q}\frac{\partial F}{\partial q}\frac{dq}{dt} + \frac{\partial}{\partial t}\frac{\partial F}{\partial q} \quad (7.47)$$

If we plug this into Eq. 7.46, we find:

$$\frac{\partial}{\partial q}\left(\frac{\partial F}{\partial q}\frac{dq}{dt} + \frac{\partial F}{\partial t}\right) \stackrel{!}{=} \frac{d}{dt}\frac{\partial F}{\partial q} \qquad \text{this is Eq. 7.46}$$

$$\downarrow \text{ Eq. 7.47}$$

$$\frac{\partial}{\partial q}\left(\frac{\partial F}{\partial q}\frac{dq}{dt} + \frac{\partial F}{\partial t}\right) \stackrel{!}{=} \frac{\partial}{\partial q}\frac{\partial F}{\partial q}\frac{dq}{dt} + \frac{\partial}{\partial t}\frac{\partial F}{\partial q}$$

$$\downarrow \text{ rearranging terms}$$

$$\frac{\partial}{\partial q}\frac{\partial F}{\partial q}\frac{dq}{dt} + \frac{\partial}{\partial q}\frac{\partial F}{\partial t} \stackrel{!}{=} \frac{\partial}{\partial q}\frac{\partial F}{\partial q}\frac{dq}{dt} + \frac{\partial}{\partial t}\frac{\partial F}{\partial q}$$

$$\downarrow \quad \frac{\partial}{\partial q}\frac{\partial F}{\partial t} = \frac{\partial}{\partial t}\frac{\partial F}{\partial q}$$

$$\frac{\partial}{\partial q}\frac{\partial F}{\partial q}\frac{dq}{dt} + \frac{\partial}{\partial t}\frac{\partial F}{\partial q} \stackrel{!}{=} \frac{\partial}{\partial q}\frac{\partial F}{\partial q}\frac{dq}{dt} + \frac{\partial}{\partial t}\frac{\partial F}{\partial q} \quad \checkmark$$

In the final step, we used the fact that partial derivatives commute, which is a property that holds in general.[43]

[43] The freedom to switch partial derivatives is known as **Schwarz's theorem**.

We've therefore successfully demonstrated that the Euler-Lagrange equation is completely unaffected by gauge transformations.

As a final comment, take note that we can also add an arbitrary constant C to the Lagrangian[44]

$$L \to L' = L + C \tag{7.48}$$

[44] This is, of course, a special case of the more general transformation in Eq. 7.39. For example, for $F(q,t) = Ct$ we have $\frac{dF}{dt} = C$.

because the Euler-Lagrange equation (Eq. 4.25) only involves derivatives of the Lagrangian:

$$\frac{\partial L'}{\partial q} = \frac{d}{dt}\left(\frac{\partial L'}{\partial \dot{q}}\right)$$

$$\downarrow \quad L' = L + C$$

$$\frac{\partial(L+C)}{\partial q} = \frac{d}{dt}\left(\frac{\partial(L+C)}{\partial \dot{q}}\right)$$

$$\downarrow \text{ rearranging terms}$$

$$\frac{\partial L}{\partial q} + \frac{\partial}{\partial q}C = \frac{d}{dt}\left(\frac{\partial L}{\partial \dot{q}}\right) + \frac{d}{dt}\frac{\partial}{\partial \dot{q}}C$$

$$\downarrow \quad C \text{ is constant}$$

$$\frac{\partial L}{\partial q} = \frac{d}{dt}\left(\frac{\partial L}{\partial \dot{q}}\right) \quad \checkmark \tag{7.49}$$

This freedom to add constants to the Lagrangian is equivalent to our freedom to shift any given potential by a constant amount[45]

[45] We use a minus sign because this allows us to see the equality to what we did previously. But the minus sign is not important because we could always use a different constant $\tilde{C} = -C$ instead and then we would have

$$V \to V' = V - C = V + \tilde{C}.$$

$$V \to V' = V - C \tag{7.50}$$

since[46]

$$L = T - V \to L' = T - (V - C) = T - V + C = L + C. \tag{7.51}$$

[46] We have this freedom because we can only measure the resulting forces but not potentials themselves. We have $F = -\frac{\partial V}{\partial q}$ and therefore a constant shift (Eq. 7.50) leaves the resulting force unchanged

$$\begin{aligned} F' &= -\frac{\partial V'}{\partial q} \\ &= -\frac{\partial (V - C)}{\partial q} \\ &= -\frac{\partial V}{\partial q} = F. \end{aligned}$$

We talked about this in Section 2.1.3.

But in the Lagrangian formalism, we have a lot more freedom. Not only we can add some constant but also the total derivative of an arbitrary function (as long as it only depends on q and t).

Moreover, not only we can add a constant to the action but we can also multiply it by a constant (non-zero) factor

$$S \to kS. \tag{7.52}$$

Such a constant factor has no influence on which path is the path of least action because the action involved in all paths is scaled equally:

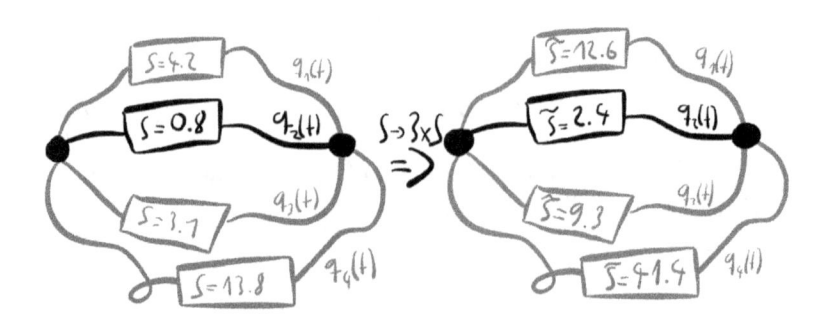

Therefore, rescaling the action by some arbitrary constant factor has no influence on the dynamics of the system.

This observation directly implies that we can also scale the Lagrangian by a constant factor

$$L \to L' = kL. \tag{7.53}$$

We can also see this by observing that the Lagrangian appears on both sides of the Euler-Lagrange equation (Eq. 4.25) and therefore, such a constant factor simply drops out:

$$\frac{\partial L'}{\partial q} = \frac{d}{dt}\left(\frac{\partial L'}{\partial \dot{q}}\right)$$

⟶ $L' = kL$, Eq. 7.53

$$\frac{\partial (kL)}{\partial q} = \frac{d}{dt}\left(\frac{\partial (kL)}{\partial \dot{q}}\right)$$

⟶ rearranging terms

$$k\frac{\partial L}{\partial q} = k\frac{d}{dt}\left(\frac{\partial L}{\partial \dot{q}}\right)$$

⟶ $\cancel{k}$

$$\frac{\partial L}{\partial q} = \frac{d}{dt}\left(\frac{\partial L}{\partial \dot{q}}\right) \checkmark \tag{7.54}$$

To summarize: there are always infinitely many Lagrangians describing the same system. The situation is somewhat analogous to how there is usually more than one word to describe a particular thing *within* a given language (synonyms).[47]

[47] Recall that the various formulations of classical mechanics are analogous to how we can describe the same thing using different languages. Here we describe the same thing in different ways using the *same* language.

More specifically:

▷ We always have the freedom to add a constant to our Lagrangian (Eq. 7.48)

$$L \to L' = L + C \tag{7.55}$$

and to multiply it by some constant factor (Eq. 7.53)

$$L \to L' = kL \tag{7.56}$$

because such transformations have no effect on the Euler-Lagrange equation.

▷ But there is even more freedom since we can add the total derivative of a general (differentiable) function F to our Lagrangian, as long as it only depends on the location q and time coordinate t (Eq. 7.39)

$$L \to L' = L + \frac{dF(q,t)}{dt}. \tag{7.57}$$

Above, we've demonstrated explicitly that any such term drops out from the Euler-Lagrange equation and therefore has no effect on the equations of motion.

▷ In general, we call such transformations of the Lagrangian itself gauge transformations.

▷ Additionally, we discussed this from a slightly different perspective. In Eq. 7.40 we demonstrated explicitly that a gauge transformation leads to an equal shift of the action for all paths. Therefore, the path of least action is the same for the original and shifted action functional. In physical terms, this means that we find exactly the same (correct) path using the original and the gauge transformed action functional.

Now, it's time to move on and talk about how all this works in the Hamiltonian formalism.

7.3 The Hamiltonian Algorithm

If we want to use the Hamiltonian formalism to describe a given system, our main task is to write down the correct Hamiltonian H. However, usually this requires a few intermediate steps.

▷ First of all, we write down the Lagrangian $L = T - V$.

▷ Then we calculate the corresponding generalized momenta $p = \frac{\partial L}{\partial \dot{q}}$.

▷ Afterwards, we solve the formulas we found in the previous step for $\dot{q}$.

▷ This allows us to eliminate $\dot{q}$ from $L = L(q, \dot{q}, t)$ and to use the formula

$$H(q, p, t) = p\dot{q}(q, p, t - L(q, \dot{q}(q, p, t), t). \quad (7.58)$$

Finally, as soon as we've successfully calculated the Hamiltonian H, we can use Hamilton's equations (Eq. 5.15)

$$\frac{dp}{dt} = -\frac{\partial H}{\partial q}$$
$$\frac{dq}{dt} = \frac{\partial H}{\partial p}, \quad (7.59)$$

to derive the equation of motion.

As a concrete example, let's discuss the falling ball a third time.

In Section 7.2, we already calculated the correct Lagrangian (Eq. 7.10)

$$L = \frac{1}{2}m(v_x^2 + v_y^2 + v_z^2) - mgz. \tag{7.60}$$

Using this Lagrangian, we can calculate the conjugate momenta:[48]

[48] We use again the notation $v_q = \dot{q}$.

$$p_x = \frac{\partial L}{\partial v_x} \stackrel{(7.60)}{=} \frac{\partial\left(\frac{1}{2}m(v_x^2 + v_y^2 + v_z^2) - mgz\right)}{\partial v_x}$$

$$= mv_x$$

$$p_y = \frac{\partial L}{\partial v_y} \stackrel{(7.60)}{=} \frac{\partial\left(\frac{1}{2}m(v_x^2 + v_y^2 + v_z^2) - mgz\right)}{\partial v_y}$$

$$= mv_y$$

$$p_z = \frac{\partial L}{\partial v_z} \stackrel{(7.60)}{=} \frac{\partial\left(\frac{1}{2}m(v_x^2 + v_y^2 + v_z^2) - mgz\right)}{\partial v_z}$$

$$= mv_z. \tag{7.61}$$

Inverting these equations yields

$$v_x = \frac{p_x}{m}, \quad v_y = \frac{p_y}{m}, \quad v_z = \frac{p_z}{m}. \tag{7.62}$$

This allows us to derive the Hamiltonian[49]

[49] Take note that the final result is indeed simply $H = T + V$. However, the method we used here to derive the Hamiltonian is more general, while $H = T + V$ is not always correct.

$$H = \vec{p}\cdot\dot{\vec{q}} - L$$

↷ Eq. 7.60

$$= (p_x v_x + p_y v_y + p_z v_z) - \left(\frac{1}{2}m(v_x^2 + v_y^2 + v_z^2) - mgz\right)$$

↷ Eq. 7.62

$$= \left(p_x \frac{p_x}{m} + p_y \frac{p_y}{m} + p_z \frac{p_z}{m}\right)$$

$$- \left(\frac{1}{2}m\left(\left(\frac{p_x}{m}\right)^2 + \left(\frac{p_y}{m}\right)^2 + \left(\frac{p_z}{m}\right)^2\right) - mgz\right)$$

↷ rearranging terms

$$= \frac{1}{2}\left(\frac{p_x^2}{m} + \frac{p_y^2}{m} + \frac{p_z^2}{m}\right) + mgz. \tag{7.63}$$

Finally, we can put this Hamiltonian into Hamilton's equations. For the x-coordinate we find

$$\frac{dp_x}{dt} = -\frac{\partial H}{\partial x}$$
↱ Eq. 7.63

$$= -\frac{\partial\left(\frac{1}{2}\left(\frac{p_x^2}{m} + \frac{p_y^2}{m} + \frac{p_z^2}{m}\right) + mgz\right)}{\partial x}$$
↱

$$= 0 \qquad (7.64)$$

$$\frac{dx}{dt} = \frac{\partial H}{\partial p_x}$$
↱ Eq. 7.63

$$= \frac{\partial\left(\frac{1}{2}\left(\frac{p_x^2}{m} + \frac{p_y^2}{m} + \frac{p_z^2}{m}\right) + mgz\right)}{\partial p_x}$$
↱

$$= \frac{p_x}{m}. \qquad (7.65)$$

We can then use Eq. 7.65 to eliminate p_x from Eq. 7.64:

$$\frac{dp_x}{dt} = 0 \qquad \text{this is Eq. 7.64}$$
↱ Eq. 7.65

$$\frac{d\left(m\frac{dx}{dt}\right)}{dt} = 0$$
↱ $m = $ const.

$$m\frac{d^2x}{dt^2} = 0. \qquad (7.66)$$

Analogously, for the y-coordinate we find

$$m\frac{d^2y}{dt^2} = 0. \qquad (7.67)$$

Moreover, for the z-coordinate we find

$$\frac{dp_z}{dt} = -\frac{\partial H}{\partial z}$$

↷ Eq. 7.63

$$= -\frac{\partial\left(\frac{1}{2}\left(\frac{p_x^2}{m} + \frac{p_y^2}{m} + \frac{p_z^2}{m}\right) + mgz\right)}{\partial z}$$

↷

$$= -mg \tag{7.68}$$

$$\frac{dz}{dt} = \frac{\partial H}{\partial p_z}$$

↷ Eq. 7.63

$$= \frac{\partial\left(\frac{1}{2}\left(\frac{p_x^2}{m} + \frac{p_y^2}{m} + \frac{p_z^2}{m}\right) + mgz\right)}{\partial p_z}$$

↷

$$= \frac{p_z}{m}. \tag{7.69}$$

We can then use Eq. 7.69 to eliminate p_z from Eq. 7.68:

$$\frac{dp_z}{dt} = -mg \qquad \text{this is Eq. 7.68}$$

↷ Eq. 7.69

$$\frac{d\left(m\frac{dz}{dt}\right)}{dt} = -mg$$

↷ $m = \text{const.}$

$$m\frac{d^2z}{dt^2} = -mg. \tag{7.70}$$

These three equations (Eq. 7.66, Eq. 7.67, Eq. 7.70) are exactly the equations of motion which we already derived using the Newtonian formalism (Eq. 7.3) and Lagrangian formalism (Eq. 7.11, Eq. 7.12, Eq. 7.13). However, the Hamiltonian way to derive them was far more cumbersome. So you'll probably not be surprised when I tell you that the real advantage of the Hamiltonian method is not a practical but mostly a conceptual one.

To understand this, we need to talk once more about coordinate and gauge transformations.

7.3.1 Canonical Transformations and Canonical Coordinates

We already discussed in Section 7.2.2 that switching coordinates is often a powerful tool. In particular, we've learned that the Euler-Lagrange equation remains valid no matter which coordinates we choose.

Now here is some good news: (almost) exactly the same statement is true for the Hamiltonian formalism and Hamilton's equations. Since we end up with the Hamiltonian formalism by modifying the Lagrangian formalism a little bit, this shouldn't be too surprising.[50]

However, we need to be careful. In the Lagrangian formalism, we describe any system using locations q in configuration space, and the coordinates we use to describe them can be chosen freely. In the Hamiltonian formalism, we describe systems in terms of the locations q *and* momenta p. So the freedom to choose coordinates for the locations q carries over to the Hamiltonian formalism.[51] But additionally, a general coordinate transformation in phase space intermingles location and momentum coordinates while a transformation in configuration space only mixes location coordinates.

In other words, in the Hamiltonian formalism we have even more freedom to choose coordinates.[52]

But we don't have absolute freedom. Although we act as if q and p are completely independent parameters, there is a close connection between them since:[53]

$$p \equiv \frac{\partial L}{\partial \dot{q}}. \qquad (7.71)$$

In particular, there is a direct connection between each location coordinate q (to be precise: its rate of change $\dot{q}$) and each momentum coordinate p. For exactly this reason, we often call p in this context *conjugate* momenta.

[50] This was shown in Chapter 5.

[51] We will show this explicitly below.

[52] Formulated differently, in the Hamiltonian formulation of mechanics, we have twice as many variables and this widens the realm of possible transformations. This is one of the advantages of the Hamiltonian formalism. In Section 11.1, we will see that it's possible to use this freedom to switch to coordinates in which the equations of motion are trivial to solve.

[53] This is the general definition of the conjugate momentum as given in Eq. 4.29.

Since there is a close connection between the two, we can't transform our location and momentum coordinates completely arbitrarily and still expect that Hamilton's equations remain valid. The interplay between the locations and momenta is precisely what determines the dynamics within the system and therefore we can't change them too wildly.[54]

In other words, there are important restrictions which our coordinate transformations $q \to Q(q,p)$ and $p \to P(q,p)$ must fulfill.[55] Good transformations respect the structure of the Hamiltonian formalism and we call them **canonical transformations**.[56]

Moreover, we call good coordinates which we can use in the Hamiltonian formalism **canonical coordinates**.[57] Only if we use canonical coordinates can we use Hamilton's equations in their usual, simple form. In other words, a canonical transformation is a switch in coordinates which does not "break" our Hamiltonian framework.

The term "canonical" is used because Hamilton's equations are often the starting point for advanced investigations in classical mechanics and are therefore so important that they are often called the canonical equations.

Our task is therefore to calculate which conditions suitable coordinates (and transformations between them) have to fulfill. Let's start by deriving the conditions canonical coordinates have to fulfill. In a second step, we will then take a closer look at canonical transformations.

We want to derive a condition on our new coordinates (Q, P) which makes sure that Hamilton's equations remain valid. In particular, this means that we now investigate which properties our new coordinates must have in order that the time evolution of the system is described by the simple equations[58]

---

[54] From a slightly different perspective, we can imagine that if we change to wildly new coordinates and then try to calculate the dynamics as usual using Hamilton's equations, we find that the system behaves completely differently. Formulated differently, if we switch to such coordinates and don't modify Hamilton's equations appropriately, we end up with new dynamical rules which no longer describe our original system appropriately.

[55] Take note that transformations in configuration space are of the form $q \to Q(q)$. The corresponding transformation of $\dot{q}$ follows automatically since $\dot{q}$ is defined as the time derivative of q.

[56] Take note that sometimes the word **contact transformation** is used instead of canonical transformation.

[57] With enough mathematical shrewdness you can, of course, use non-canonical coordinates. But this requires a modification of the Hamiltonian formalism.

[58] We are only interested in coordinates in which Hamilton's equations remain valid. What we have here are Hamilton's equations (Eq. 5.15):

$$\frac{dp}{dt} = -\frac{\partial H}{\partial q}$$
$$\frac{dq}{dt} = \frac{\partial H}{\partial p},$$

in terms of the new coordinates. The important thing is that the form of the equations is exactly the same and no new terms are necessary on the left-hand or right-hand sides.

$$\frac{dQ}{dt} = \frac{\partial \tilde{H}}{\partial P}$$
$$\frac{dP}{dt} = -\frac{\partial \tilde{H}}{\partial Q}, \qquad (7.72)$$

where $\tilde{H}(Q,P) = H\big(q(Q,P), p(Q,P)\big)$ is the original Hamiltonian rewritten in terms of the new coordinates.[59]

In general, we have some formulas which relate the old coordinates (q,p) to the new coordinates (Q,P):[60]

$$Q = Q(q,p), \qquad P = P(q,p). \qquad (7.73)$$

In Section 5.1.1, we learned that the time-evolution of *any* phase space function is described by Hamilton's general equation of motion (Eq. 5.36), which we recite here for convenience:

$$\frac{d}{dt}F = \{F, H\}. \qquad (7.74)$$

Since our new coordinates (Q,P) are defined as functions of the old coordinates (q,p) (Eq. 7.73), we have

$$\frac{d}{dt}Q = \{Q, H\}_{q,p} = \frac{\partial Q}{\partial q}\frac{\partial H}{\partial p} - \frac{\partial Q}{\partial p}\frac{\partial H}{\partial q}$$
$$\frac{d}{dt}P = \{P, H\}_{q,p} = \frac{\partial P}{\partial q}\frac{\partial H}{\partial p} - \frac{\partial P}{\partial p}\frac{\partial H}{\partial q}. \qquad (7.75)$$

Here we've added subscripts to the Poisson brackets to remind us that we need to evaluate them in terms of the old coordinates (q,p).[61] We are not finished, because we still have expressions involving the old coordinates and the original Hamiltonian H on the right-hand side.

To change this, we can invert the formulas in Eq. 7.73.[62] This yields formulas which allow us to calculate what the Hamiltonian looks like in terms of the new coordinates:

$$H(q,p) \xrightarrow{\text{Eq. 7.73}} H\big(q(Q,P), p(Q,P)\big) \equiv \tilde{H}(Q,P). \qquad (7.76)$$

This allows us to evaluate the partial derivatives that appear in Eq. 7.75 using the chain rule:[63]

---

[59] Once more it is important to keep in mind that, in general, we have $\tilde{H}(Q,P) \neq H(Q,P)$. We discussed this subtlety in Section 5.1 and you can find a few more comments in Appendix D.

[60] In words, this means that our new coordinates are defined as some function of the old coordinates. For example, if we switch from Cartesian coordinates to polar coordinates, we have

$$r(x,y) = \sqrt{x^2 + y^2}$$
$$\phi(x,y) = \arccos\left(\frac{x}{\sqrt{x^2+y^2}}\right).$$

[61]
$$\{A,B\}_{q,p} \equiv \frac{\partial A}{\partial q}\frac{\partial B}{\partial p} - \frac{\partial A}{\partial p}\frac{\partial B}{\partial q}.$$

This is the definition of the Poisson bracket as given in Eq. 5.33.

[62] For example, for the switch from Cartesian to polar coordinates this would mean that we use

$$x = r\cos\phi$$
$$y = r\sin\phi.$$

[63] Recall that our goal is to calculate the equations of motion for our new coordinates Q and P which no longer depend on the old coordinates. See Eq. 7.72. The chain rule is explained in Appendix A.4.

$$\frac{\partial H\big(q(Q,P),p(Q,P)\big)}{\partial q} \stackrel{(7.76)}{=} \frac{\partial \tilde{H}(Q,P)}{\partial q}$$

⟩ chain rule

$$= \frac{\partial \tilde{H}}{\partial Q}\frac{\partial Q}{\partial q} + \frac{\partial \tilde{H}}{\partial P}\frac{\partial P}{\partial q}$$

$$\frac{\partial H\big(q(Q,P),p(Q,P)\big)}{\partial p} \stackrel{(7.76)}{=} \frac{\partial \tilde{H}(Q,P)}{\partial p}$$

⟩ chain rule

$$= \frac{\partial \tilde{H}}{\partial Q}\frac{\partial Q}{\partial p} + \frac{\partial \tilde{H}}{\partial P}\frac{\partial P}{\partial p}. \tag{7.77}$$

Substituting these expression into the first line in Eq. 7.75 yields

$$\frac{d}{dt}Q = \frac{\partial Q}{\partial q}\frac{\partial H}{\partial p} - \frac{\partial Q}{\partial p}\frac{\partial H}{\partial q}$$

⟩ Eq. 7.77

$$= \frac{\partial Q}{\partial q}\left(\frac{\partial \tilde{H}}{\partial Q}\frac{\partial Q}{\partial p} + \frac{\partial \tilde{H}}{\partial P}\frac{\partial P}{\partial p}\right) - \frac{\partial Q}{\partial p}\left(\frac{\partial \tilde{H}}{\partial Q}\frac{\partial Q}{\partial q} + \frac{\partial \tilde{H}}{\partial P}\frac{\partial P}{\partial q}\right)$$

⟩ $\cancel{\dfrac{\partial Q}{\partial q}\dfrac{\partial \tilde{H}}{\partial Q}\dfrac{\partial Q}{\partial p}}$

$$= \frac{\partial Q}{\partial q}\frac{\partial \tilde{H}}{\partial P}\frac{\partial P}{\partial p} - \frac{\partial Q}{\partial p}\frac{\partial \tilde{H}}{\partial P}\frac{\partial P}{\partial q}$$

⟩ rearranging terms

$$= \frac{\partial \tilde{H}}{\partial P}\left(\frac{\partial Q}{\partial q}\frac{\partial P}{\partial p} - \frac{\partial Q}{\partial p}\frac{\partial P}{\partial q}\right)$$

⟩ definition of { , }, Eq. 5.33

$$= \frac{\partial \tilde{H}}{\partial P}\{Q,P\}_{q,p}. \tag{7.78}$$

Analogously, for the second line in Eq. 7.75 we find

$$\frac{d}{dt}P = \frac{\partial P}{\partial q}\frac{\partial H}{\partial p} - \frac{\partial P}{\partial p}\frac{\partial H}{\partial q}$$

⟩ Eq. 7.77

$$= \frac{\partial P}{\partial q}\left(\frac{\partial \tilde{H}}{\partial Q}\frac{\partial Q}{\partial p} + \frac{\partial \tilde{H}}{\partial P}\frac{\partial P}{\partial p}\right) - \frac{\partial P}{\partial p}\left(\frac{\partial \tilde{H}}{\partial Q}\frac{\partial Q}{\partial q} + \frac{\partial \tilde{H}}{\partial P}\frac{\partial P}{\partial q}\right)$$

⟩ $\cancel{\dfrac{\partial P}{\partial q}\dfrac{\partial \tilde{H}}{\partial P}\dfrac{\partial P}{\partial p}}$

$$= \frac{\partial P}{\partial q}\frac{\partial \tilde{H}}{\partial Q}\frac{\partial Q}{\partial p} - \frac{\partial P}{\partial p}\frac{\partial \tilde{H}}{\partial Q}\frac{\partial Q}{\partial q}$$

⟩ rearranging terms

$$= \frac{\partial \tilde{H}}{\partial Q}\left(\frac{\partial P}{\partial q}\frac{\partial Q}{\partial p} - \frac{\partial P}{\partial p}\frac{\partial Q}{\partial q}\right)$$

⟩ rearranging terms

$$= -\frac{\partial \tilde{H}}{\partial Q}\left(\frac{\partial Q}{\partial q}\frac{\partial P}{\partial p} - \frac{\partial Q}{\partial p}\frac{\partial P}{\partial q}\right)$$

⟩ definition of { , }, Eq. 5.33

$$= -\frac{\partial \tilde{H}}{\partial Q}\{Q,P\}_{q,p}. \tag{7.79}$$

[64] For your convenience: Eq. 7.72 reads
$$\frac{dQ}{dt} = \frac{\partial \tilde{H}}{\partial P}$$
$$\frac{dP}{dt} = -\frac{\partial \tilde{H}}{\partial Q}.$$

[65] Strictly speaking, we have
$$\{Q,P\}_{q,p} = k,$$
where k is some arbitrary non-zero number, which would work, too, because we can always absorb such a constant in our new Hamiltonian
$$\tilde{\tilde{H}} \equiv k\tilde{H}.$$
Therefore, such a constant would not affect the form of the equations. However, it is conventional to *choose* $k = 1$ because usually we are not interested in such a change of scale.

[66] These are indeed extremely useful coordinates, at least when we want to describe the harmonic oscillator. We will discuss this in Section 8.5.

Now compare this to Eq. 7.72, which is what we want our equations of motion to look like.[64]

We can see that Eq. 7.78/Eq. 7.79 and Eq. 7.72 are equal if[65]

$$\boxed{\{Q,P\}_{q,p} = 1.} \qquad (7.80)$$

We can therefore conclude that only if our new coordinates (Q, P) fulfill this condition will Hamilton's equations remain valid. Therefore, this is the condition good, new coordinates must fulfill. Formulated differently, Eq. 7.80 is the **defining condition of canonical coordinates**. Any set of coordinates which fulfills this condition is canonical. For this reason, we call Eq. 7.80 the **canonical Poisson bracket relation**.

So whenever someone hands us some wild new coordinates like, for example,[66]

$$Q = \sqrt{\frac{2p}{m\omega}} \sin q$$
$$P = \sqrt{2m\omega p} \cos q \qquad (7.81)$$

we can check immediately if this is a valid choice by calculating

$$\{Q,P\}_{q,p} = \frac{\partial Q}{\partial q}\frac{\partial P}{\partial p} - \frac{\partial Q}{\partial p}\frac{\partial P}{\partial q}$$

$$= \frac{\partial\left(\sqrt{\frac{2p}{m\omega}}\sin q\right)}{\partial q}\frac{\partial\left(\sqrt{2m\omega p}\cos q\right)}{\partial p} - \frac{\partial\left(\sqrt{\frac{2p}{m\omega}}\sin q\right)}{\partial p}\frac{\partial\left(\sqrt{2m\omega p}\cos q\right)}{\partial q}$$

$$= \sqrt{\frac{2p}{m\omega}}\cos q \sqrt{\frac{m\omega}{2p}}\cos q - \frac{1}{\sqrt{2m\omega p}}\sin q \sqrt{2m\omega p}(-\sin q)$$

$$= \cos^2 q + \sin^2 q$$

$$= 1 \quad \checkmark \qquad (7.82)$$

This tells us that the new coordinates in Eq. 7.81 are indeed canonical and therefore are a valid choice.

ALGORITHMS AND TOOLS   157

For an alternative perspective on why the condition in Eq. 7.80 is so important, we can investigate how Hamilton's general equation (Eq. 5.36) changes under phase space transformations.[67] Our goal is to show that as long as $\{Q,P\}_{q,p} = 1$, we have

$$\frac{d\tilde{F}}{dt} = \{\tilde{F}, \tilde{H}\}_{Q,P}, \qquad (7.83)$$

[67] Reminder: Eq. 5.36 reads
$$\frac{dF}{dt} = \{F, H\}_{q,p}.$$

where $\tilde{F}$ and $\tilde{H}$ are the functions that we get if we substitute q, p with their expressions in terms of Q and P, i.e., $q(Q,P), p(P,Q)$:

$$\tilde{F}(Q,P) \equiv F\big(q(Q,P), p(Q,P)\big)$$
$$\tilde{H}(Q,P) \equiv H\big(q(Q,P), p(Q,P)\big). \qquad (7.84)$$

If we can demonstrate this, we have proven that Hamilton's general equation is valid for any choice of coordinates that fulfill $\{Q,P\}_{q,p} = 1$, i.e., all canonical coordinates.

So let's check this explicitly:

$$\frac{dF}{dt} = \{F, H\}_{q,p}$$

⟩ substitute old coordinates

$$\therefore \quad \frac{dF\big(q(Q,P), p(Q,P)\big)}{dt} = \{F\big(q(Q,P), p(Q,P)\big), H\big(q(Q,P), p(Q,P)\big)\}_{q,p}$$

⟩ Eq. 7.84

$$\therefore \quad \frac{d\tilde{F}}{dt} = \{\tilde{F}, \tilde{H}\}_{q,p}$$

⟩ def. of $\{\,,\,\}$, Eq. 5.34

$$\therefore \quad \frac{d\tilde{F}}{dt} = \frac{\partial \tilde{F}}{\partial q}\frac{\partial \tilde{H}}{\partial p} - \frac{\partial \tilde{F}}{\partial p}\frac{\partial \tilde{H}}{\partial q}$$

⟩ Eq. 7.77

$$\therefore \quad \frac{d\tilde{F}}{dt} = \left(\frac{\partial \tilde{F}}{\partial Q}\frac{\partial Q}{\partial q} + \frac{\partial \tilde{F}}{\partial P}\frac{\partial P}{\partial q}\right)\left(\frac{\partial \tilde{H}}{\partial Q}\frac{\partial Q}{\partial p} + \frac{\partial \tilde{H}}{\partial P}\frac{\partial P}{\partial p}\right)$$
$$- \left(\frac{\partial \tilde{F}}{\partial Q}\frac{\partial Q}{\partial p} + \frac{\partial \tilde{F}}{\partial P}\frac{\partial P}{\partial p}\right)\left(\frac{\partial \tilde{H}}{\partial Q}\frac{\partial Q}{\partial q} + \frac{\partial \tilde{H}}{\partial P}\frac{\partial P}{\partial q}\right)$$

⟩ expanding the products

$$\therefore \quad \frac{d\tilde{F}}{dt} = \cancel{\frac{\partial \tilde{F}}{\partial Q}\frac{\partial Q}{\partial q}\frac{\partial \tilde{H}}{\partial Q}\frac{\partial Q}{\partial p}} + \frac{\partial \tilde{F}}{\partial Q}\frac{\partial Q}{\partial q}\frac{\partial \tilde{H}}{\partial P}\frac{\partial P}{\partial p}$$
$$+ \frac{\partial \tilde{F}}{\partial P}\frac{\partial P}{\partial q}\frac{\partial \tilde{H}}{\partial Q}\frac{\partial Q}{\partial p} + \cancel{\frac{\partial \tilde{F}}{\partial P}\frac{\partial P}{\partial q}\frac{\partial \tilde{H}}{\partial P}\frac{\partial P}{\partial p}}$$
$$- \cancel{\frac{\partial \tilde{F}}{\partial Q}\frac{\partial Q}{\partial p}\frac{\partial \tilde{H}}{\partial Q}\frac{\partial Q}{\partial q}} - \frac{\partial \tilde{F}}{\partial Q}\frac{\partial Q}{\partial p}\frac{\partial \tilde{H}}{\partial P}\frac{\partial P}{\partial q}$$
$$- \frac{\partial \tilde{F}}{\partial P}\frac{\partial P}{\partial p}\frac{\partial \tilde{H}}{\partial Q}\frac{\partial Q}{\partial q} - \cancel{\frac{\partial \tilde{F}}{\partial P}\frac{\partial P}{\partial p}\frac{\partial \tilde{H}}{\partial P}\frac{\partial P}{\partial q}}$$

$$\therefore \quad \frac{d\tilde{F}}{dt} = \frac{\partial \tilde{F}}{\partial Q}\frac{\partial Q}{\partial q}\frac{\partial \tilde{H}}{\partial P}\frac{\partial P}{\partial p} + \frac{\partial \tilde{F}}{\partial P}\frac{\partial P}{\partial q}\frac{\partial \tilde{H}}{\partial Q}\frac{\partial Q}{\partial p}$$
$$- \frac{\partial \tilde{F}}{\partial Q}\frac{\partial Q}{\partial p}\frac{\partial \tilde{H}}{\partial P}\frac{\partial P}{\partial q} - \frac{\partial \tilde{F}}{\partial P}\frac{\partial P}{\partial p}\frac{\partial \tilde{H}}{\partial Q}\frac{\partial Q}{\partial q}$$

⟩ sharp observation

$$\therefore \quad \frac{d\tilde{F}}{dt} = \left(\frac{\partial \tilde{F}}{\partial Q}\frac{\partial \tilde{H}}{\partial P} - \frac{\partial \tilde{F}}{\partial P}\frac{\partial \tilde{H}}{\partial Q} \right)\left(\frac{\partial Q}{\partial q}\frac{\partial P}{\partial p} - \frac{\partial Q}{\partial p}\frac{\partial P}{\partial q} \right)$$

⟩ def. of $\{,\}_{q,p}$ and $\{,\}_{Q,P}$

$$\therefore \quad \frac{d\tilde{F}}{dt} = \{\tilde{F}, \tilde{H}\}_{Q,P} \{Q, P\}_{q,p} \qquad (7.85)$$

Therefore, Hamilton's general equation (Eq. 5.36) remains valid for all new coordinates for which $\{Q, P\}_{q,p} = 1$ (Eq. 7.80) holds.

Now that we've established how we can identify canonical coordinates, we want to understand the transformations between them a little better.

In other words, we have talked so far about canonical *coordinates* and now we want to talk about canonical *transformations*. Of course, these two notions are directly connected and each new set of canonical coordinates Q, P comes necessarily with a canonical transformation that tells us how the old coordinates are related to the new ones $Q = Q(q, p)$, $P = P(q, p)$.

But it makes sense to study canonical transformations in a bit more detail because we have not yet developed any understanding of how we can use them.

There are three approaches that we will talk about in the following sections. Each approach will reveal something important about canonical transformations.

▷ First, to get some rough understanding for what canonical transformations are, we translate the transformations we already know (point transformations, gauge transformations) into the language of the Hamiltonian formalism.[68]

[68] We discussed point transformations in Section 7.2.2 and gauge transformations in Section 7.2.3.

We will discuss how this works in Section 7.3.2.

▷ A second approach is to study (infinitesimally) small canonical transformations. Studying (infinitesimally) small transformations is a standard approach in physics because understanding them is much easier than understanding large ones.

Moreover, large transformations can be generated by combining many small ones. Often, a small set of small transformations is sufficient to understand all possible large transformations.[69]

[69] In mathematical terms, this means that we study the corresponding Lie algebra instead of the Lie group itself.

Infinitesimal transformations are especially important in the context of Noether's theorem which establishes a connection between conserved quantities and symmetries. Noether's theorem is the topic of Chapter 10, and in Section 7.3.3, we will talk about infinitesimal canonical transformations.

▷ Finally, there is a completely general method for constructing canonical transformations which is especially important for practical applications. This method is based on so-called generating functions and allows us to find exactly those canonical transformations which simplify a given problem.

We will talk about generating functions and how they help us to categorize canonical transformations in Section 7.3.4.

We will start by studying transformations that we already know. In particular, we will see that point transformations in configuration space correspond to a particular type of canonical transformation in phase space.

Moreover, the second type of allowed transformation that we discovered in the Lagrangian formalism (gauge transformations) corresponds to another type of canonical transformations:[70]

[70] We discussed point transformations in Section 7.2.2 and gauge transformations in Section 7.2.3.

All this is summarized in the following diagram:

configuration space:

point transformation	gauge transformation
$q \to Q = Q(q)$	$q \to Q = q$
$\dot{q} \to \dot{Q} = \frac{\partial Q}{\partial q}\dot{q}$	$\dot{q} \to \dot{Q} = \dot{q}$
$L(q,\dot{q},t) \to \tilde{L} = \tilde{L}(Q,\dot{Q},t)$	$L \to L' = L + \frac{dF(q,t)}{dt}$

phase space:

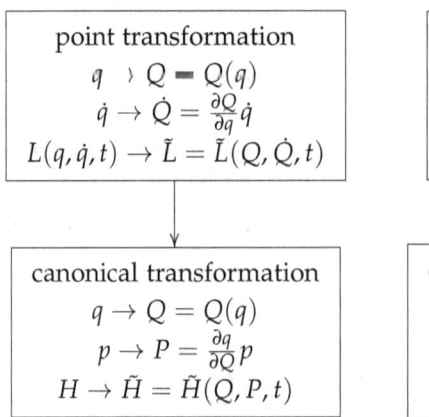

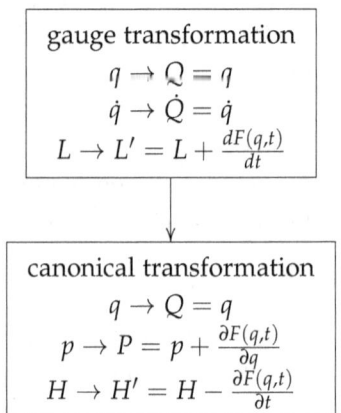

canonical transformation	canonical transformation
$q \to Q = Q(q)$	$q \to Q = q$
$p \to P = \frac{\partial q}{\partial Q}p$	$p \to P = p + \frac{\partial F(q,t)}{\partial q}$
$H \to \tilde{H} = \tilde{H}(Q,P,t)$	$H \to H' = H - \frac{\partial F(q,t)}{\partial t}$

7.3.2 Canonical Point and Gauge Transformations

First of all, let's check that point transformations in configuration space do indeed correspond to canonical transformations in phase space. In addition, this allows us to understand exactly how a change of coordinates (e.g., rectangular coordinates to polar coordinates) affects our phase space description. In particular, we see that such a change in coordinates modifies the Hamiltonian and the canonical momenta directly. Moreover, this allows us to understand that point transformations are one important class of canonical transformations.[71]

A point transformation in configuration space reads

$$q \to Q = Q(q). \tag{7.86}$$

In words, this means that our new coordinates Q are defined as some function of the old ones.[72]

If we now want to translate a point transformation into a phase space transformation, we need to be careful. Naively, we could think that since a point transformation only influences the locations q, we have for the momentum simply $p \to P = p$. But this is wrong, as we will see in a moment.

First of all, take note that a point transformation (Eq 7.86) di-

---

[71] As mentioned above, we'll talk about a second class of canonical transformations afterwards and about canonical transformations in more general terms in Section 7.3.4.

[72] For example, when we switch to polar coordinates, we have

$$r(x,y) = \sqrt{x^2 + y^2}.$$

rectly modifies the velocities, too:

$$\dot{q} \to \dot{Q} = \dot{Q}(q) = \frac{d}{dt}Q(q) = \frac{\partial Q(q)}{\partial q}\frac{dq}{dt} = \frac{\partial Q(q)}{\partial q}\dot{q}. \quad (7.87)$$

This tells us[73]

[73] We will need this result in a moment.

$$\frac{\partial \dot{Q}}{\partial \dot{q}} \stackrel{(7.87)}{=} \frac{\partial}{\partial \dot{q}}\left(\frac{\partial Q(q)}{\partial q}\dot{q}\right) = \frac{\partial Q(q)}{\partial q}. \quad (7.88)$$

This, in turn, implies that the momentum p is also directly modified whenever we perform a point transformation

$$p = \frac{\partial L}{\partial \dot{q}} \to P = \frac{\partial \tilde{L}(Q,\dot{Q}))}{\partial \dot{Q}}$$

$\circlearrowright \quad \tilde{L}(Q,\dot{Q}) \equiv L\big(q(Q),\dot{q}(Q,\dot{Q})\big)$

$$= \frac{\partial L(q(Q),\dot{q}(Q,\dot{Q}))}{\partial \dot{Q}}$$

$\circlearrowright \quad$ chain rule

$$= \frac{\partial L(q,\dot{q})}{\partial \dot{q}}\frac{\partial \dot{q}}{\partial \dot{Q}}$$

$\circlearrowright \quad$ Eq. 7.88 and $p \equiv \dfrac{\partial L(q,\dot{q})}{\partial \dot{q}}$

$$= p\frac{\partial q}{\partial Q}. \quad (7.89)$$

This tells us that the new momentum P is not simply the old one p. Instead, there is a factor $\frac{\partial q}{\partial Q}$ which directly depends on the point transformation $q \to Q(q)$.

Now, with Eq 7.86 and Eq 7.89 in hand, we can check that point transformations really correspond to canonical transformations in phase space.

In Section 7.3.1, we discovered that canonical coordinates are characterized by the condition (Eq 7.80)

$$\{Q,P\}_{q,p} = 1, \quad (7.90)$$

where $\{\,,\,\}_{q,p}$ denotes the Poisson bracket evaluated with respect to the old coordinates. Therefore, to check whether a point transformation leads to a canonical phase space transformation, we have to evaluate:[74]

[74] To get to the second line, we use that the defining feature of a point transformation is that $q \to Q(q)$. In words, this means that our new location coordinates do not depend on the momentum p.

$$\{Q,P\}_{q,p} = \frac{\partial Q}{\partial q}\frac{\partial P}{\partial p} - \frac{\partial Q}{\partial p}\frac{\partial P}{\partial q}$$

↪ $\frac{\partial Q}{\partial p} = 0$

$$= \frac{\partial Q}{\partial q}\frac{\partial P}{\partial p}$$

↪ Eq. 7.89

$$= \frac{\partial Q}{\partial q}\frac{\partial}{\partial p}\left(p\frac{\partial q}{\partial Q}\right)$$

↪ $\frac{\partial p}{\partial p} = 1$

$$= \frac{\partial Q}{\partial q}\frac{\partial q}{\partial Q}$$

↪ $\frac{\partial q}{\partial Q} = \frac{1}{\frac{\partial Q}{\partial q}}$

$$= 1 \checkmark \qquad (7.91)$$

Therefore, we can conclude, as promised, that point transformations indeed correspond to canonical phase space transformations. From a slightly different perspective, we can see this by investigating how the Hamiltonian changes under a point transformation:

$$H(q,p) \stackrel{(5.12)}{\equiv} p\dot{q} - \tilde{L}(q,\dot{q})$$
$$\Rightarrow \tilde{H}(Q,P) = p(Q,P)\dot{q}(Q,\dot{Q},P) - L\Big(q(Q,P),\dot{q}(Q,\dot{Q},P)\Big)$$

↪ Eq. 7.87, Eq. 7.89

$$= P\frac{\partial Q}{\partial q}\dot{Q}\frac{\partial q}{\partial Q} - L\Big(q(Q,P),\dot{q}(Q,\dot{Q},P)\Big)$$

↪ $\frac{\partial Q}{\partial q}\frac{\partial q}{\partial Q} = 1$

$$= P\dot{Q} - L\Big(q(Q,P),\dot{q}(Q,\dot{Q},P)\Big). \qquad (7.92)$$

We can rearrange the terms to calculate the corresponding transformed "Hamiltonian-Lagrangian"

$$L\Big(q(Q,P),\dot{q}(Q,\dot{Q},P)\Big) = P\dot{Q} - \tilde{H}(Q,P). \qquad (7.93)$$

With this Lagrangian in hand, we can follow exactly the same steps as in Section 5.1 to calculate the resulting equation of motion using the principle of least action.

Since this Hamiltonian has exactly the same form as the untransformed Hamiltonian $L_H = p\dot{q} - H$ (Eq. 5.19), this yields

Hamilton's equations once more, but with a Hamiltonian $\tilde{H}$.[75]

Now secondly, let's check what gauge transformations look like in the Hamiltonian formalism.

The first main observation is that the definition of the conjugate momentum $p = \frac{\partial L}{\partial \dot{q}}$ (Eq. 4.29) directly involves the Lagrangian L. Moreover, recall that in Section 7.2.3, we've discovered that there isn't one unique Lagrangian for a given system. Instead, there are infinitely many since we can always add the total time derivative of an arbitrary function $F = F(q,t)$:[76]

$$L \to L' = L + \frac{dF}{dt}. \quad (7.94)$$

If we combine these two puzzle pieces, we reach the conclusion that each such choice leads to a different conjugate momentum (Eq. 4.29):

$$p = \frac{\partial L}{\partial \dot{q}} \quad \to \quad P = \frac{\partial L'}{\partial \dot{q}}$$

$$= \frac{\partial\left(L + \frac{dF(q,t)}{dt}\right)}{\partial \dot{q}} \quad \circlearrowleft \text{ Eq. 7.94}$$

$$= \frac{\partial L}{\partial \dot{q}} + \frac{\partial}{\partial \dot{q}} \frac{dF(q,t)}{dt} \quad \circlearrowleft \text{ rearranging terms}$$

$$= \frac{\partial L}{\partial \dot{q}} + \frac{\partial}{\partial \dot{q}}\left(\frac{\partial F}{\partial q}\frac{dq}{dt} + \frac{\partial F}{\partial t}\right) \quad \circlearrowleft \frac{d}{dt}F(q,t) = \frac{\partial F}{\partial q}\frac{dq}{dt} + \frac{\partial F}{\partial t}$$

$$= \frac{\partial L}{\partial \dot{q}} + \frac{\partial F}{\partial q} \quad \circlearrowleft \frac{\partial}{\partial \dot{q}}\frac{\partial F}{\partial t} = 0 \text{ and } \frac{\partial}{\partial \dot{q}}\frac{dq}{dt} = 1$$

$$= p + \frac{\partial F}{\partial q}. \quad \circlearrowleft \frac{\partial L}{\partial \dot{q}} = p \quad (7.95)$$

Moreover, the location coordinates remain completely unaffected by a gauge transformation:[77]

$$q \to Q(q) = q. \quad (7.96)$$

Using these formulas we can check whether a gauge transformation indeed corresponds to a canonical phase space transformation:

[75] Specifically, we find
$$\frac{dQ}{dt} = \frac{\partial \tilde{H}}{\partial P}$$
$$\frac{dP}{dt} = -\frac{\partial \tilde{H}}{\partial Q}.$$

[76] This is how we defined a gauge transformation in Eq. 7.39.

[77] This is, after all, one of the defining features of a gauge transformation. While a point transformation modifies the location coordinates (plus conjugate momentum coordinates), a gauge transformation only modifies the Lagrangian and, as a result, also the conjugate momenta.

$$\{Q,P\} = \frac{\partial Q}{\partial q}\frac{\partial P}{\partial p} - \frac{\partial Q}{\partial p}\frac{\partial P}{\partial q}$$
⟳ Eq. 7.95, Eq. 7.96

$$= \frac{\partial q}{\partial q}\frac{\partial \left(p + \frac{\partial F}{\partial q}\right)}{\partial p} - \frac{\partial q}{\partial p}\frac{\partial \left(p + \frac{\partial F}{\partial q}\right)}{\partial q}$$
⟳ $\frac{\partial q}{\partial q} = 1, \frac{\partial q}{\partial p} = 0$

$$= \frac{\partial \left(p + \frac{\partial F}{\partial q}\right)}{\partial p}$$
⟳ rearranging terms

$$= \frac{\partial p}{\partial p} + \frac{\partial}{\partial p}\frac{\partial F}{\partial q}$$
⟳ $\frac{\partial}{\partial p}\frac{\partial F}{\partial q} = 0$ since $F = F(q,t)$

$$= \frac{\partial p}{\partial p}$$
⟳

$$= 1 \checkmark \tag{7.97}$$

So we can conclude that the gauge transformations (which we discovered in Section 7.2.3) are indeed canonical transformations. Once more we can understand this from a slightly different perspective by investigating how the Hamiltonian changes under a gauge transformation:

$$H(q,p) \stackrel{(5.12)}{\equiv} p\dot{q} - L$$
⟳ Eq. 7.94, Eq. 7.95, Eq. 7.96

$$= (P - \frac{\partial F}{\partial Q})\dot{Q} - (L' - \frac{dF}{dt})$$
⟳ rearranging

$$= P\dot{Q} - L' - \frac{\partial F}{\partial Q}\dot{Q} + \frac{dF}{dt}$$
⟳ $H' \equiv P\dot{Q} - L'$

$$= H'(Q,P) - \frac{\partial F}{\partial Q}\dot{Q} + \frac{dF}{dt}$$
⟳ $\frac{dF}{dt} = \frac{\partial F}{\partial Q}\frac{dQ}{dt} + \frac{\partial F}{\partial t}$

$$= H'(Q,P) - \frac{\partial F}{\partial Q}\dot{Q} + \left(\frac{\partial F}{\partial Q}\frac{dQ}{dt} + \frac{\partial F}{\partial t}\right)$$
⟳ $\frac{\partial F}{\partial Q}\frac{dQ}{dt}$ cancels

$$= H'(Q,P) + \frac{\partial F}{\partial t} \tag{7.98}$$

Therefore, we learn here that while the Lagrangian changes by a total derivative under a gauge transformation:

$$L \to L' \stackrel{(7.94)}{=} L + \frac{dF}{dt}, \qquad (7.99)$$

the Hamiltonian changes by a partial derivative

$$H \to H' \stackrel{(7.98)}{=} H - \frac{\partial F}{\partial t}. \qquad (7.100)$$

In particular, this implies that our Hamiltonian is only modified at all if F depends explicitly on t.[78]

Moreover, we can rearrange the terms in

$$H' \equiv P\dot{Q} - L' \qquad (7.101)$$

to calculate the corresponding transformed "Hamiltonian Lagrangian"

$$L' = P\dot{Q} - H'. \qquad (7.102)$$

With this Lagrangian in hand, we can follow exactly the same steps as in Section 5.1 to calculate the resulting equation of motion using the principle of least action. And since the form of this Hamiltonian only differs from the form of the untransformed Hamiltonian Lagrangian $L = p\dot{q} - H$ (Eq. 5.19) by the total derivative $\frac{dF}{dt}$ (c.f. Eq. 7.94), this yields Hamilton's equation once more but this time with Hamiltonian H'.[79]

Using these results, we can check *explicitly* that Hamilton's equations are unchanged under gauge transformations.[80]

We can demonstrate the invariance of Hamilton's first equation (Eq. 5.15)

$$\frac{dq}{dt} = \frac{\partial H}{\partial p} \qquad (7.103)$$

[78] Maybe you find it puzzling that we can shift the Hamiltonian since it usually represents the total energy. To understand this, recall that we can always shift any potential by a constant amount because the resulting force remains unchanged. This, in turn, implies that we can shift the potential energy of an object analogously. (In physical terms, this is possible because only potential energy differences are important and not the absolute value of the potential energy itself.) And therefore, we can also shift the Hamiltonian.

[79] Recall that in Section 7.2.3, we discovered that a term of the form $\frac{dF}{dt}$ in the Lagrangian has no influence on the resulting equations of motion.

[80] We've already shown this above. But still it is instructive to see how it works out if we simply calculate everything by brute force. Nevertheless, as usual, feel free to skip the following calculation because we are not really learning something new.

without major problems:

$$\frac{dQ}{dt} = \frac{\partial H'}{\partial P}$$
⤳ Eq. 7.100

$$\therefore \quad \frac{dQ}{dt} = \frac{\partial \left(H - \frac{\partial F}{\partial t} \right)}{\partial P}$$
⤳ rearranging terms

$$\therefore \quad \frac{dQ}{dt} = \frac{\partial H}{\partial P} - \frac{\partial}{\partial P}\left(\frac{\partial F}{\partial t}\right)$$
⤳ $F = F(q, t)$ does not depend on P.

$$\therefore \quad \frac{dQ}{dt} = \frac{\partial H}{\partial P}$$
⤳ chain rule

$$\therefore \quad \frac{dQ}{dt} = \frac{\partial H}{\partial p}\frac{\partial p}{\partial P}$$
⤳ $Q = q$, Eq. 7.96

$$\therefore \quad \frac{dq}{dt} = \frac{\partial H}{\partial p}\frac{\partial p}{\partial P}$$
⤳ $P = p + \frac{\partial F}{\partial q}$, Eq. 7.95

$$\therefore \quad \frac{dq}{dt} = \frac{\partial H}{\partial p}\frac{\partial}{\partial P}\left(P - \frac{\partial F}{\partial q}\right)$$
⤳ $\frac{\partial P}{\partial P} = 1$, $\frac{\partial}{\partial P}\frac{\partial F(q,t)}{\partial q} = 0$

$$\therefore \quad \frac{dq}{dt} = \frac{\partial H}{\partial p} \checkmark \tag{7.104}$$

Demonstrating the invariance of Hamilton's second equation (Eq. 5.15)

$$\frac{dp}{dt} = -\frac{\partial H}{\partial q} \tag{7.105}$$

is a bit more difficult.

After the gauge transformation, the equation reads

$$\frac{dP}{dt} = -\frac{\partial H'}{\partial Q}$$
⤳ Eq. 7.100

$$\therefore \quad \frac{dP}{dt} = -\frac{\partial \left(H - \frac{\partial F}{\partial t} \right)}{\partial Q}$$
⤳ rearranging terms

$$\therefore \quad \frac{dP}{dt} = -\frac{\partial H}{\partial Q} + \frac{\partial}{\partial Q}\left(\frac{\partial F}{\partial t}\right), \tag{7.106}$$

and we need to be *very* careful when we evaluate the first term on the right-hand side. The key observation is that $P = p + \frac{\partial F}{\partial q}$

(Eq. 7.95) implies that p can also be a function of Q since $F = F(Q,t)$ depends on Q.[81] Therefore:

$$\begin{aligned}
\frac{\partial H'(Q,P)}{\partial Q} &= \frac{\partial H\big(q(Q), p(Q,P), t\big)}{\partial Q} \\
&= \frac{\partial H(q,p,t)}{\partial q}\frac{\partial q(Q)}{\partial Q} + \frac{\partial H(q,p,t)}{\partial p}\frac{\partial p(Q)}{\partial Q} & & \circlearrowright \text{ chain rule} \\
&= \frac{\partial H}{\partial q}\frac{\partial Q}{\partial Q} + \frac{\partial H}{\partial p}\frac{\partial}{\partial Q}\left(P - \frac{\partial F}{\partial q}\right) & & \circlearrowright \text{ Eq. 7.96, Eq. 7.95} \\
&= \frac{\partial H}{\partial q} - \frac{\partial H}{\partial p}\frac{\partial^2 F}{\partial q^2} & & \circlearrowright \frac{\partial Q}{\partial Q} = 1 \text{ and } \frac{\partial P}{\partial Q} = 0 \\
&= \frac{\partial H}{\partial q} - \dot{q}\frac{\partial^2 F}{\partial q^2}. & & \circlearrowright \text{ Eq. 7.103} \\
& & & \quad (7.107)
\end{aligned}$$

Moreover, for the term on the left-hand side, we find

$$\begin{aligned}
\frac{dP}{dt} &\stackrel{(7.95)}{=} \frac{d}{dt}\left(p + \frac{\partial F}{\partial q}\right) \\
&= \frac{d}{dt}p + \frac{d}{dt}\frac{\partial F}{\partial q} & & \circlearrowright \text{ rearranging terms} \\
&= \frac{d}{dt}p + \frac{d}{dt}G & & \circlearrowright G \equiv \frac{\partial F}{\partial q} \text{ (definition)} \\
&= \frac{d}{dt}p + \frac{\partial G}{\partial q}\frac{dq}{dt} + \frac{\partial G}{\partial t} & & \circlearrowright \text{ evaluating total derivative} \\
&= \frac{d}{dt}p + \left(\frac{\partial}{\partial q}\frac{\partial F}{\partial q}\right)\frac{dq}{dt} + \frac{\partial}{\partial t}\frac{\partial F}{\partial q} & & \circlearrowright \text{ substituting back} \\
&= \frac{d}{dt}p + \frac{\partial^2 F}{\partial q^2}\dot{q} + \frac{\partial}{\partial t}\frac{\partial F}{\partial q}. & & \circlearrowright \text{ simplifying the notation} \\
& & & \quad (7.108)
\end{aligned}$$

Using these two results we can rewrite our gauge transformed second Hamilton equation (Eq. 7.106) as follows:

[81] Maybe you are now a bit scared because you would've made a mistake here. So it will probably make you feel better when I tell you that a few years ago there was a little controversy because several professional physicists missed this subtlety. To learn more about this story, see

Z. K. Silagadze. Gauge transformations are canonical transformations, redux, 2014

$$\frac{dP}{dt} = -\frac{\partial H}{\partial Q} + \frac{\partial}{\partial Q}\frac{\partial F}{\partial t}$$

⤷ Eq. 7.107 and Eq. 7.108

$$\frac{d}{dt}p + \frac{\partial^2 F}{\partial q^2}\dot{q} + \frac{\partial}{\partial t}\frac{\partial F}{\partial q} = -\frac{\partial H}{\partial q} + \dot{q}\frac{\partial^2 F}{\partial q^2} + \frac{\partial}{\partial Q}\frac{\partial F}{\partial t}$$

⤷ Eq. 7.96

$$\frac{d}{dt}p + \frac{\partial^2 F}{\partial q^2}\dot{q} + \frac{\partial}{\partial t}\frac{\partial F}{\partial q} = -\frac{\partial H}{\partial q} + \dot{q}\frac{\partial^2 F}{\partial q^2} + \frac{\partial}{\partial q}\frac{\partial F}{\partial t}$$

⤷ $\frac{\partial}{\partial q}\frac{\partial}{\partial t} = \frac{\partial}{\partial t}\frac{\partial}{\partial q}$

$$\frac{d}{dt}p + \frac{\partial^2 F}{\partial q^2}\dot{q} + \frac{\partial}{\partial t}\frac{\partial F}{\partial q} = -\frac{\partial H}{\partial q} + \dot{q}\frac{\partial^2 F}{\partial q^2} + \frac{\partial}{\partial t}\frac{\partial F}{\partial q}$$

⤷ ~~$\frac{\partial}{\partial t}\frac{\partial F}{\partial q}$~~ and ~~$\frac{\partial^2 F}{\partial q^2}\dot{q}$~~

$$\frac{d}{dt}p = -\frac{\partial H}{\partial q} \checkmark \qquad (7.109)$$

So once more we've successfully shown that Hamilton's equations are unchanged by gauge transformations.

The following diagram summarizes what we've discovered in this section:

configuration space:

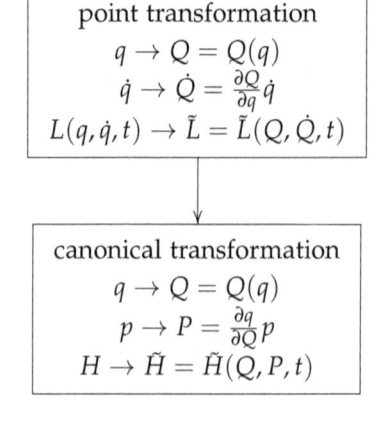

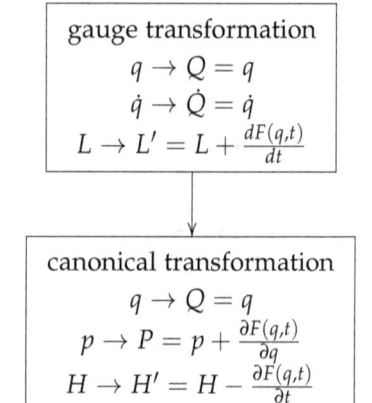

point transformation	gauge transformation
$q \to Q = Q(q)$	$q \to Q = q$
$\dot{q} \to \dot{Q} = \frac{\partial Q}{\partial q}\dot{q}$	$\dot{q} \to \dot{Q} = \dot{q}$
$L(q,\dot{q},t) \to \tilde{L} = \tilde{L}(Q,\dot{Q},t)$	$L \to L' = L + \frac{dF(q,t)}{dt}$

phase space:

canonical transformation	canonical transformation
$q \to Q = Q(q)$	$q \to Q = q$
$p \to P = \frac{\partial q}{\partial Q}p$	$p \to P = p + \frac{\partial F(q,t)}{\partial q}$
$H \to \tilde{H} = \tilde{H}(Q,P,t)$	$H \to H' = H - \frac{\partial F(q,t)}{\partial t}$

Next, let's talk about tiny canonical transformations. This allows us to develop a much deeper understanding of the underlying structure.

7.3.3 Infinitesimal Canonical Transformation

Mathematically, we can write a tiny transformation of our phase space coordinates as[82]

$$q \to Q = q + \epsilon D(q,p)$$
$$p \to P = p + \epsilon E(q,p), \quad (7.110)$$

where ϵ is an infinitesimally small number.[83] In words, this means that our new coordinates Q, P are defined as the old ones plus a little change.

In the previous sections, we've already learned that we can't transform q and p completely independently. The relationship between q and p contains important information about the dynamics within the system. Therefore, we need to be careful to keep it intact.

Thus, let's see what the defining condition in Eq. 7.80

$$\{Q,P\}_{q,p} = 1 \quad (7.111)$$

implies for our new objects $D(q,p)$, $E(q,p)$ which represent the canonical transformation here:[84]

[82] Take note that if we include more than one dimension, we have

$$q_i \to Q_i = q_i + \epsilon D_i(q_j, p_j)$$
$$p_i \to P_i = p_i + \epsilon E_i(q_j, p_j).$$

By choosing the functions D_i and E_i appropriately, not only can we write translations like this, but we can also write, for example, rotations of our phase space coordinate system.

[83] In some sense, an infinitesimal number is a very small quantity which is greater than zero yet less than any positive number. Formulated differently, an infinitesimal number is as small as possible but not yet zero. This implies for an infinitesimal number ϵ that we have $\epsilon^2 = 0$.

[84] We will see below that this is a smart thing to investigate.

$$\{Q,P\}_{q,p} = \frac{\partial Q}{\partial q}\frac{\partial P}{\partial p} - \frac{\partial Q}{\partial p}\frac{\partial P}{\partial q}$$

↷ Eq. 7.110

$$= \frac{\partial(q+\epsilon D)}{\partial q}\frac{\partial(p+\epsilon E)}{\partial p} - \frac{\partial(q+\epsilon D)}{\partial p}\frac{\partial(p+\epsilon E)}{\partial q}$$

↷ $\frac{\partial q}{\partial p}=0, \frac{\partial q}{\partial q}=1, \frac{\partial p}{\partial q}=0, \frac{\partial p}{\partial p}=1$

$$= \left(1+\epsilon\frac{\partial D}{\partial q}\right)\left(1+\epsilon\frac{\partial E}{\partial p}\right) - \left(0+\epsilon\frac{\partial D}{\partial p}\right)\left(0+\epsilon\frac{\partial E}{\partial q}\right)$$

↷

$$= 1 + \epsilon\frac{\partial D}{\partial q} + \epsilon\frac{\partial E}{\partial p} + \epsilon^2\frac{\partial D}{\partial q}\frac{\partial E}{\partial p} - \epsilon^2\frac{\partial D}{\partial p}\frac{\partial E}{\partial q}$$

↷ $\epsilon \ll 1 \Rightarrow \epsilon^2 \approx 0$

$$= 1 + \epsilon\frac{\partial D}{\partial q} + \epsilon\frac{\partial E}{\partial p}. \quad (7.112)$$

If the transformation in Eq. 7.110 is indeed canonical, $\{Q,P\}_{q,p} = 1$ must hold and therefore by comparison with Eq. 7.112, we can conclude that

$$\frac{\partial D}{\partial q} + \frac{\partial E}{\partial p} \stackrel{!}{=} 0. \quad (7.113)$$

This implies that there must be a connection between D and E.[85] Specifically, we can conclude that Eq. 7.113 is fulfilled if

[85] This is hardly surprising because whenever we mix location and momentum coordinates using a canonical transformation, we must be careful how we mix them. This is how we ended up with the condition in Eq. 7.80 ($\{Q,P\}_{q,p} = 1$) in Section 7.3.1.

$$D = \frac{\partial G}{\partial p}, \qquad E = -\frac{\partial G}{\partial q}, \qquad (7.114)$$

where $G = G(q,p)$ is some arbitrary function, because

$$0 \stackrel{!}{=} \frac{\partial D}{\partial q} + \frac{\partial E}{\partial p}$$

$\circlearrowright$ Eq. 7.114

$$= \frac{\partial}{\partial q}\frac{\partial G}{\partial p} - \frac{\partial}{\partial p}\frac{\partial G}{\partial q}$$

$\circlearrowright \frac{\partial}{\partial q}\frac{\partial}{\partial p} = \frac{\partial}{\partial p}\frac{\partial}{\partial q}$

$$= \frac{\partial}{\partial p}\frac{\partial G}{\partial q} - \frac{\partial}{\partial p}\frac{\partial G}{\partial q}$$

$\circlearrowright$

$$= 0 \quad \checkmark \qquad (7.115)$$

Motivated by this observation, it is conventional to say that the new object G *generates* the transformation. Moreover, G is called the **generator** of the canonical transformation.

Our infinitesimal transformation in Eq. 7.110 therefore reads

$$q \to Q = q + \epsilon D(q,p) \stackrel{\text{Eq. 7.114}}{=} q + \epsilon \frac{\partial G}{\partial p}$$

$$p \to P = p + \epsilon E(q,p) \stackrel{\text{Eq. 7.114}}{=} p - \epsilon \frac{\partial G}{\partial q}. \qquad (7.116)$$

This result allows us to understand canonical transformations from a completely new perspective.

We already learned in Section 5.1.1 and Section 7.3.1 that the Poisson bracket is, in some sense, the natural product in phase space.[86] Therefore, if we consider an abstract phase space transformation

[86] For instance, we've learned that the Poisson bracket allows us to calculate how functions evolve in time (Eq. 5.36) and to determine which coordinates are canonical (Eq. 7.80).

$$q \to Q = T \circ q, \qquad p \to P = T \circ p, \qquad (7.117)$$

it's not a huge stretch to expect that in more concrete terms such a transformation reads

$$q \to Q = T \circ q, \qquad p \to P = T \circ p$$
$$= \{q, T\}, \qquad \qquad = \{p, T\}.$$

$$(7.118)$$

This is exactly what we've discovered above! To see this, we let our generator G act on the coordinates q, p via the Poisson bracket. To that end, we define an infinitesimal transformation as a linear combination of the identity transformation $I \circ X = X$ (which does nothing) and the generator action:

$$T_{\text{inf}} \circ X \equiv (I + \epsilon G) \circ X = X + \epsilon G \circ X. \qquad (7.119)$$

We can then calculate:

$$
\begin{aligned}
q \to Q &= T_{\text{inf}} \circ q, & p \to P &= T_{\text{inf}} \circ p \\
&= q + \epsilon G \circ q, & &= p + \epsilon G \circ p \\
&= q + \epsilon \{q, G\}, & &= p + \epsilon \{p, G\} \\
&= q + \epsilon \frac{\partial q}{\partial q}\frac{\partial G}{\partial p} - \epsilon \frac{\partial q}{\partial p}\frac{\partial G}{\partial q}, & &= p + \epsilon \frac{\partial p}{\partial q}\frac{\partial G}{\partial p} - \epsilon \frac{\partial p}{\partial p}\frac{\partial G}{\partial q} \\
&= q + \epsilon \frac{\partial q}{\partial q}\frac{\partial G}{\partial p}, & &= p - \epsilon \frac{\partial p}{\partial p}\frac{\partial G}{\partial q} \\
&= q + \epsilon \frac{\partial G}{\partial p}, & &= p - \epsilon \frac{\partial G}{\partial q}.
\end{aligned}
$$
(7.120)

↷ Eq. 7.119

↷

↷ Eq. 5.33

↷ $\frac{\partial q}{\partial p} = 0, \frac{\partial p}{\partial q} = 0$

↷ $\frac{\partial q}{\partial q} = 1, \frac{\partial p}{\partial p} = 1$

This is exactly what we discussed in Eq. 7.116. Therefore, our generator G really acts on the phase space coordinates via the Poisson bracket.

In words, this tells us that (infinitesimal) canonical transformations are not really mysterious after all. We can transform our phase space coordinates using any phase space function $G(q, p)$. We only have to make sure that G acts on our coordinates via the appropriate phase space product, i.e., the Poisson bracket. This alone makes sure that we end up with a canonical transformation.

We will talk about the generators G in more detail in Chapter 10. The most important discovery will be that some of the generators are actually old acquaintances.[87]

[87] To spoil the surprise: we will discover that the momentum p generates spatial translations, the Hamiltonian H generates temporal translations and angular momentum generates rotations.

In the previous sections, we've explored canonical transformation by translating the transformations that we discovered in the Lagrangian formalism into the language of the Hamiltonian formalism and by studying infinitesimal canonical transformations. However, there is also an alternative perspective which allows us to understand and describe canonical transformations in more general terms. This is what we'll talk about in the next section.

7.3.4 Generating Functions

In this section, we'll talk about canonical transformations from yet another perspective. Most importantly, we will discover a new method that allows us to construct transformations between canonical coordinates systematically. This allows us to use canonical transformations as a powerful practical tool.[88]

[88] We will discover in Section 11.1 that by performing a particularly clever canonical transformation, we can often solve the problem at hand much easier.

In the previous sections we've discovered that we can understand infinitesimal canonical transformations using a single function G. Therefore, it probably won't surprise you when I proclaim that we can understand finite canonical transformations using a single function too. In other words, the main idea in this section is analogous to what we discussed in the previous section. But this time we don't restrict ourselves to infinitesimal transformations.

We assume that there is some function F which encodes all of the information about one particular canonical transformation. Starting from such a **generating function**, we can derive what the corresponding canonical transformation looks like.[89]

[89] Take note that: generating function $\neq$ generator.

To understand how this is possible, we first need to talk about what we really need to specify a canonical transformation. One possibility are formulas which tell us how each new coordinate Q, P is related to the old coordinates q, p:[90]

$$Q = Q(q, p, t), \qquad P = P(q, p, t). \qquad (7.121)$$

[90] As usual, Q, P and q, p represent symbolically our $2N$ new and $2N$ old coordinates. This means that we really have here $2N$ such formulas
$Q_i = Q_i(q_j, p_j, t), \qquad P_i = P_i(q_j, p_j, t)$.

But it would be equally sufficient to specify the inverse formu-

las
$$q = q(Q, P, t), \qquad p = p(Q, P, t). \qquad (7.122)$$
In addition, relations of the form
$$p = p(q, Q, t), \qquad P = P(q, Q, t). \qquad (7.123)$$
work equally well because we can invert the formulas of the form $P = P(q, Q, t)$ to find out how the old locations q are related to the new coordinates, i.e., to find $q = q(Q, P, t)$. Then we can use these formulas to substitute for q in $p = p(q, Q, t)$ to get the formulas that tell us how the old momentum coordinates are related to the new ones
$$p = p(q, Q, t) = p\big(q(Q, P, t), Q, t\big) = p(Q, P, t).$$
So in other words, Eq. 7.123 is simply a little bit more indirect way to relate the old to the new coordinates.[91]

With this in mind, I now claim that a canonical transformation can be specified completely using a single function $F = F(q, Q, t)$ through the formulas:[92]
$$P = \frac{\partial F}{\partial Q}$$
$$p = -\frac{\partial F}{\partial q}. \qquad (7.124)$$

Since F is a function of q, Q, and t, we know that $\frac{\partial F}{\partial Q}$ will be a function of q, Q, and t too. Analogously, the second line in Eq. 7.124 is an equation of the form $p = p(q, Q, t)$. We can then invert this equation to get an expression of the form $Q = Q(q, p, t)$. So in other words, given some function F, the second line tells us what our new location coordinates look like in terms of the original location and momentum coordinates. The first line yields a formula of the form $P = P(q, Q, t)$. But we already know from the second line what Q looks like in terms of the original coordinates. Therefore, we can use this expression to get a formula that specifies the new momentum coordinates P solely in terms of the original phase space coordinates $P = P(q, p, t)$.

In this sense, the formulas in Eq. 7.124 are sufficient to establish relations of the form given in Eq. 7.123. Therefore, Eq. 7.124

[91] You might rightfully wonder, why do we consider such an awkward way to specify our new coordinates? Be assured that there is a good reason which we will learn below. For the moment, all we care about is that relations of this form are equally sufficient.

[92] For the moment, simply assume that someone really smart came up with these formulas that allow us to specify a canonical transformation completely in terms of a single function. We specify the new coordinates rather indirectly because only in this way can we make sure that we actually end up with canonical coordinates. We will discuss how these two formulas can actually be derived below.

tells us everything that we need to know about how our old coordinates are related to the new ones.

However, the main point we are interested in is that by using the formulas in Eq. 7.124, we end up not just with any coordinates but with *canonical* coordinates for *any* function $F = F(q, Q, t)$. In other words, Eq. 7.124 allows us to write down canonical coordinates and canonical transformations en masse.

Of course, we need to check such a grand claim.[93] As usual, we do this by checking that the canonical Poisson bracket relation remains valid for any choice of $F = F(q, Q, t)$.[94] But before we can evaluate the canonical Poisson bracket relation explicitly, we need two formulas which follow directly from Eq. 7.124. First, we note that if we take the partial derivative of P with respect to the old momentum coordinates p, we find[95]

$$\left.\frac{\partial P}{\partial p}\right|_q = \left.\frac{\partial P}{\partial Q}\right|_q \left.\frac{\partial Q}{\partial p}\right|_q \qquad (7.125)$$

because p appears only in $Q = Q(q, p)$. Secondly, when we take the partial derivative of P with respect to the old location coordinates q, we find

$$\left.\frac{\partial P}{\partial q}\right|_p = \left.\frac{\partial P}{\partial Q}\right|_q \left.\frac{\partial Q}{\partial q}\right|_p + \left.\frac{\partial P}{\partial q}\right|_Q$$

↷ rearranging terms

$$\left.\frac{\partial P}{\partial q}\right|_p - \left.\frac{\partial P}{\partial Q}\right|_q \left.\frac{\partial Q}{\partial q}\right|_p = \left.\frac{\partial P}{\partial q}\right|_Q \qquad (7.126)$$

because $P = \frac{\partial F(q,Q,t)}{\partial Q}$ and therefore, q can appear directly in P and in $Q = Q(q, p)$.

Using these two formulas, we can check that the coordinates we get by using Eq. 7.124 are indeed canonical for any function

---

[93] Feel free to skip the following somewhat lengthy calculations on a first encounter.

[94] This is the defining property of canonical coordinates that we discovered in Section 7.3.1.

[95] Here $|_x$ denotes that we hold x fixed when we calculate the partial derivative. We spell this out explicitly here because otherwise, it's really easy to get confused.

$F = F(q, Q, t)$:

$$\{Q, P\}_{q,p} = \frac{\partial Q}{\partial q}\bigg|_p \frac{\partial P}{\partial p}\bigg|_q - \frac{\partial Q}{\partial p}\bigg|_q \frac{\partial P}{\partial q}\bigg|_p$$

↪ Eq. 7.125

$$= \frac{\partial Q}{\partial q}\bigg|_p \frac{\partial P}{\partial Q}\bigg|_q \frac{\partial Q}{\partial p}\bigg|_q - \frac{\partial Q}{\partial p}\bigg|_q \frac{\partial P}{\partial q}\bigg|_p$$

↪ rearranging terms

$$= -\frac{\partial Q}{\partial p}\bigg|_q \left(\frac{\partial P}{\partial q}\bigg|_p - \frac{\partial Q}{\partial q}\bigg|_p \frac{\partial P}{\partial Q}\bigg|_q\right)$$

↪ Eq. 7.126

$$= -\frac{\partial Q}{\partial p}\bigg|_q \left(\frac{\partial P}{\partial q}\bigg|_Q\right)$$

↪ Eq. 7.124

$$= -\frac{\partial Q}{\partial p}\bigg|_q \left(\frac{\partial}{\partial q}\bigg|_Q \frac{\partial F}{\partial Q}\bigg|_q\right)$$

↪ $\frac{\partial}{\partial x}\frac{\partial}{\partial y} = \frac{\partial}{\partial y}\frac{\partial}{\partial x}$

$$= -\frac{\partial Q}{\partial p}\bigg|_q \left(\frac{\partial}{\partial Q}\bigg|_q \frac{\partial F}{\partial q}\bigg|_Q\right)$$

↪ Eq. 7.124

$$= -\frac{\partial Q}{\partial p}\bigg|_q \left(\frac{\partial (-p)}{\partial Q}\bigg|_q\right)$$

↪

$$= 1 \checkmark \qquad\qquad (7.127)$$

Thus, as promised, the new coordinates we get through Eq. 7.124 are indeed canonical for any function $F = F(q, Q, t)$.

As a concrete example, let's calculate the canonical transformation generated by

$$F(q, Q) = qQ. \qquad (7.128)$$

Using Eq. 7.124, we find

$$P = \frac{\partial F}{\partial Q} = \frac{\partial(qQ)}{\partial Q} = q$$

$$p = -\frac{\partial F}{\partial q} = -\frac{\partial(qQ)}{\partial q} = -Q. \qquad (7.129)$$

This means that the canonical transformation generated by $F(q,Q) = qQ$ flips the roles of the momentum and location coordinates.

Now, you are probably wondering why this method works.

Here's why. First of all, recall that we are able to derive Hamilton's equations using the principle of least action.[96] Hamilton's equations follow as a condition on the path of least action for the Hamiltonian action (Eq 5.20)

[96] We discussed this in Chapter 5.

$$S = \int_{t_i}^{t_f} L\, dt = \int_{t_i}^{t_f} \left(p\dot{q} - H\right) dt, \qquad (7.130)$$

where

$$L \equiv p\dot{q} - H \qquad (7.131)$$

is the "Hamiltonian Lagrangian".

The defining feature of a canonical transformation is that the form of Hamilton's equations remains unchanged. This implies that after the transformation $q, p \to Q, P$, the least action principle with an analogous Hamiltonian action

$$\tilde{S} = \int_{t_i}^{t_f} L'\, dt = \int_{t_i}^{t_f} dt \left(P\dot{Q} - H'\right) \qquad (7.132)$$

must still be valid, where

$$L' \equiv P\dot{Q} - H' \qquad (7.133)$$

is the transformed "Hamiltonian Lagrangian".[97]

[97] We have already discussed this at the end of Section 7.3.2. In particular, see Eq. 7.102.

The original Lagrangian L and the transformed Lagrangian differ by the total derivative of an arbitrary function $F = F(q, Q, t)$:[98]

$$L' - L = \frac{dF}{dt}. \quad (7.134)$$

In general, the total time derivative of a function $F = F(q, Q, t)$ reads:[99]

$$\frac{dF(q, Q, t)}{dt} = \frac{\partial F}{\partial q}\frac{\partial q}{\partial t} + \frac{\partial F}{\partial Q}\frac{\partial Q}{\partial t} + \frac{\partial F}{\partial t}$$

$$= \frac{\partial F}{\partial q}\dot{q} + \frac{\partial F}{\partial Q}\dot{Q} + \frac{\partial F}{\partial t}. \quad (7.135)$$

[98] See Eq. 7.94. Moreover, recall that this is the case because only the initial and final locations are fixed and therefore, this total derivative term yields zero after the integration.

[99] The total derivative of a general function is discussed in Appendix A.3.

Using this, we can rewrite Eq. 7.134 as follows:

$$L' - L = \frac{dF}{dt}$$

$\circlearrowright$ Eq. 7.102, Eq. 5.12, Eq. 7.135

$$\left(P\dot{Q} - H'\right) - \left(p\dot{q} - H\right) = \frac{\partial F}{\partial q}\dot{q} + \frac{\partial F}{\partial Q}\dot{Q} + \frac{\partial F}{\partial t}$$

$\circlearrowright$ rearranging

$$\left(P - \frac{\partial F}{\partial Q}\right)\dot{Q} - \left(p + \frac{\partial F}{\partial q}\right)\dot{q} = H' - H + \frac{\partial F}{\partial t}. \quad (7.136)$$

Now we can use the fact that none of the functions in Eq. 7.136 depend explicitly on $\dot{q}$. Therefore, if we calculate the partial derivative with respect to $\dot{q}$, we find:

$$\frac{\partial}{\partial \dot{q}}\left(P - \frac{\partial F}{\partial Q}\right)\dot{Q} - \frac{\partial}{\partial \dot{q}}\left(p + \frac{\partial F}{\partial q}\right)\dot{q} = \frac{\partial}{\partial \dot{q}}\left(H' - H + \frac{\partial F}{\partial t}\right)$$

$$-\left(p + \frac{\partial F}{\partial q}\right) = 0. \quad (7.137)$$

Analogously, since none of the functions depend explicitly on $\dot{Q}$, we find that

$$\frac{\partial}{\partial \dot{Q}}\left(P - \frac{\partial F}{\partial Q}\right)\dot{Q} - \frac{\partial}{\partial \dot{Q}}\left(p + \frac{\partial F}{\partial q}\right)\dot{q} = \frac{\partial}{\partial \dot{Q}}\left(H' - H + \frac{\partial F}{\partial t}\right)$$

$$\left(P - \frac{\partial F}{\partial Q}\right) = 0. \quad (7.138)$$

Therefore, we can conclude:

$$P = \frac{\partial F}{\partial Q}$$
$$p = -\frac{\partial F}{\partial q}. \quad (7.139)$$

This is exactly Eq. 7.124.

We ended up with this somewhat indirect relationship between the generating function F and the new coordinates because only this way can we make sure that we end up with canonical coordinates. We achieved this result because the least action principle holds for the same path expressed in different coordinates.

Moreover, we can also calculate how the Hamiltonian gets modified by the canonical transformation generated by F. If we use Eq. 7.139 in Eq. 7.136, we find:

$$\left(P - \frac{\partial F}{\partial Q}\right)\dot{Q} - \left(p + \frac{\partial F}{\partial q}\right)\dot{q} = H' - H + \frac{\partial F}{\partial t} \quad \text{this is Eq. 7.136}$$

$$\downarrow \text{Eq. 7.139}$$

$$(0)\dot{Q} - (0)\dot{q} = H' - H + \frac{\partial F}{\partial t}$$

$$0 = H' - H + \frac{\partial F}{\partial t}. \tag{7.140}$$

This tells us again that the new Hamiltonian H' is related to the old one H by the partial time derivative of the generating function:[100]

$$H' = H - \frac{\partial F}{\partial t}. \tag{7.141}$$

[100] This is exactly what we discovered already in Eq. 7.100.

To summarize, the algorithm to produce a canonical transformation goes like this:

1. Write down a specific generating function $F = F(q, Q, t)$.
2. Use Eq. 7.139 to find the formulas which express the new coordinates (Q, P) in terms of the old ones (q, p).
3. Use Eq. 7.141 to calculate the new Hamiltonian $H'(Q, P, t)$.

Take note that we can also reverse this procedure. This means that we can start with a given canonical transformation

$$Q = Q(q, p), \quad P = P(q, p)$$

and then look for the function which generates it. Specifically, we can use the equations which relate the new coordinates to the old ones to find formulas which relate p and P to q, Q and t. We can then use these formulas to eliminate p and P from Eq. 7.139. This is a system of differential equations that we need to solve for $F = F(q, Q, t)$. If we find a solution, we've successfully identified the function which generates the canonical transformation we started with. Therefore, from a slightly different perspective, we can argue that as soon as we've found a suitable generating function for a given transformation, we can conclude that the transformation is indeed canonical. In other words, searching for a suitable generating function F is another method to check whether a given transformation is canonical.[101]

[101] Alternatively, as discussed above, we can check if the canonical Poisson bracket relation $\{Q, P\}_{q,p} = 1$ is fulfilled by the new coordinates Q, P.

Now, it's finally time to talk about concrete systems.

8
The Harmonic Oscillator

Sidney Coleman once remarked that *"the career of a young theoretical physicist consists of treating the harmonic oscillator in ever-increasing levels of abstraction."* And Michael Peskin famously wrote that *"physics is that subset of human experience which can be reduced to coupled harmonic oscillators"*.[1]

So it certainly makes sense to study the harmonic oscillator in some detail. But first, what is a harmonic oscillator?

In fact, we call all systems with a potential of the form $V = cx^2$, where c is some constant, a harmonic oscillator. The potential of the harmonic oscillator looks like this:

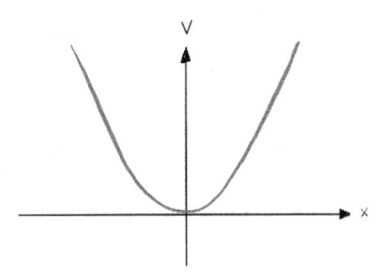

[1] In quantum mechanics, the quantum harmonic oscillator is one of the best examples to understand what quantization really means and how the quantum framework works. In classical field theory, we can use that a field is simply a bunch of harmonic oscillators coupled together (think: like a mattress). If we then turn these harmonic oscillators into quantum harmonic oscillators, we end up with quantum field theory.

An example of such a system is an object attached to a spring.

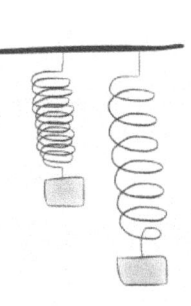

We can therefore visualize the meaning of the potential as follows:

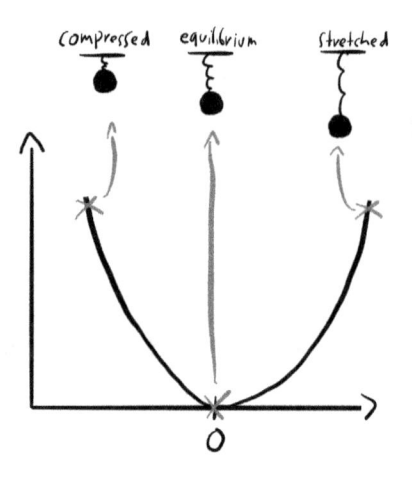

Now, why are harmonic oscillators so important?

It turns out that, in the first approximation, *lots* of potentials are extremely similar to the harmonic potential. This means that the first term of the Taylor expansion[2] of many potentials is exactly the harmonic potential:

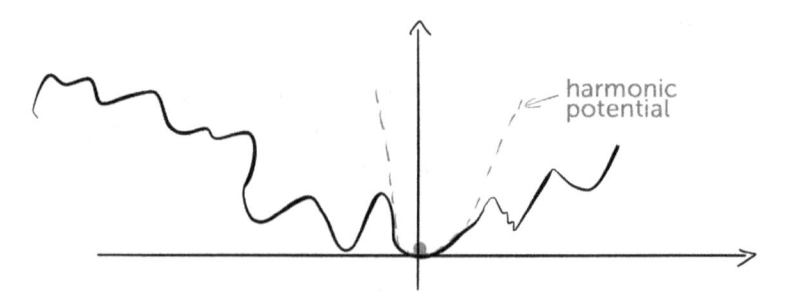

[2] For the basic idea behind the Taylor expansion see Appendix F.

For example, the Taylor expansion of a much more complicated function like $\cos x$ is

$$\cos x = 1 - \frac{x^2}{2} + \dots.$$

So for small x, the potential can be approximated by $1 - \frac{x^2}{2}$, which is exactly the potential of a harmonic oscillator shifted by +1. A concrete physical example is a pendulum which is described by the potential $V = 1 - \cos x$:[3]

[3] We will talk about the pendulum in Chapter 9.

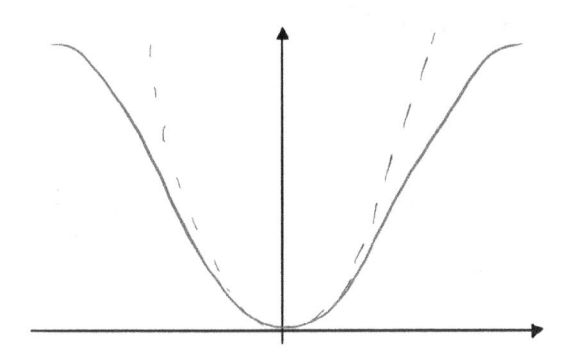

Thus, by studying the harmonic potential closely, we can learn a lot about many other systems (at least as long as we are dealing with small excitations/low energies).

The potential of the harmonic oscillator is usually written as[4]

$$V(x) = \frac{1}{2}kx^2, \quad (8.1)$$

where k is the **spring constant** which characterizes the strength of the spring.

[4] The factor $\frac{1}{2}$ is included because, as we will see below, it makes the formula for the resulting force simpler.

In the following sections, we will derive the correct equation of motion for the harmonic oscillator using the Newtonian, Lagrangian, and Hamiltonian algorithm:

$$m\frac{d^2x}{dt^2} = -kx$$

Newtonian Algorithm — Lagrangian Algorithm — Hamiltonian Algorithm

$$F = -kx \qquad L = \tfrac{1}{2}m\dot{x}^2 - \tfrac{1}{2}kx^2 \qquad H = \tfrac{1}{2}\tfrac{p^2}{m} + \tfrac{1}{2}kx^2$$

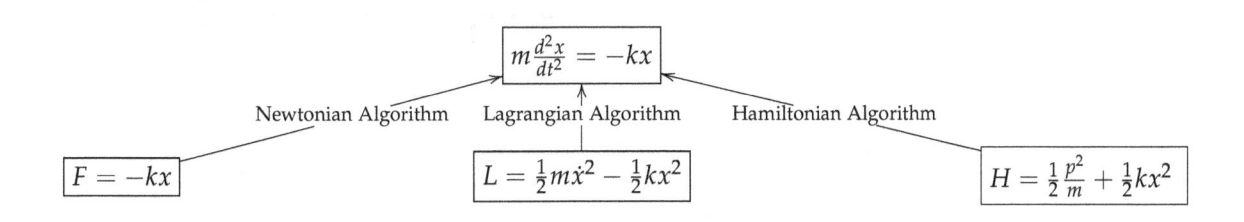

Afterwards, in Section 8.4 we will discuss how we can solve this equation of motion.

8.1 Newtonian Description

In the Newtonian formulation of classical mechanics, our first task is always to determine the forces that act on each object. Here we are dealing with just one object (the object attached to the spring) moving in one dimension (up and down) and there is only one force which is exerted by the spring the object is attached to.

This force can be described by the law

$$F(x) = -kx, \tag{8.2}$$

where k is the spring constant and x the distance that the object is away from its equilibrium position at $x = 0$.[5] This empirical law is known as **Hooke's law**. In words, it tells us that the force exerted by the spring gets larger if we move the object farther away from its equilibrium position in a linear manner.

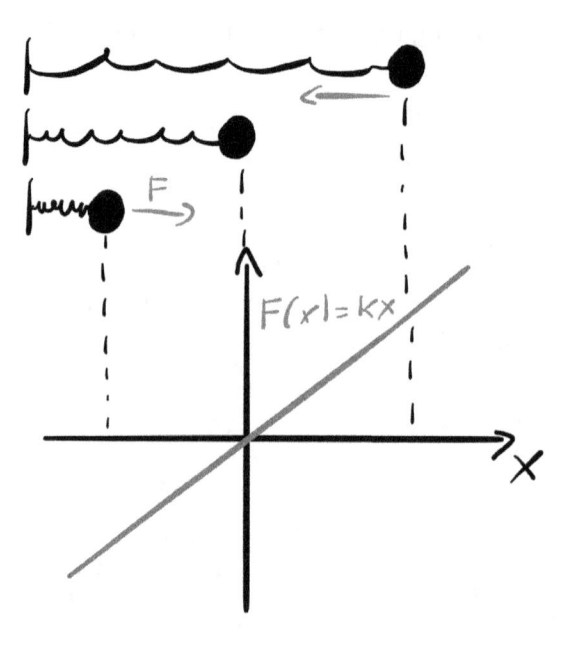

[5] We choose the coordinate system in such a way that the equilibrium position is at $x = 0$.

So it takes twice as much force to stretch our spring twice as far.

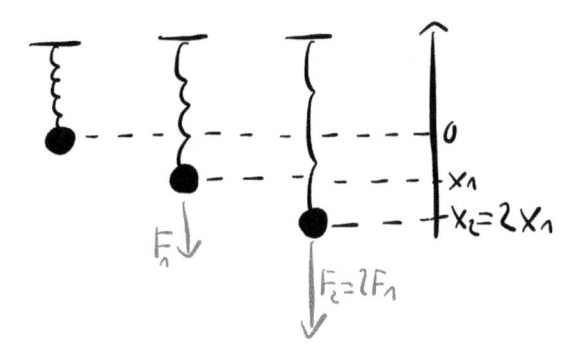

Using the general formula that relates a potential $V(x)$ to the corresponding force

$$F = -\frac{\partial V(x)}{\partial x} \quad \text{(this is Eq. 2.16)}$$

we can check that the force law in Eq. 8.2 is indeed consistent with the formula for the harmonic oscillator potential given above:

$$F = -\frac{\partial V(x)}{\partial x}$$

$$\circlearrowright \quad V(x) = \frac{1}{2}kx^2, \text{ Eq. 8.1}$$

$$= -\frac{\partial(\frac{1}{2}kx^2)}{\partial x}$$

$$\circlearrowright \quad \frac{\partial x^2}{\partial x} = 2x$$

$$= -kx \quad \checkmark$$

With this formula in hand, we can immediately write down the equation of motion for the harmonic oscillator using Newton's second law (Eq. 3.1)[6]

[6] We discussed equations of motion in general terms in Section 2.2.

$$\frac{d}{dt}p = F \quad \text{this is Newton's second law (Eq. 3.1)}$$

$$\circlearrowright \quad F(x) = -kx, \text{ Eq. 8.2}$$

$$\frac{d}{dt}p = -kx$$

$$\circlearrowright \quad p = m\frac{dx}{dt}, \text{ Eq. 2.7}$$

$$\frac{d}{dt}\left(m\frac{dx}{dt}\right) = -kx$$

$$\circlearrowright \quad \text{assuming mass } m \text{ is constant: } \frac{d}{dt}\left(m\frac{dx}{dt}\right) = m\frac{d}{dt}\left(\frac{dx}{dt}\right)$$

$$m\frac{d^2x}{dt^2} = -kx. \quad (8.3)$$

This is the equation of motion for an object attached to a spring.

Maybe you wonder why the gravitational force plays no role here. To understand this, let's include it in the equation of motion

$$m\frac{d^2x}{dt^2} = F_{\text{spring}} + F_{\text{gravity}} = -kx - mg. \qquad (8.4)$$

Now, we can simply shift our coordinate system

$$x \to x' = x + mg/k \qquad (8.5)$$

and then find

$$m\frac{d^2x}{dt^2} = -kx - mg$$

$$\to \quad m\frac{d^2(x' - mg/k)}{dt^2} = -k(x' - mg/k) - mg \qquad \circlearrowright\ x = x' - mg/k$$

$$m\frac{d^2x'}{dt^2} = -kx' + mg - mg \qquad \circlearrowright\ \frac{d^2(mg/k)}{dt^2} = 0$$

$$m\frac{d^2x'}{dt^2} = -kx'. \qquad \circlearrowright\ \cancel{mg} \qquad (8.6)$$

This is exactly the equation of motion that we already derived above. In words, this means that the gravitational force indeed has an effect, but not on the dynamics of the system. The only thing the gravitational force does is to shift the equilibrium position. And that's why we can ignore it here. We can always shift our coordinate system such that the equilibrium position is at the origin of the coordinate system (see Eq. 8.5). Intuitively, this happens because the gravitational force $F_{\text{gravity}} = -mg$ is constant. When our object is at the equilibrium position, the gravitational force is balanced by the spring force. (Otherwise the object would fall to the floor.) But the gravitational force remains constant when the object is displaced from the equilibrium position. Therefore, it yields no contribution to the restoring force which pushes the object back to the equilibrium position, and we can conclude that it has no further effect on the movement of the object.

Next, let's see how all this works in the Lagrangian formalism.

8.2 Lagrangian Description

In the Lagrangian formulation of classical mechanics, our first task is always to write down the correct Lagrangian $L = T - V$. The potential energy of the object attached to a spring can be described by (Eq. 8.1)

$$V(x) = \frac{1}{2}kx^2. \qquad (8.7)$$

The corresponding Lagrangian therefore reads

$$L = T - V = \frac{1}{2}m\dot{x}^2 - \frac{1}{2}kx^2. \qquad (8.8)$$

We can then calculate the correct equation of motion by putting this Lagrangian into the Euler-Lagrange equation (Eq. 4.25):

$$\frac{\partial L}{\partial x} = \frac{d}{dt}\left(\frac{\partial L}{\partial \dot{x}}\right)$$

$\circlearrowright$ $L = \frac{1}{2}m\dot{x}^2 - \frac{1}{2}kx^2$, Eq. 8.8

$$\frac{\partial\left(\frac{1}{2}m\dot{x}^2 - \frac{1}{2}kx^2\right)}{\partial x} = \frac{d}{dt}\left(\frac{\partial\left(\frac{1}{2}m\dot{x}^2 - \frac{1}{2}kx^2\right)}{\partial \dot{x}}\right)$$

$\circlearrowright$ $\frac{\partial x}{\partial \dot{x}} = 0$ and $\frac{\partial \dot{x}}{\partial x} = 0$

$$\frac{\partial\left(-\frac{1}{2}kx^2\right)}{\partial x} = \frac{d}{dt}\left(\frac{\partial\left(\frac{1}{2}m\dot{x}^2\right)}{\partial \dot{x}}\right)$$

$\circlearrowright$ $\frac{\partial x^2}{\partial x} = 2x$ and $\frac{\partial \dot{x}^2}{\partial \dot{x}} = 2\dot{x}$

$$-kx = m\frac{d}{dt}\dot{x}$$

$\circlearrowright$ $\dot{x} \equiv \frac{dx}{dt}$

$$-kx = m\frac{d^2x}{dt^2}. \qquad (8.9)$$

This is exactly the equation of motion that we already derived using the Newtonian formalism (Eq. 8.3).

Next, let's derive this equation of motion one more time, using the Hamiltonian formalism.

8.3 Hamiltonian Description

In the Hamiltonian formalism our first task is always to determine the Hamiltonian H, which is defined as (Eq. 5.18)

$$H = p\dot{x} - L, \tag{8.10}$$

where p denotes the generalized momentum (Eq. 4.29)

$$p = \frac{\partial L}{\partial \dot{x}}. \tag{8.11}$$

Using the Lagrangian that we've discovered in the previous section, we find for the canonical momentum:

$$p = \frac{\partial L}{\partial \dot{x}}$$

$$= \frac{\partial \left(\frac{1}{2}m\dot{x}^2 - \frac{1}{2}kx^2\right)}{\partial \dot{x}} \qquad \circlearrowright \quad L = \frac{1}{2}m\dot{x}^2 - \frac{1}{2}kx^2,\ \text{Eq. 8.8}$$

$$= m\dot{x}. \qquad \circlearrowright \quad \frac{\partial x}{\partial \dot{x}} = 0 \text{ and } \frac{\partial \dot{x}^2}{\partial \dot{x}} = 2\dot{x} \tag{8.12}$$

This tells us that

$$\dot{x} = \frac{p}{m}. \tag{8.13}$$

And for the Hamiltonian we find:

$$H = p\dot{x} - L$$

$$= p\left(\frac{p}{m}\right) - L \qquad \circlearrowright \quad \dot{x} = \frac{p}{m},\ \text{Eq. 8.13}$$

$$= \frac{p^2}{m} - \left(\frac{1}{2}m\dot{x}^2 - \frac{1}{2}kx^2\right) \qquad \circlearrowright \quad L = \frac{1}{2}m\dot{x}^2 - \frac{1}{2}kx^2,\ \text{Eq. 8.8}$$

$$= \frac{p^2}{m} - \left(\frac{1}{2}m\left(\frac{p}{m}\right)^2 - \frac{1}{2}kx^2\right) \qquad \circlearrowright \quad \dot{x} = \frac{p}{m},\ \text{Eq. 8.13}$$

$$= \frac{1}{2}\frac{p^2}{m} + \frac{1}{2}kx^2. \qquad \circlearrowright \quad \text{rearranging terms} \tag{8.14}$$

We can see here that for the harmonic oscillator, the Hamiltonian function H represents the total energy $E = T + V$.

Next, to derive the equation of motion, we put this result into Hamilton's first equation (Eq. 5.15)

$$\frac{dp}{dt} = -\frac{\partial H}{\partial x}$$

↳ Eq. 8.14

$$= -\frac{\partial \left(\frac{1}{2}\frac{p^2}{m} + \frac{1}{2}kx^2\right)}{\partial x}$$

↳

$$= -kx. \qquad (8.15)$$

Moreover, Hamilton's second equation (Eq. 5.15) tells us

$$\frac{dx}{dt} = \frac{\partial H}{\partial p}$$

↳ $H = \frac{1}{2}\frac{p^2}{m} + \frac{1}{2}kx^2$, Eq. 8.14

$$= \frac{\partial \left(\frac{1}{2}\frac{p^2}{m} + \frac{1}{2}kx^2\right)}{\partial p}$$

↳

$$= \frac{p}{m}. \qquad (8.16)$$

Next, we can take the derivative of Eq. 8.16[7]

[7] We will see in a moment why this is a smart idea.

$$\frac{dx}{dt} = \frac{p}{m} \qquad \text{this is Eq. 8.16}$$

↳ $\frac{d}{dt}$

$$\frac{d^2x}{dt^2} = \frac{\frac{dp}{dt}}{m}$$

↳ $\times m$

$$m\frac{d^2x}{dt^2} = \frac{dp}{dt} \qquad (8.17)$$

and use this result to eliminate p from Eq. 8.15:

$$\frac{dp}{dt} = -kx \qquad \text{this is Eq. 8.15}$$

↳ Eq. 8.17

$$m\frac{d^2x}{dt^2} = -kx. \qquad (8.18)$$

So again, we find the correct equation of motion, but this time we derived it using Hamilton's equations.

8.4 Solving the Equation of Motion

Now, let's talk about the equation we derived three times (Eq. 8.3)

$$m\frac{d^2x}{dt^2} = -kx \qquad (8.19)$$

in a bit more detail.

In Section 2.2, we discussed already that, in general, solutions of this equation read (Eq. 2.30)

$$x(t) = A\cos(\omega t) + B\sin(\omega t). \qquad (8.20)$$

Let's recall the main points. We get solutions like this because Eq. 8.3 tells us that we need a function which yields itself if we differentiate it twice. The trigonometric functions $\cos(\omega t)$ and $\sin(\omega t)$ have exactly this property.

We can check[8]

[8] Completely analogously, we can check that $x(t) = B\sin(\omega t)$ and $x(t) = A\cos(\omega t) + B\sin(\omega t)$ solve the equation of motion.

$$m\frac{d^2x}{dt^2} = -kx$$

$$\circlearrowright \text{ ansatz } x(t) = A\cos(\omega t)$$

$$m\frac{d^2\big(A\cos(\omega t)\big)}{dt^2} = -kA\cos(\omega t)$$

$$\circlearrowright \frac{d^2\cos(\omega t)}{dt^2} = -\omega^2\cos(\omega t)$$

$$m\big(-A\omega^2\cos(\omega t)\big) = -kA\cos(\omega t)$$

$$\circlearrowright \text{ rearranging terms and } \cancel{A}$$

$$m\omega^2\cos(\omega t) = k\cos(\omega t). \qquad (8.21)$$

We can see that $x(t) = A\cos(\omega t)$ indeed solves our equation of motion provided that

$$\omega = \sqrt{\frac{k}{m}} \qquad (8.22)$$

because then

$$mw^2 \cos(wt) = k\cos(wt) \quad \text{this is the last line in Eq. 8.21}$$
$$\Rightarrow w = \sqrt{\frac{k}{m}}$$

$$m\left(\sqrt{\frac{k}{m}}\right)^2 \cos(wt) = k\cos(wt)$$

$$m\frac{k}{m}\cos(wt) = k\cos(wt) \quad \Rightarrow$$

$$\Rightarrow \frac{m}{m} = 1$$

$$k\cos(wt) = k\cos(wt) \checkmark \tag{8.23}$$

Intuitively, these solutions describe a periodic up-and-down movement of the object attached to the spring. The constant A is the amplitude of the oscillation and depends on how far we displace the object from its equilibrium position at $t = 0$:

$$x(0) = A\cos(w0) = A. \tag{8.24}$$

If we don't displace the object at all, we have $A = 0$ and therefore no oscillation at all. In other words, the object simply remains at its equilibrium position.

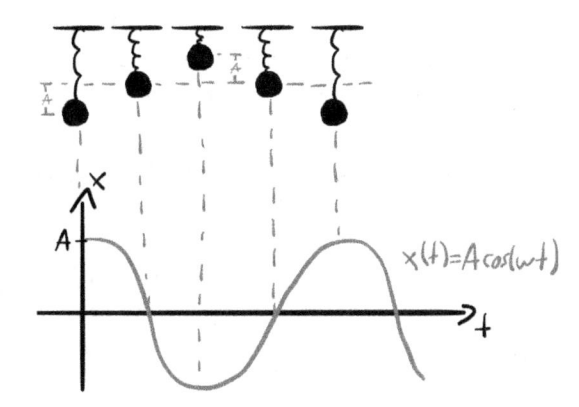

The constant w describes the frequency of the oscillation which is how often per unit time our object passes a certain location (e.g., the maximum position). In other words, w tells us how many oscillations there are *per second*. This is a feature of the system and does not depend on initial conditions. We can see this because Eq. 8.22 tells us that w only depends on the mass of the object m and the spring constant k. This may be a bit surprising because it means, no matter how far we displace the object at $t = 0$, the time the object needs to move one full up-and-down cycle is always the same.

Moreover, if we consider the full solution (Eq. 2.30), there is an additional constant B. This constant encodes information about the initial velocity of the object (i.e., if we release the object from its displaced position at rest or give it some initial push):[9]

[9] The constant B plays no role for the initial position because $\sin(0) = 0$.

$$\frac{d}{dt}x(t) = \frac{d}{dt}\Big(A\cos(\omega t) + B\sin(\omega t)\Big)$$

↓ $\frac{d}{dt}\cos(\omega t) = -\omega \sin(\omega t)$, $\frac{d}{dt}\sin(\omega t) = \omega \cos(\omega t)$

$$= -A\omega \sin(\omega t) + B\omega \cos(\omega t)$$

↓ initial condition

$$\therefore \frac{dx(0)}{dt} = -A\omega \sin(\omega 0) + B\omega \cos(\omega 0)$$

↓ $\cos(0) = 1$ and $\sin(0) = 0$

$$= B\omega \quad \Rightarrow \quad B = \frac{\frac{dx(0)}{dt}}{\omega} , \tag{8.25}$$

where $\frac{dx(0)}{dt}$ is the initial velocity.

Maybe you are wondering how we can derive the solution of the equation of motion instead of guessing it. One possible clever approach works as follows.[10]

[10] Of course, if you're not interested in a mathematical derivation of the solution, feel free to skip the following paragraphs.

First of all, we multiply the equation of motion (Eq. 8.3) by the velocity $\dot{x}$

$$m\frac{d^2x}{dt^2} = -kx \qquad \text{this is Eq. 8.3}$$

↓ ×$\dot{x}$

$$m\dot{x}\frac{d^2x}{dt^2} = -k\dot{x}x . \tag{8.26}$$

This is clever because we can now rewrite the left-hand side using the chain rule $\left(\frac{d}{dt}f(t)^2 = 2f(t)\frac{df(t)}{dt}\right)$

$$\dot{x}\frac{d^2x}{dt^2} = \dot{x}\frac{d\dot{x}}{dt} = \frac{1}{2}\frac{d}{dt}\dot{x}^2 . \tag{8.27}$$

Moreover, we can rewrite the right-hand side analogously using

$$x\dot{x} = x\frac{dx}{dt} = \frac{1}{2}\frac{d}{dt}x^2 . \tag{8.28}$$

Eq. 8.26 then reads

$$m\dot{x}\frac{d^2x}{dt^2} = -k\dot{x}x$$

⤳ Eq. 8.27 and Eq. 8.28

$$\frac{d}{dt}\left(\frac{1}{2}m\dot{x}^2\right) = -\frac{d}{dt}\left(\frac{1}{2}kx^2\right)$$

⤳ rearranging terms

$$\frac{d}{dt}\left(\frac{1}{2}m\dot{x}^2 + \frac{1}{2}kx^2\right) = 0. \tag{8.29}$$

This means that the expression between the parentheses on the left-hand side is constant in time. As already discussed in Section 2.1.3, we usually call this conserved quantity the total energy

$$E = \frac{1}{2}m\dot{x}^2 + \frac{1}{2}kx^2 \equiv T + V, \tag{8.30}$$

where $T = \frac{1}{2}m\dot{x}^2$ is the kinetic energy and $V = \frac{1}{2}kx^2$ the potential energy.

Equipped with this observation, we can use the separation of variables method, which we already discussed in Section 2.2. We now rewrite Eq. 8.30 as follows:

$$E = \frac{1}{2}m\dot{x}^2 + \frac{1}{2}kx^2$$

⤳ rearranging terms

$$E - \frac{1}{2}kx^2 = \frac{1}{2}m\dot{x}^2$$

⤳ $\times \frac{2}{m}$

$$\frac{2E - kx^2}{m} = \dot{x}^2$$

⤳ $\sqrt{}$

$$\sqrt{\frac{2E - kx^2}{m}} = \dot{x}$$

⤳ $\dot{x} = \frac{dx}{dt}$

$$\sqrt{\frac{2E - kx^2}{m}} = \frac{dx}{dt}$$

⤳ rearranging terms

$$dt = \sqrt{\frac{m}{2E - kx^2}}\,dx$$

⤳ integrating

$$\int dt = \int \sqrt{\frac{m}{2E - kx^2}}\,dx$$

⤳

$$t = \int \sqrt{\frac{m}{2E - kx^2}}\,dx. \tag{8.31}$$

We've therefore "reduced" the problem of solving the equation of motion to the problem of solving the integral on the right-hand side in Eq. 8.31.

Solving integrals is, like solving differential equations, somewhat of an art. The integral here can be solved by making the substitution[11]

$$x \equiv \sqrt{\frac{2E}{k}} \sin(\phi), \tag{8.32}$$

which implies

$$\frac{dx}{d\phi} = \sqrt{\frac{2E}{k}} \cos(\phi)$$

⤷ multiplying by $d\phi$

$$dx = \sqrt{\frac{2E}{k}} \cos(\phi) d\phi. \tag{8.33}$$

This is helpful because the term under the square root simplifies dramatically.

$$t = \int \sqrt{\frac{m}{2E - kx^2}} dx$$

this is the last line in Eq. 8.31

⤷ substitution, Eq. 8.32, Eq. 8.33

$$= \int \sqrt{\frac{m}{2E - k\left(\sqrt{\frac{2E}{k}}\sin(\phi)\right)^2}} \left(\sqrt{\frac{2E}{k}} \cos(\phi) d\phi\right)$$

⤷ $\cancel{2E}$

$$= \sqrt{\frac{m}{k}} \int \sqrt{\frac{1}{1 - \sin^2(\phi)}} \cos(\phi) d\phi$$

⤷ $1 - \sin^2(\phi) = \cos^2(\phi)$

$$= \sqrt{\frac{m}{k}} \int \sqrt{\frac{1}{\cos^2(\phi)}} \cos(\phi) d\phi$$

⤷ $\cancel{\cos(\phi)}$

$$= \sqrt{\frac{m}{k}} \int d\phi$$

⤷

$$= \sqrt{\frac{m}{k}} (\phi + \phi_0), \tag{8.34}$$

where ϕ_0 is the constant of integration. Solving for ϕ yields

$$\phi = \sqrt{\frac{k}{m}} t - \phi_0. \tag{8.35}$$

That's our solution. All we now need to do is put this result

---

[11] The main idea which motivates this substitution is the trigonometric identity $\cos^2(\phi) + \sin^2(\phi) = 1$, as we will see below.

into Eq. 8.32:

$$x = \sqrt{\frac{2E}{k}} \sin(\phi) \qquad \text{this is Eq. 8.32}$$

$$\circlearrowright \text{ Eq. 8.35}$$

$$= \sqrt{\frac{2E}{k}} \sin\left(\sqrt{\frac{k}{m}}t - \phi_0\right). \qquad (8.36)$$

We can rewrite this result using the trigonometric identity

$$\sin(a - b) = \sin(a)\cos(b) - \cos(a)\sin(b) \qquad (8.37)$$

to bring it into the form given above:

$$x = \sqrt{\frac{2E}{k}} \sin\left(\sqrt{\frac{k}{m}}t - \phi_0\right)$$

$$\circlearrowright \text{ Eq. 8.37}$$

$$= \sqrt{\frac{2E}{k}} \left(\sin\left(\sqrt{\frac{k}{m}}t\right)\cos(\phi_0) - \cos\left(\sqrt{\frac{k}{m}}t\right)\sin(\phi_0)\right)$$

$$\circlearrowright \text{ definitions}$$

$$\equiv B \sin\left(\sqrt{\frac{k}{m}}t\right) + A \cos\left(\sqrt{\frac{k}{m}}t\right)$$

$$\circlearrowright \; \omega \equiv \sqrt{\frac{k}{m}} \quad \text{(Eq. 8.22)}$$

$$\equiv B \sin(\omega t) + A \cos(\omega t).$$

This is exactly the general solution that we already wrote down in Eq. 2.30.

Now, before we move on and discuss how we can describe the pendulum, we will discuss a second method to solve the equations of motion of the harmonic oscillator. The idea behind this second method is to use a canonical transformation $(q, p) \to (Q, P)$ such that the equation of motion becomes much easier to solve when rewritten in terms of the new phase space coordinates (Q, P).

Let's see how this works.

8.5 Solving the Harmonic Oscillator Using a Canonical Transformation

First of all, we assume someone hands us the following generating function[12]

$$F(q, Q) = -\frac{1}{2} m\omega q^2 \cot(Q), \qquad (8.38)$$

where $\omega \equiv \sqrt{\frac{k}{m}}$ and $\cot(x)$ denotes the cotangent function: $\cot(x) \equiv \frac{1}{\tan(x)} = \frac{\cos(x)}{\sin(x)}$.

[12] We talked about generating functions in Section 7.3.4. Moreover, we will discuss in Section 11.1 how, in general, we can find clever generating functions like this which make a given problem much simpler.

We can then use Eq. 7.124 to determine the corresponding canonical transformation:

$$P \stackrel{7.124}{=} \frac{\partial F}{\partial Q}$$

$\circlearrowleft$ Eq. 8.38

$$= \frac{\partial \left(-\frac{1}{2} m\omega q^2 \cot(Q)\right)}{\partial Q}$$

$\circlearrowleft$ $\frac{\partial \cot(x)}{\partial x} = -\frac{1}{\sin^2(x)}$

$$= \frac{m\omega q^2}{2 \sin^2(Q)} \qquad (8.39)$$

$$p \stackrel{7.124}{=} -\frac{\partial F}{\partial q}$$

$\circlearrowleft$ Eq. 8.38

$$= -\frac{\partial \left(-\frac{1}{2} m\omega q^2 \cot(Q)\right)}{\partial q}$$

$\circlearrowleft$ $\frac{\partial q^2}{\partial q} = 2q$

$$= m\omega q \cot(Q). \qquad (8.40)$$

These two formulas implicitly define the new coordinates Q, P in terms of the old ones q, p. Our next task is to use these formulas to rewrite the Hamiltonian (Eq. 8.14)

$$H = \frac{1}{2} \frac{p^2}{m} + \frac{1}{2} kq^2$$

$\circlearrowleft$ $\omega \equiv \sqrt{\frac{k}{m}}$

$$= \frac{1}{2m}(p^2 + m^2\omega^2 q^2) \qquad (8.41)$$

in terms of the new coordinates.

This requires that we solve Eq. 8.39 and Eq. 8.40 for q and p.
First of all, we can solve Eq. 8.39 for q:

$$P = \frac{m\omega q^2}{2\sin^2(Q)}$$

↪ rearranging terms

$$\frac{2}{m\omega}P\sin^2(Q) = q^2$$

↪ √

$$\sqrt{\frac{2}{m\omega}P}\sin(Q) = q. \qquad (8.42)$$

Putting this into Eq. 8.40 yields

$$p = m\omega q \cot(Q)$$

↪ Eq. 8.42

$$= m\omega \left(\sqrt{\frac{2}{m\omega}P}\sin(Q)\right)\cot(Q)$$

↪ $\cot(Q) \equiv \frac{\cos(Q)}{\sin(Q)}$

$$= \sqrt{2m\omega P}\cos(Q). \qquad (8.43)$$

Using these two results, we can rewrite the Hamiltonian (Eq. 8.41) in terms of the new coordinates:

$$H\big(q(Q,P), p(Q,P)\big) \stackrel{(8.41)}{=} \frac{1}{2m}\left(\big(p(Q,P)\big)^2 + m^2\omega^2\big(q(Q,P)\big)^2\right)$$

↪ Eq. 8.42 and Eq. 8.43

$$= \frac{1}{2m}\left(\big(\sqrt{2m\omega P}\cos(Q)\big)^2 + m^2\omega^2\left(\sqrt{\frac{2}{m\omega}P}\sin(Q)\right)^2\right)$$

↪

$$= \frac{1}{2m}\left(2m\omega P\cos^2(Q) + m^2\omega^2\left(\frac{2}{m\omega}P\sin^2(Q)\right)\right)$$

↪ $\frac{m^2\omega^2}{m\omega} = m\omega$

$$= \frac{1}{2m}\Big(2m\omega P\cos^2(Q) + 2m\omega P\sin^2(Q)\Big)$$

↪ $2m$ and factor out $P\omega$

$$= P\omega\Big(\cos^2(Q) + \sin^2(Q)\Big)$$

↪ $\cos^2(Q) + \sin^2(Q) = 1$

$$= P\omega \equiv \tilde{H}(Q,P). \qquad (8.44)$$

We can see that the Hamiltonian looks much simpler in terms of the new coordinates.

Let's see what Hamilton's equations tell us if we use this new Hamiltonian. Hamilton's first equation (Eq. 5.15) reads

$$\frac{dP}{dt} = -\frac{\partial \tilde{H}}{\partial Q}$$

↰ Eq. 8.44

$$= -\frac{\partial (P\omega)}{\partial Q}$$

↰

$$= 0. \qquad (8.45)$$

This means that P is a constant.

Moreover, Hamilton's second equation (Eq. 5.15) tells us that

$$\frac{dQ}{dt} = \frac{\partial \tilde{H}}{\partial P}$$

↰ (Eq. 8.44)

$$= \frac{\partial (P\omega)}{\partial P}$$

↰

$$= \omega. \qquad (8.46)$$

We can immediately integrate this equation to determine how Q changes in time:

$$Q = \omega t + Q_0, \qquad (8.47)$$

where Q_0 is a constant of integration.

Now, as a final step, we can substitute these results into Eq. 8.42 and Eq. 8.43 to determine the solution in terms of our original coordinates:

$$q \stackrel{(8.42)}{=} \sqrt{\frac{2}{m\omega}} P \sin(Q)$$

↰ Eq. 8.47

$$= \sqrt{\frac{2}{m\omega} P} \sin(\omega t + Q_0)$$

$$p \stackrel{(8.43)}{=} \sqrt{2m\omega P} \cos(Q)$$

↰ Eq. 8.47

$$= \sqrt{2m\omega P} \cos(\omega t + Q_0), \qquad (8.48)$$

where P is a constant that we need to determine using specific initial conditions. This solution is exactly what we found in the previous method using a more conventional approach.[13]

Let's now move on and talk about the pendulum, which is a bit more complicated because to describe it, we need to take a constraint into account.

[13] Reminder: in Eq. 8.36 we found:

$$q = \sqrt{\frac{2E}{k}} \sin\left(\sqrt{\frac{k}{m}} t - \phi_0 \right)$$

$$= \sqrt{\frac{2E}{m\omega^2}} \sin(\omega t - \phi_0),$$

where we used $\omega \equiv \sqrt{\frac{k}{m}}$ to get to the second line. To see the equivalence, take note that the new momentum coordinate P (which we found out is just a constant) is exactly equal to the total energy E of the oscillator divided by the frequency ω: $P = \frac{E}{\omega}$.

9

The Pendulum

Right after the harmonic oscillator, the second most important system in physics is the pendulum. In the context of classical mechanics, it is especially useful to understand why the Lagrangian formalism is helpful.

A pendulum consists of a bob of mass m which is suspended from some fixed ceiling by a string of length l.[1] We want to describe how the bob swings back and forth under the influence of the Earth's gravitational field.[2]

[1] By the way, "pendulus" is Latin and means "hanging".

[2] Take note that we will consider only the mathematical abstraction of a pendulum and ignore several effects which play a role for a real-world pendulum. For example, we ignore that the string has a mass and we assume that the string cannot be bent. So in other words, instead of a bob attached to a string, we consider a bob attached to a massless rigid rod. Moreover, we assume that the pendulum swings in a perfect plane and that we can ignore air resistance.

Using Cartesian coordinates, we need (x, y) to describe the position of the bob. However, x and y are not completely independent because our bob is not moving around freely but always remains attached to the string of length l. Mathematically, this means that we have the constraint

$$x^2 + y^2 = l^2. \tag{9.1}$$

Therefore, our pendulum really only has one degree of freedom. When x changes, the variation of y is completely fixed by the condition in Eq. 9.1. Therefore, we could eliminate one of the coordinates using Eq. 9.1:

$$x^2 + y^2 = l^2$$

$$\therefore \quad y = \sqrt{l^2 - x^2}. \tag{9.2}$$

Using this formula, we can express everything in terms of x.

However, a much more convenient method is to use the angle ϕ between the pendulum and its equilibrium position to describe the configuration of the pendulum. Each value of the angle ϕ corresponds to one specific position of the bob. This is another way to see that there is just one degree of freedom.

The Cartesian coordinates are related to ϕ by

$$x = l \sin(\phi)$$
$$y = l \cos(\phi). \tag{9.3}$$

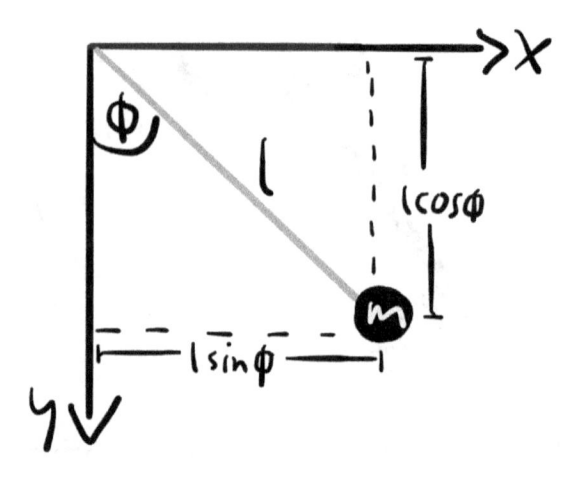

From a slightly different perspective, we can see that by using ϕ as our new coordinate we are "hard coding" the constraint in Eq. 9.1 into our description since:

$$
\begin{aligned}
x^2 + y^2 &= l^2 && \quad \text{↷ } x = l\sin(\phi),\ y = l\cos(\phi) \\
\Big(l\sin(\phi)\Big)^2 + \Big(l\cos(\phi)\Big)^2 &= l^2 && \quad \text{↷} \\
l^2\Big(\sin^2(\phi) + \cos^2(\phi)\Big) &= l^2 && \quad \text{↷ } \sin^2(\phi) + \cos^2(\phi) = 1 \\
l^2 &= l^2 \quad \checkmark
\end{aligned}
\tag{9.4}
$$

Now, let's see how we can describe the pendulum in the various frameworks. Again we will derive the correct equation of motion using the Newtonian, Lagrangian, and Hamiltonian algorithms:

$$\frac{d^2\phi}{dt^2} = -\frac{g}{l}\sin(\phi)$$

Newtonian Algorithm Lagrangian Algorithm Hamiltonian Algorithm

$$\vec{F} = \begin{pmatrix} 0 \\ mg \end{pmatrix} + \begin{pmatrix} -t\sin(\phi) \\ -t\cos(\phi) \end{pmatrix} \qquad L = \frac{1}{2}ml^2\dot{\phi}^2 + mgl\cos(\phi) \qquad H = \frac{1}{2}\frac{p^2}{ml^2} - mgl\cos(\phi)$$

And afterwards, in Section 9.4, we will discuss how we can solve this equation of motion.

9.1 Newtonian Description

As usual, the first question that we need to answer in the Newtonian framework is: which forces act on our object in question?

While for the harmonic oscillator system discussed in the previous chapter, the gravitational force didn't play an important role, for the pendulum it is the main actor. The gravitational force $\vec{F}_G$ is what pushes the pendulum back to the equilibrium position and therefore causes it to swing back and forth. However, gravity is not the only force acting on our bob. If it was, the bob would simply fall to the ground. This does not happen because the string our bob is attached to exerts a tension force $\vec{F}_T$.[3]

[3] In other words, the string provides a constant centripetal force.

Our next goal is to write down formulas which describe these two forces.

The gravitational force always points downward and therefore reads[4]

[4] As shown in the images above, we choose our coordinate system such that the y-axis points downward. Therefore, the gravitational force points in the positive y-direction.

$$\vec{F}_G = \begin{pmatrix} 0 \\ mg \end{pmatrix}. \qquad (9.5)$$

What can we say about the tension force $\vec{F}_T$?

Well, we know that it always points radially toward the suspension of the pendulum. Mathematically, this means that

$$\vec{F}_T = \begin{pmatrix} -t\sin(\phi) \\ -t\cos(\phi) \end{pmatrix}, \qquad (9.6)$$

where t describes the magnitude of the tension force which we still need to figure out.[5]

[5] We can do this by using the constraint that our bob remains attached to the string of fixed length l.

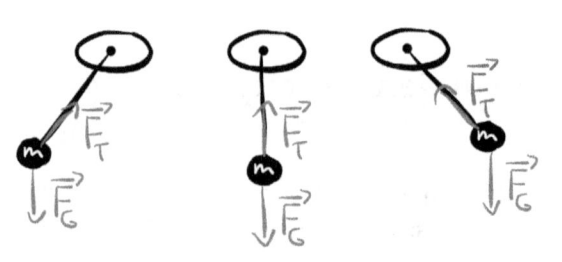

Equipped with Eq. 9.5 and Eq. 9.6 we can write down the equation of motion using Newton's second law (Eq. 3.1):

$$\frac{d}{dt}\vec{p} = \vec{F}$$

$$\frac{d}{dt}\vec{p} = \vec{F}_G + \vec{F}_T$$

⟩ Eq. 9.5 and Eq. 9.6

$$\frac{d}{dt}\vec{p} = \begin{pmatrix} 0 \\ mg \end{pmatrix} + \begin{pmatrix} -t\sin(\phi) \\ -t\cos(\phi) \end{pmatrix}$$

⟩ $p = \begin{pmatrix} m\dot{x} \\ m\dot{y} \end{pmatrix}$

$$\frac{d}{dt}\begin{pmatrix} m\dot{x} \\ m\dot{y} \end{pmatrix} = \begin{pmatrix} 0 \\ mg \end{pmatrix} + \begin{pmatrix} -t\sin(\phi) \\ -t\cos(\phi) \end{pmatrix}$$

$$\begin{pmatrix} m\ddot{x} \\ m\ddot{y} \end{pmatrix} = \begin{pmatrix} 0 \\ mg \end{pmatrix} + \begin{pmatrix} -t\sin(\phi) \\ -t\cos(\phi) \end{pmatrix}. \tag{9.7}$$

The first row in the last line tells us:

$$m\ddot{x} = -t\sin(\phi)$$

⟩ Eq. 9.3

$$= -t\frac{x}{l} \tag{9.8}$$

and the second row tells us:

$$m\ddot{y} = mg - t\cos(\phi)$$

⟩ Eq. 9.3

$$= mg - t\frac{y}{l}. \tag{9.9}$$

By switching to a description of the pendulum in terms of the angle ϕ, we can use these two equations to determine the magnitude of the tension force t and derive the equation of motion for ϕ. But first of all, we need

$$x = l\sin(\phi)$$

⟩ chain rule

$$\therefore \quad \dot{x} = l\cos(\phi)\dot{\phi}$$

⟩ product rule

$$\therefore \quad \ddot{x} = -l\sin(\phi)\dot{\phi}^2 + l\cos(\phi)\ddot{\phi}$$

$$y = l\cos(\phi)$$

⟩ chain rule

$$\therefore \quad \dot{y} = -l\sin(\phi)\dot{\phi}$$

⟩ product rule

$$\therefore \quad \ddot{y} = -l\cos(\phi)\dot{\phi}^2 - l\sin(\phi)\ddot{\phi}. \tag{9.10}$$

Putting this into our equations of motion (Eq. 9.8, Eq. 9.9) yields

$$m\ddot{x} = -t\sin(\phi) \quad \text{Eq. 9.10}$$

$$m\Big(-l\sin(\phi)\dot{\phi}^2 + l\cos(\phi)\ddot{\phi}\Big) = -t\sin(\phi) \tag{9.11}$$

$$m\ddot{y} = mg - t\cos(\phi) \quad \text{Eq. 9.10}$$

$$m\Big(-l\cos(\phi)\dot{\phi}^2 - l\sin(\phi)\ddot{\phi}\Big) = mg - t\cos(\phi). \tag{9.12}$$

These two equations can be combined cleverly such that we find[6]

$$\ddot{\phi} = -\frac{g}{l}\sin(\phi)$$

$$t = ml\dot{\phi}^2 + mg\cos(\phi). \tag{9.13}$$

The first equation here is the correct equation of motion that we need to solve for $\phi(t)$ as soon as initial conditions are specified. This resulting function $\phi(t)$ describes how our pendulum swings. The second equation tells us exactly how large the tension force is. However, often we are not really interested in things like the tension force explicitly.[7] Usually we only care about how the object in question moves. That's where the Lagrangian formalism comes in handy.

But first, let's see how the formulas in Eq. 9.13 follow from Eq. 9.11 and Eq. 9.12.[8]

Multiplying Eq. 9.11 by $\sin(\phi)$ and Eq. 9.12 by $\cos(\phi)$ and then taking their sum:

$$\Big(\text{Eq. 9.11} \times \sin\phi\Big) + \Big(\text{Eq. 9.12} \times \cos\phi\Big)$$

yields

$$m(l\ddot{\phi}\cos\phi\sin\phi - l\dot{\phi}^2\sin^2\phi) + m(-l\ddot{\phi}\sin\phi\cos\phi - l\dot{\phi}^2\cos^2\phi) = -t\sin^2\phi - t\cos^2\phi + mg\cos\phi$$

$$\therefore \quad -ml\dot{\phi}^2 = -t + mg\cos\phi. \quad \text{$\cos^2\phi + \sin^2\phi = 1$}$$

Analogously, multiplying Eq. 9.11 by $\cos(\phi)$ and Eq. 9.12 by $\sin(\phi)$ and then taking their difference:

$$\Big(\text{Eq. 9.11} \times \cos\phi\Big) - \Big(\text{Eq. 9.12} \times \sin\phi\Big)$$

---

[6] We will discuss how this can be done in a moment.

[7] The explicit formula for the tension force is useful, for example, when we want to calculate how fast the pendulum can spin before it breaks.

[8] As usual, if you're not interested in mathematical details, feel free to skip the following paragraphs.

yields

$$m(l\ddot{\phi}\cos^2\phi - l\dot{\phi}^2\sin\phi\cos\phi) - m(-l\ddot{\phi}\sin^2\phi - l\dot{\phi}^2\cos\phi\sin\phi) = -t\sin\phi\cos\phi + t\cos\phi\sin\phi - mg\sin\phi$$

$$\therefore\ ml\ddot{\phi} = -mg\sin\phi.$$

⟩ $\cos^2\phi + \sin^2\phi = 1$

Next, let's discuss how all this becomes much easier when we use the Lagrangian formalism.

9.2 Lagrangian Description

The main new feature of the pendulum compared to the harmonic oscillator is that we have to take a constraint (Eq. 9.1)[9]

$$x^2 + y^2 = l^2 \quad (9.14)$$

into account.

We can do this by using the method of Lagrange multipliers.[10] So, to describe the pendulum, we use the Lagrangian for an object moving freely in two dimensions in the Earth's gravitational field

$$L_{\text{free}} = T - V = \frac{1}{2}m(\dot{x}^2 + \dot{y}^2) + mgy \quad (9.15)$$

and add the Lagrange multiplier term:

$$L_{\text{con}} = \frac{1}{2}\lambda(x^2 + y^2 - l^2) \quad (9.16)$$

which yields

$$L_{\text{pendulum}} = L_{\text{free}} + L_{\text{con}}$$

$$= \frac{1}{2}m(\dot{x}^2 + \dot{y}^2) + mgy + \frac{1}{2}\lambda(x^2 + y^2 - l^2), \quad (9.17)$$

⟩

where λ is our Lagrange multiplier. This Lagrange multiplier represents the pendulum constraint (Eq. 9.1) in the Lagrangian formalism.

[9] Strictly speaking, this is not quite true because our object attached to a spring was not allowed to move around freely either. We only allowed up-and-down-movements but no swinging etc. In other words, our harmonic oscillator was restricted to one dimension which means mathematically that we introduced the constraints

$$y = 0$$

and

$$z = 0.$$

Analogously, here we only allow that the pendulum swings in two dimensions, which means we have the additional constraint $z = 0$.

[10] We discussed Lagrange multipliers in Section 7.2.1. Moreover, the Lagrange multiplier method is discussed in general terms in Appendix C.

Equipped with this Lagrangian for the pendulum, we can directly calculate the equations of motion by using the Euler-Lagrange equation (Eq. 4.25)

$$\frac{\partial L}{\partial q} = \frac{d}{dt}\left(\frac{\partial L}{\partial \dot{q}}\right). \tag{9.18}$$

For $q = x$, we find

$$\frac{\partial L}{\partial x} = \frac{d}{dt}\left(\frac{\partial L}{\partial \dot{x}}\right)$$

$$\frac{\partial\left(\frac{1}{2}m(\dot{x}^2 + \dot{y}^2) + mgy + \frac{1}{2}\lambda(x^2 + y^2 - l^2)\right)}{\partial x} = \frac{d}{dt}\left(\frac{\partial\left(\frac{1}{2}m(\dot{x}^2 + \dot{y}^2) + mgy + \frac{1}{2}\lambda(x^2 + y^2 - l^2)\right)}{\partial \dot{x}}\right) \quad \circlearrowleft \text{ Eq. 9.17}$$

$$\lambda x = \frac{d}{dt}(m\dot{x})$$

$$\lambda x = m\ddot{x}. \tag{9.19}$$

And for $q = y$, we find

$$\frac{\partial L}{\partial y} = \frac{d}{dt}\left(\frac{\partial L}{\partial \dot{y}}\right)$$

$$\frac{\partial\left(\frac{1}{2}m(\dot{x}^2 + \dot{y}^2) + mgy + \frac{1}{2}\lambda(x^2 + y^2 - l^2)\right)}{\partial y} = \frac{d}{dt}\left(\frac{\partial\left(\frac{1}{2}m(\dot{x}^2 + \dot{y}^2) + mgy + \frac{1}{2}\lambda(x^2 + y^2 - l^2)\right)}{\partial \dot{y}}\right) \quad \circlearrowleft \text{ Eq. 9.17}$$

$$mg + \lambda y = m\ddot{y}. \tag{9.20}$$

These are the correct equations of motion for our pendulum that we have already derived using the Newtonian formalism (Eq. 9.8, Eq. 9.9). By comparing our result here with our previous result, we can conclude that $\lambda = -\frac{t}{l}$, which means that our Lagrange multiplier λ is directly proportional to the tension t.

Moreover, as a consistency check, we can evaluate the Euler-

Lagrange equation for $q = \lambda$:

$$\frac{\partial L}{\partial \lambda} = \frac{d}{dt}\left(\frac{\partial L}{\partial \dot{\lambda}}\right)$$

⤷ Eq. 9.17

$$\frac{\partial\left(\frac{1}{2}m(\dot{x}^2+\dot{y}^2)+mgy+\frac{1}{2}\lambda(x^2+y^2-l^2)\right)}{\partial \lambda} = \frac{d}{dt}\left(\frac{\partial\left(\frac{1}{2}m(\dot{x}^2+\dot{y}^2)+mgy+\frac{1}{2}\lambda(x^2+y^2-l^2)\right)}{\partial \dot{\lambda}}\right)$$

⤷

⤷

$$\frac{1}{2}(x^2+y^2-l^2) = 0$$

$$x^2+y^2-l^2 = 0. \qquad (9.21)$$

This is exactly our pendulum constraint (Eq. 9.1).

There is another cool thing we can do using the Lagrangian formalism. First of all, we need to recall that the Euler-Lagrange equation is valid for any choice of coordinates.[11] In addition, we are often not really interested in constraint forces like the tension force F_T but only want to know how the object in question moves. If this is the case, we can use our freedom to choose suitable coordinates to make the Lagrangian description even simpler.

[11] We discussed this in Section 7.2.1.

In fact, we've seen this already at the beginning of this chapter. When we switch from our Cartesian coordinates (x, y) to a description in terms of the angle ϕ, the constraint becomes trivially true.[12] In particular, this means that our Lagrange multiplier term (Eq. 9.16) becomes as simple as it gets:

[12] This was shown in Eq. 9.4.

$$L_{\text{con}} = \frac{1}{2}\lambda(x^2+y^2-l^2)$$

⤷ Eq. 9.3

$$= \frac{1}{2}\lambda\left((l\sin\phi)^2+(l\cos\phi)^2-l^2\right)$$

⤷

$$= \frac{1}{2}\lambda\left(l^2(\sin^2\phi+\cos^2\phi)-l^2\right)$$

⤷ $\sin^2\phi+\cos^2\phi = 1$

$$= \frac{1}{2}\lambda\left(l^2-l^2\right)$$

⤷

$$= 0. \qquad (9.22)$$

In terms of ϕ, the full pendulum Lagrangian (Eq. 9.17) therefore reads:

$$\begin{aligned}
L_{\text{pendulum}} &= \frac{1}{2}m(\dot{x}^2 + \dot{y}^2) + mgy + \frac{1}{2}\lambda(x^2 + y^2 - l^2) \\
&\quad \circlearrowright \text{ Eq. 9.10 and Eq. 9.22} \\
&= \frac{1}{2}m\left(\left(l\cos(\phi)\dot{\phi}\right)^2 + \left(-l\sin(\phi)\dot{\phi}\right)^2\right) + mg\left(l\cos(\phi)\right) \\
&\quad \circlearrowright \\
&= \frac{1}{2}ml^2\dot{\phi}^2\left(\cos^2(\phi) + \sin^2(\phi)\right) + mgl\cos(\phi) \\
&\quad \circlearrowright \; \sin^2\phi + \cos^2\phi = 1 \\
&= \frac{1}{2}ml^2\dot{\phi}^2 + mgl\cos(\phi).
\end{aligned} \quad (9.23)$$

If we now use the Euler-Lagrange equation for $q = \phi$, we find

$$\frac{\partial L}{\partial \phi} = \frac{d}{dt}\left(\frac{\partial L}{\partial \dot{\phi}}\right)$$

$\circlearrowright$ Eq. 9.17

$$\frac{\partial\left(\frac{1}{2}ml^2\dot{\phi}^2 + mgl\cos(\phi)\right)}{\partial \phi} = \frac{d}{dt}\left(\frac{\partial\left(\frac{1}{2}ml^2\dot{\phi}^2 + mgl\cos(\phi)\right)}{\partial \dot{\phi}}\right)$$

$\circlearrowright$

$$-mgl\sin(\phi) = ml^2\frac{d}{dt}(\dot{\phi})$$

$\circlearrowright$

$$-mg\sin(\phi) = ml\ddot{\phi}. \quad (9.24)$$

This is exactly the equation of motion that we already derived in Eq. 9.13. However, the derivation using the Lagrangian formalism is much simpler. In the Newtonian framework, we have to perform lots of intermediate steps. In particular, we have to find the tension force exerted by the string and need to think carefully about the directions of the various forces. Moreover, we've discovered that the most natural description of the pendulum is in terms of the angle ϕ. But Newton's second law is a relationship between vectors in a Cartesian coordinate system, and while it is possible to switch to different coordinates, this is quite cumbersome.

In contrast, in the Lagrangian framework, we only have to figure out the potential and kinetic energy in terms of the convenient coordinate ϕ. Since the Euler-Lagrange equation is valid for any choice of coordinates, we can then directly calculate the equation of motion without thinking about the tension force and the directions of the forces.

In general, in the Lagrangian formalism, as soon as we have found suitable coordinates which make the constraints trivially true, we don't have to care about the constraints at all. This happens because constraints are implemented in the Lagrangian framework using Lagrange multipliers and, for a suitable choice of coordinates, the Lagrange multiplier terms vanish. Therefore, as soon as we include these additional terms, we no longer need to think about the constraints.

Next, let's talk about the Hamiltonian description of a pendulum.

9.3 Hamiltonian Description

As usual in the Hamiltonian framework, our first task is to calculate the Hamiltonian H and the generalized momentum p. Using the pendulum Lagrangian we derived in the previous section (Eq. 9.23) and the general definition of the generalized momentum p (Eq. 4.29), we can calculate

$$p = \frac{\partial L}{\partial \dot\phi}$$

$$= \frac{\partial \left(\frac{1}{2} m l^2 \dot\phi^2 + mgl\cos(\phi)\right)}{\partial \dot\phi}$$

$$= ml^2 \dot\phi. \tag{9.25}$$

This tells us

$$\dot\phi = \frac{p}{ml^2}. \tag{9.26}$$

With this result in hand, we can derive the Hamiltonian using the general definition in terms of the Lagrangian (Eq. 5.18)[13]

[13] Take note that this Hamiltonian has the usual form $H = T + V$ where $V = -mgl\cos(\phi)$ is the bob's potential energy in the Earth's gravitational field and $T = \frac{1}{2}ml^2\dot\phi^2$ is its kinetic energy.

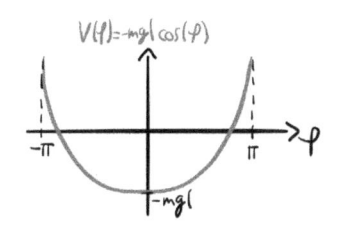

$$H = p\dot{\phi} - L$$
⤳ Eq. 9.26
$$= \frac{p^2}{ml^2} - L$$
⤳ Eq. 9.23
$$= \frac{p^2}{ml^2} - \left(\frac{1}{2}ml^2\dot{\phi}^2 + mgl\cos(\phi)\right)$$
⤳ Eq. 9.26
$$= \frac{p^2}{ml^2} - \left(\frac{1}{2}ml^2\left(\frac{p}{ml^2}\right)^2 + mgl\cos(\phi)\right)$$
⤳
$$= \frac{1}{2}\frac{p^2}{ml^2} - mgl\cos(\phi). \tag{9.27}$$

Then, we can use Hamilton's first equation to calculate

$$\frac{dp}{dt} = -\frac{\partial H}{\partial \phi}$$
⤳ Eq. 9.27
$$= -\frac{\partial\left(\frac{1}{2}\frac{p^2}{ml^2} - mgl\cos(\phi)\right)}{\partial \phi}$$
⤳ $\frac{\partial \cos(\phi)}{\partial \phi} = -\sin(\phi)$
$$= -mgl\sin(\phi). \tag{9.28}$$

And Hamilton's second equation tells us

$$\frac{d\phi}{dt} = \frac{\partial H}{\partial p}$$
⤳ Eq. 9.27
$$= \frac{\partial\left(\frac{1}{2}\frac{p^2}{ml^2} - mgl\cos(\phi)\right)}{\partial p}$$
⤳
$$= \frac{p}{ml^2}. \tag{9.29}$$

[14] We will see in a moment why this is a clever idea.

We can then take the derivative of Eq. 9.29[14]

$$\frac{d\phi}{dt} = \frac{p}{ml^2}$$
this is Eq. 9.29
⤳ $\frac{d}{dt}$
$$\frac{d^2\phi}{dt^2} = \frac{\frac{dp}{dt}}{ml^2}$$
⤳ $\times ml^2$
$$ml^2\frac{d^2\phi}{dt^2} = \frac{dp}{dt} \tag{9.30}$$

and use this result to eliminate p from Eq. 9.28:

$$\frac{dp}{dt} = -mgl\sin(\phi) \quad \text{this is Eq. 9.28}$$

$$\downarrow \text{Eq. 9.30}$$

$$ml^2 \frac{d^2\phi}{dt^2} = -mgl\sin(\phi)$$

$$\downarrow \not{m}\not{l}$$

$$\frac{d^2\phi}{dt^2} = -\frac{g}{l}\sin(\phi). \tag{9.31}$$

Once more, this is the correct equation of motion that we have already derived above (Eq. 9.13).

We can therefore see that the Hamiltonian formalism does not offer any further simplifications. It's merely a different way of doing the same thing.

9.4 Solving the Equation of Motion

Now, it's finally time to talk about the equation

$$\frac{d^2\phi}{dt^2} = -\frac{g}{l}\sin(\phi) \tag{9.32}$$

that we derived three times in the previous sections in a bit more detail.

Although this equation seems really simple, it's extremely difficult to solve.[15]

[15] To spoil the surprise, we can't write down a solution of Eq. 9.32 in terms of elementary functions.

As a first step, we can use the same trick that we already used for the equation of motion of the harmonic oscillator.

So first of all, we multiply the equation of motion by the velocity $\dot\phi$

$$\frac{d^2\phi}{dt^2} = -\frac{g}{l}\sin(\phi)$$

$$\rightarrow \quad \dot\phi\frac{d^2\phi}{dt^2} = -\dot\phi\frac{g}{l}\sin(\phi). \tag{9.33}$$

This is clever because we can now rewrite the left-hand side using the chain rule $\left(\frac{d}{dt} f(t)^2 = 2f(t) \frac{df(t)}{dt} \right)$:

$$\dot{\phi} \frac{d^2\phi}{dt^2} = \dot{\phi} \frac{d\dot{\phi}}{dt} = \frac{1}{2} \frac{d}{dt} \dot{\phi}^2. \tag{9.34}$$

Moreover, we can rewrite the right-hand side using

$$\dot{\phi} \sin(\phi) = -\frac{d}{dt} \cos(\phi). \tag{9.35}$$

Eq. 9.33 then reads

$$\dot{\phi} \frac{d^2\phi}{dt^2} = -\dot{\phi} \frac{g}{l} \sin(\phi)$$

↪ Eq. 9.34 and Eq. 9.35

$$\frac{1}{2} \frac{d}{dt} \dot{\phi}^2 = \frac{g}{l} \frac{d}{dt} \cos(\phi)$$

↪ rearranging terms

$$\frac{d}{dt} \left(\frac{1}{2} \dot{\phi}^2 - \frac{g}{l} \cos(\phi) \right) = 0. \tag{9.36}$$

In words, this tells us that the quantity between the big parentheses on the left-hand side is constant in time.[16]

Moreover, we can integrate Eq. 9.36 which yields

$$\frac{d}{dt} \left(\frac{1}{2} \dot{\phi}^2 - \frac{g}{l} \cos(\phi) \right) = 0$$

↪ integrating

$$\frac{1}{2} \dot{\phi}^2 - \frac{g}{l} \cos(\phi) = C, \tag{9.37}$$

where C is an integration constant.[17]

Next, by taking the square root and separating the variables, we

[16] If we multiply the quantity we find here by ml^2, we get exactly the Hamiltonian which describes the total energy.

[17] C is directly proportional to the energy of the pendulum: $C = \frac{E}{ml^2}$. This follows because if we multiply the quantity on the left-hand side by ml^2, we get exactly the total energy.

find

$$\frac{1}{2}\dot{\phi}^2 - \frac{g}{l}\cos(\phi) = C$$

$$\dot{\phi}^2 = 2C + \frac{2g}{l}\cos(\phi) \quad \circlearrowright \text{ rearranging terms}$$

$$\dot{\phi} = \sqrt{2C + \frac{2g}{l}\cos(\phi)} \quad \circlearrowright \checkmark$$

$$\frac{d\phi}{dt} = \sqrt{2C + \frac{2g}{l}\cos(\phi)} \quad \circlearrowright \dot{\phi} = \frac{d\phi}{dt}$$

$$\frac{d\phi}{\sqrt{2C + \frac{2g}{l}\cos(\phi)}} = dt \quad \circlearrowright \text{ separating variables}$$

$$\int \frac{d\phi}{\sqrt{2C + \frac{2g}{l}\cos(\phi)}} = t. \quad \circlearrowright \int \quad (9.38)$$

This type of integral is known as an elliptic integral and there is no simple solution. But we can solve it, for example, using a tool like Mathematica or Matlab.

To get at least some insight, we can evaluate the equation of motion in the limit that the pendulum only swings a little. Mathematically, this means $\phi \ll 1$. The equation of motion simplifies dramatically in this limit which follows when we Taylor expand $\sin(\phi)$:[18]

$$\sin(\phi) = \phi - \frac{\phi^3}{6} + \frac{\phi^5}{120} + \ldots \quad (9.39)$$

As long as the pendulum only swings a little ($\phi \ll 1$), we can safely ignore all higher order terms ($\phi^3 \approx 0, \phi^5 \approx 0$, etc.)[19] and the equation of motion (Eq. 9.32) therefore reads:

$$\frac{d^2\phi}{dt^2} = -\frac{g}{l}\sin(\phi)$$

$$= -\frac{g}{l}\left(\phi - \frac{\phi^3}{6} + \frac{\phi^5}{120} + \ldots\right) \quad \circlearrowright \text{ Taylor expansion (Eq. 9.39)}$$

$$\approx -\frac{g}{l}\phi. \quad \circlearrowright \phi \ll 1 \quad (9.40)$$

[18] The Taylor expansion of $\sin(\phi)$ is discussed in Appendix F.

[19] This follows because for any positive number smaller than 1, higher orders are much smaller. For example, for $\phi = 0.1$, we find $\phi^3 = 0.001$ and $\phi^5 = 0.00001$.

This means that for small excitations of the pendulum, we can describe it using solutions of the simpler equation of motion

$$\frac{d^2\phi}{dt^2} = -\frac{g}{l}\phi. \tag{9.41}$$

But this is exactly the equation of motion of the harmonic oscillator (Eq. 8.3)! In words, this means that as long as the pendulum only swings a little, its movement is analogous to the movement of the harmonic oscillator. Therefore, in the small angle limit $\phi \ll 1$, we can again use the solutions that we discovered in Section 8.4.

Now that we've developed some understanding of how the different formulations of classical mechanics work in practice, it's time to move on to the good stuff.

Part III
Get an Understanding of Classical Mechanics You Can Be Proud Of

"There will come a time when you believe everything is finished. That will be the beginning."

Louis L'Amour

PS: You can discuss the content of Part III with other readers and give feedback at www.nononsensebooks.com/cm/bonus.

In this final part, we'll elevate our understanding of classical mechanics to a completely new level. In particular, we'll talk about three major topics: Noether's theorem, additional formulations of classical mechanics and the origin of classical mechanics.

Let me outline shortly why these topics are interesting.

▷ Noether's theorem is one of the most beautiful insights in all of physics. It allows us to understand that there is a deep connection between symmetries and conserved quantities. In particular, it allows us to understand why the concepts of momentum, energy and angular momentum are so important in all fundamental theories of physics.

▷ Alternative formulations like Hamilton-Jacobi mechanics and Koopman-von Neumann mechanics allow us to tackle problems in completely new ways and help us to reveal deep insights about the general structure of classical mechanics. In addition, we will talk about statistical mechanics. We need statistical mechanics whenever we are dealing with uncertainty, which is necessarily the case when we want to describe *lots* of objects at once. Moreover, a solid understanding of statistical mechanics is essential to understand the difference between classical mechanics and quantum mechanics.

▷ Finally, to really understand classical mechanics, we need to talk about its origin. This means that we need to understand how exactly classical mechanics is related to more fundamental theories like quantum mechanics and special relativity. By discussing these connections, we will finally understand *why* the least action principle works and why the Lagrangian has the form it has in classical mechanics ($L = T - V$).

▷ In addition, we'll talk about books you should read to learn more about specific topics.

Without further ado, let's dive in.

10

Noether's Theorem

In a nutshell, Noether's theorem tells us that for each symmetry of a given system, there must be a conserved quantity:[1]

$$\boxed{\text{Symmetry}} \xrightarrow{\text{Noether's theorem}} \boxed{\text{Conserved Quantity}}$$

In particular:

▷ If the system possesses rotational symmetry, we know immediately that angular momentum is conserved. In other words, this means that if we can rotate our system without changing anything, angular momentum is conserved.

▷ If the system is symmetric with respect to spatial translations $x \to x + \epsilon$, we know immediately that momentum is conserved. This means that if we can change the position of the whole system and nothing changes, momentum is conserved.

▷ If the system is symmetric with respect to temporal translations $t \to t + \epsilon$, we know immediately that energy is conserved. Formulated differently, if the system behaved yesterday exactly as it does today, energy is conserved.

Before we can really understand Noether's theorem, we need to understand what a symmetry is. Additionally, we need to

[1] There is also a converse Noether theorem which tells us how we can find the symmetries related to specific conserved quantities:

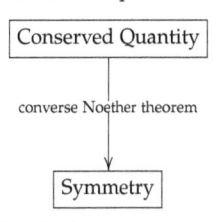

$$\boxed{\text{Conserved Quantity}}$$
$$\downarrow \text{converse Noether theorem}$$
$$\boxed{\text{Symmetry}}$$

We will talk about the converse Noether theorem in Section 10.3.2.

find a way to describe in mathematical terms that a physical system possesses a specific symmetry. So let's first talk about symmetries in quite general terms.

10.1 Symmetries

Imagine a friend stands in front of you and holds a perfectly round red ball in her hand. Then you close your eyes, your friend performs a transformation of the ball, and afterward you open your eyes again. If she rotates the ball while your eyes are closed, it is impossible for you to determine whether or not she did anything at all. Hence, rotations are symmetries of the ball.

In contrast, if she holds a cube, only very special rotations can be done without you noticing it. In general, all transformations which, in principle, change something but lead to an indistinguishable result are symmetries. Formulated differently, a symmetry takes us from one state to a different one which happens to have the same properties.[2]

[2] In contrast, a redundancy takes us from one description of a state to another description of the same state. We will talk a bit more about this distinction below.

10.1.1 Symmetries of Physical Systems

It's important to take note that with the definition given in the previous section, symmetries are *observable* properties of objects or systems. In other words, we can find out experimentally whether a given system or object possesses a specific symmetry.[3]

[3] In particular, it is important to not get confused by coordinate transformations which are always possible without physical consequences. This is discussed in a bit more detail in Appendix E.

This point can be confusing especially when it comes to symmetries of physical systems. To understand it, it's really helpful to think about **subsystems**. A subsystem is a sufficiently isolated part of the universe. And when we are interested in the symmetries of a physical system, we usually investigate how a subsystem behaves under specific transformations.[4]

For example, let's imagine there is a physicist inside a boat who is unable to look outside. Here, the boat is our subsystem. Moreover, let's assume that the physicist performs a specific physics experiment. To find out whether the physics governing the behavior of the objects in the experiment possesses, say, rotational symmetry, we rotate the boat as a whole (including the physicist and his experiment). If the physicist can't find any difference in the behavior of the experiment before and after the rotation, the laws governing the experiment are indeed rotational symmetric.[5]

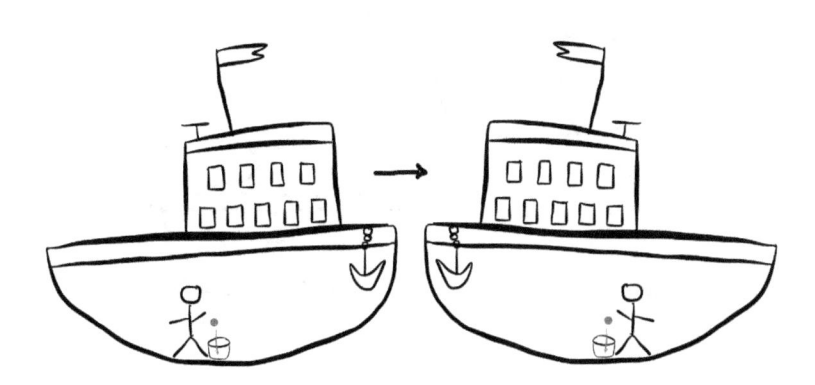

[4] The concept of a subsystem is crucial because there is no way we could ever detect a rotation of the whole universe. Hence, to establish that symmetries are indeed observable properties, we need to talk about subsystems. Moreover, in physics we almost always talk about subsystems even when this is not explicitly stated and we take the limit $r \to \infty$. (An exception is cosmology which sometimes deals with the universe as a whole.) This may seem pedantic at this point, but I promise that we can avoid a lot of confusion by thinking in terms of subsystems.

[5] Take note that the physicist is only allowed to compare the behaviour of the experiment before and after but not during the rotation. In more technical terms, we say that we perform the rotation adiabatically. This means that the transformation happens so slowly and gently that it doesn't disturb the objects (including the physicist) within the subsystem.

Analogously, we could move the boat as a whole to a new location and then let the physicist repeat the experiment. If he still can't find any difference, we say the physics governing the experiment possesses spatial translational symmetry.[6] If we let the physicist repeat the experiment after waiting for some time and there is no difference, we say the physics governing the experiment possesses temporal translational symmetry.[7]

[6] A spatial translation is a shift to a new location $x \to x + \epsilon$.

[7] A temporal translation is a shift to a new point in time $t \to t + \epsilon$.

The crucial point in all these examples is that, in principle, it would be possible to detect a difference, for example, by bringing the subsystem in contact with the outside world. Formulated differently, the ship and the rotated ship are clearly two distinct states if we look at them from the outside. But from inside the boat, it's impossible to find any difference.[8]

Now, how can we describe all this in mathematical terms?

The physicist inside the boat describes the experiment using a concrete equation of motion. Moreover, to compare the results of the experiment before and after the transformation, we can imagine that he prepares it in exactly the same initial configuration.[9] Given a specific initial configuration, he can solve the equation of motion and then compare the resulting solution with the actual behavior of the system.

[8] Again this may seem pedantic, but I emphasize this point because otherwise it's really easy to get confused by the difference between the invariance under a pure coordinate transformation (i.e., a redundancy) and a real symmetry.

[9] If he uses different initial configurations, he can't expect to see the same behavior.

If he is able to describe the experiment before and after the transformation using exactly the same solution:

| Solution of E.O.M. | ——Transformation——→ | Solution of E.O.M. |

we say the transformation is a symmetry of the solution.

From a slightly different perspective, we can imagine that our physicist repeats the experiment several times using different initial configurations before and after the transformation.

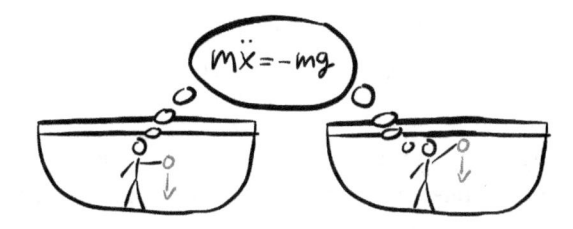

[Figure: two physicists in bowls with thought bubble $m\ddot{x} = -mg$]

If the equation of motion he deduces this way before and after the transformation are exactly the same:

| Equation of Motion | —Transformation→ | Equation of Motion |

we say the transformation is a symmetry of the equation of motion. Intuitively, this means that there is no way he can find any difference in the laws which govern the behavior of the objects in the experiment. This is really what we mean by a symmetry in physics.

And finally, we can imagine that the physicist tries to describe the system using, say, the Lagrangian formalism.

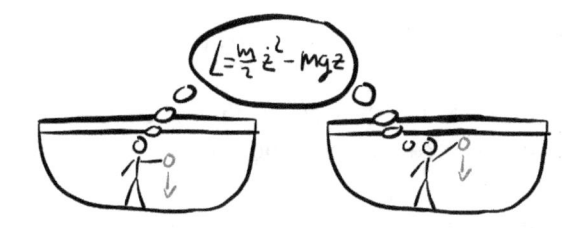

[Figure: two physicists in bowls with thought bubble $L = \frac{m}{2}\dot{z}^2 - mgz$]

If he finds that before and after the transformation the same Lagrangian describes the experiment accurately:

| Lagrangian | —Transformation→ | Lagrangian |

we say the transformation is a symmetry of the Lagrangian.

With this in mind, let's talk about Noether's theorem.

10.2 Noether's Theorem Intuitively

To get a rough feeling for why there is a connection between conserved quantities and symmetries, it's helpful to understand when quantities are not conserved and when a system is not symmetric.[10]

Again, it's instructive to think about a physicist in a subsystem. But this time let's assume that our subsystem is a spaceship.

Moreover, let's assume that the experiment our physicist carries out is that he releases a ball and observes its trajectory.

Since we want to discuss a system in which a specific quantity, say, momentum is not conserved and which is not translationally symmetric, we assume that there is a potential $U(x)$ which directly influences the ball. We assume that our physicist can describe this potential in his specific coordinate system using:[11]

$$U(x) = ax^2, \qquad (10.1)$$

where a is a constant that describes the potential strength.

Moreover, since we are in a spaceship there is no gravitational pull or any other force acting on the ball.

Let's assume that the physicist inside the box uses a coordinate system with its origin exactly at the location of his shoulder such that the ball's initial position is $x(0) = 1$ m. This detail will be important in a moment. In particular, it will be important that our physicist uses (at least from his perspective) the same coordinate system before and after the transformation.

In this setup, our physicist quickly notices that the momentum of the ball is not conserved. Even if he releases the ball without giving it a push, it will start moving as a result of the potential $U(x)$:

---

[10] As discussed in the previous section, the symmetry of physical systems is a quite different concept from the symmetry of concrete geometric shapes. In particular, a system can posses a symmetry even though a concrete configuration is not invariant under the corresponding transformation. Formulated differently, there is a distinction between the symmetries of physical laws and the symmetries of things.

[11] Take note that the total momentum is always conserved. But, for the sake of argument, we imagine that the potential $U(x)$ arises from some unknown background structure and simply absorbs the momentum of the ball and then leaks it to some object or structure outside the spaceship. This means that the momentum inside the spaceship is not conserved. But nevertheless the total momentum in the whole universe is conserved. In general, a quantity like energy or momentum is only not conserved if the subsystem we consider is not really isolated and therefore some energy or momentum can leak out through interactions with the environment.

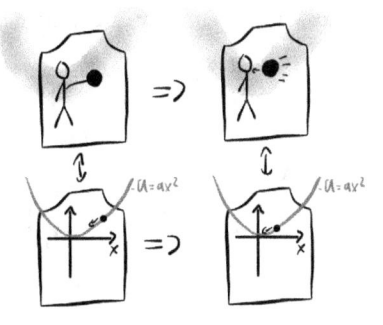

The force resulting from $U(x)$ is

$$F(x) = -\frac{dU(x)}{dx} = -\frac{d(ax^2)}{dx} = -2ax. \quad (10.2)$$

Therefore Newton's second law (Eq. 3.1) tells us

$$\frac{dp}{dt} = F(x) = -2ax. \quad (10.3)$$

If momentum were conserved, we would have $\frac{dp}{dt} = 0$, but Eq. 10.3 tells us that this is not the case here.

Now, what about translational symmetry?

To answer this question, we shift the spaceship 3 meters. Since we assume that the potential is due to some unspecified background structure, it stays where it is.

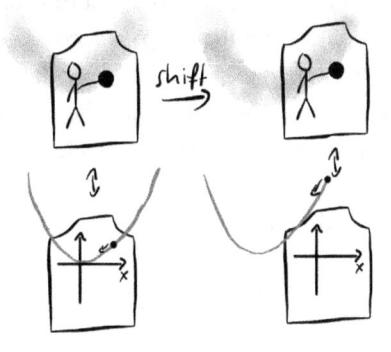

Therefore, the physicist inside the spaceship quickly notices that we performed a shift because now, he has to use the formula

$$\tilde{U}(x) = a(x + 3 \text{ m})^2 \quad (10.4)$$

to get the correct equation of motion describing the ball. The resulting force is now

$$\tilde{F}(x) = -\frac{d\tilde{U}(x)}{dx} = -\frac{d(a(x+3 \text{ m})^2)}{dx}$$
$$= -\frac{d(ax^2 + xa6 \text{ m} + a9 \text{ m}^2)}{dx}$$
$$= -2ax - a6 \text{ m} \qquad (10.5)$$

and the correct equation of motion after the shift therefore reads

$$\frac{dp}{dt} = \tilde{F}(x) = -2ax - a6 \text{ m}. \qquad (10.6)$$

So here we have a system which is not invariant under spatial translations and in which momentum is not conserved. But this, of course, does not prove by any means that there really is a direct connection. Nevertheless, it gives us a first hint of why they could be connected.

Next, let's contrast the situation discussed above, with a new one involving a system that *is* invariant under spatial translations. To construct such a system, all we have to do is assume that there is no longer any background structure which introduced the spatial inhomogeneity in the example above. To get a non-trivial situation, let's assume that there is again a potential $V(x)$, but this time it originates from a second object which is also located within the spaceship. For simplicity, let's assume this object is located exactly at the shoulder of our physicist, i.e., directly at the origin of the coordinate system he is using.

So our new potential reads

$$V(x) = bx^2, \qquad (10.7)$$

where b is again a constant that specifies the potential strength. If we now shift the spaceship by 3 meters, it's impossible for the physicist inside the spaceship to find out that anything changed because now the potential is shifted too.

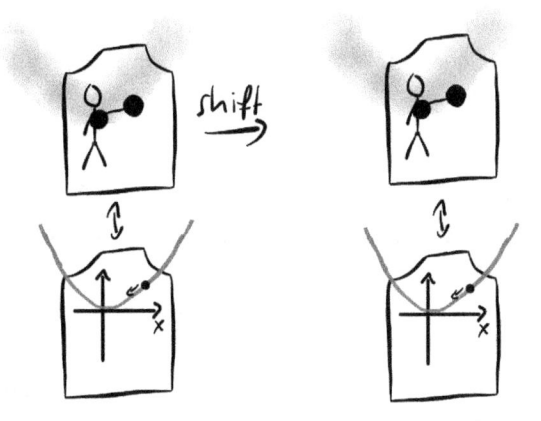

Therefore, after the shift, the potential still reads[12]

$$V(x) = bx^2,\quad (10.8)$$

and the equation of motion remains unchanged.

Now, what about the conservation of momentum in this second situation?

If our physicist releases the ball without pushing it, it will start moving again. Therefore, the momentum of this ball is still not conserved. However, the source of the potential (the second ball) is now located within the spaceship too. Thus, our physicist will observe that as soon as the ball he releases starts moving in one direction, the second ball will move in the opposite direction. If he calculates the total momentum of the two balls, he will find that it is conserved.

To summarize:

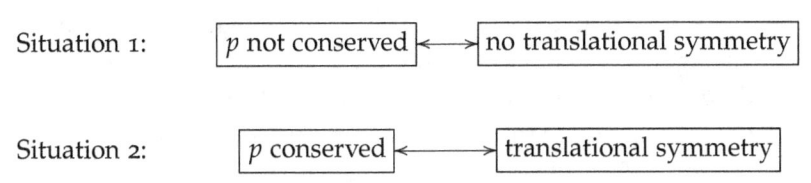

Situation 1: p not conserved ⟷ no translational symmetry

Situation 2: p conserved ⟷ translational symmetry

So hopefully it now seems a bit more plausible that there is a connection between conservation of momentum and translational symmetry and, more generally, between conserved

---

[12] From our outside perspective, the old x is related to the new one by $X = x - 3$ m. But this also implies that the location of the second object becomes $X_0 = 0 - 3$. (This is how the origin of the original coordinate system the physicist inside the spaceship uses is related to the origin of the new coordinate system after the shift, from our outside perspective.) Therefore, the potential reads after the shift

$$V(x(X)) = b(X+3-3)^2 = bX^2.$$

[13] PS: It's really fun to think about similar thought experiments which connect conservation of energy with temporal translational symmetry and the conservation of angular momentum with rotational symmetry.

quantities and symmetries.[13] If not, don't worry because in the following section, we'll discuss this connection a lot more systematically.

10.3 Noether's Theorem in the Hamiltonian Formalism

To derive Noether's theorem in the context of the Hamiltonian formalism, we only have to recall a few things we learned in previous chapters.

In Section 7.3.1, we've learned that Hamilton's equations have the same *form* no matter which canonical transformation $q, p \to Q, P$ we perform. Hamilton's equations before the transformation

$$\frac{dp}{dt} = -\frac{\partial H}{\partial q}, \quad \frac{dq}{dt} = \frac{\partial H}{\partial p}, \tag{10.9}$$

read afterwards

$$\frac{dP}{dt} = -\frac{\partial \tilde{H}}{\partial Q}, \quad \frac{dQ}{dt} = \frac{\partial \tilde{H}}{\partial P}. \tag{10.10}$$

This is always true. But when it comes to Noether's theorem, we only care about symmetries and about general canonical transformations. Mathematically, we are only dealing with a symmetry if[14]

[14] This condition is discussed in a bit more detail in Appendix D.

$$\tilde{H}(Q, P) \stackrel{!}{=} H(Q, P) \tag{10.11}$$

where $\tilde{H}(Q, P) = H(q(Q, P), p(Q, P))$ because only then will Hamilton's equations not only have the same form but are actually equivalent before and after the transformation. Formulated differently, only when Eq. 10.11 holds are Hamilton's equations invariant (and not just covariant) under the transformation $q, p \to Q, P$. And, as discussed in detail in Section 10.1 this is exactly our criterion for a symmetry.

To understand this, let's again consider the situation discussed in the previous section. The Hamiltonian for our system is

$$H = T + U = \frac{1}{2}m\dot{x}^2 + ax^2, \quad (10.12)$$

which reads after the shift $x \to X = x - s$:[15]

$$\tilde{H} = H(x(X)) = \frac{1}{2}m\dot{X}^2 + a(X+s)^2. \quad (10.13)$$

This is not equal to $H(X) = \frac{1}{2}m\dot{X}^2 + aX^2$ and therefore the shift $x \to X = x - s$ is not a symmetry.

In contrast, if there is no potential, the Hamiltonian reads[16]

$$H = T = \frac{1}{2}m\dot{x}^2. \quad (10.14)$$

This Hamiltonian reads after the shift $x \to X = x - s$:

$$\tilde{H} = H(x(X)) = T = \frac{1}{2}m\dot{X}^2. \quad (10.15)$$

This is equal to $H(X) = \frac{1}{2}m\dot{X}^2$ and therefore the shift $x \to X = x - s$ is indeed a symmetry.

Next, we need to recall that we've learned in Section 7.3.3 that we can describe infinitesimal canonical transformations using so-called generators $G(q, p)$ (Eq. 7.116):

$$q \to Q = q + \epsilon \frac{\partial G}{\partial p}$$
$$p \to P = p - \epsilon \frac{\partial G}{\partial q}. \quad (10.16)$$

By using these transformation rules, we can rewrite the invariance condition in Eq. 10.11 as follows:

[15] Take note that we have $x(X) = X + s$ and $\dot{X} = \frac{d}{dt}(x - s) = \dot{x}$ because s is a constant.

[16] Alternatively, recall the second situation described above. Here the key difference was that the potential is also shifted because the location of the second object which causes the potential is actively moved from $x = 0$ to $x = s$:

$$V(x) = ax^2$$
$$\to \tilde{V}(X) = a\big(x(X) - s\big)^2$$
$$= a(X + s - s)^2$$
$$= aX^2 = V(X)$$

and therefore the Hamiltonian is indeed invariant:

$$\tilde{H} = H(x(X))$$
$$= \frac{1}{2}m\dot{X}^2 + aX^2$$
$$= H(X).$$

$$\tilde{H}(Q,P) \stackrel{!}{=} H(Q,P) \qquad \circlearrowright \ \tilde{H}(Q,P) \equiv H(q(Q,P), p(Q,P))$$

$$H\big(q(Q,P), p(Q,P)\big) \stackrel{!}{=} H(Q,P) \qquad \circlearrowright \ \text{Eq. 10.16}$$

$$H\big(q(Q,P), p(Q,P)\big) \stackrel{!}{=} H\left(q + \epsilon\frac{\partial G}{\partial p}, p - \epsilon\frac{\partial G}{\partial q}\right)$$

$$\qquad \circlearrowright \ \text{Taylor expansion, c.f. Eq. 4.20}$$

$$H\big(q(Q,P), p(Q,P)\big) \stackrel{!}{=} H(q,p) + \epsilon\frac{\partial H}{\partial q}\frac{\partial G}{\partial p} - \epsilon\frac{\partial H}{\partial p}\frac{\partial G}{\partial q} + \ldots$$

$$\qquad \circlearrowright \ \cancel{H(q,p)}$$

$$0 \stackrel{!}{=} \epsilon\frac{\partial H}{\partial q}\frac{\partial G}{\partial p} - \epsilon\frac{\partial H}{\partial p}\frac{\partial G}{\partial q} + \ldots$$

$$\qquad \circlearrowright \ \text{def. of } \{\,,\,\}, \text{Eq. 5.34}$$

$$0 \stackrel{!}{=} \{H, G\} \qquad (10.17)$$

To get to the last line, we used the fact that for an infinitesimal transformation ($\epsilon \ll 1$) all higher order terms vanish ($\epsilon^2 = 0, \epsilon^3 = 0, \ldots$). In words, the final result here tells us that the canonical transformation generated by G is a symmetry if the Poisson bracket of G with the Hamiltonian vanishes. This is how we can check if the transformation generated by G is a symmetry.

Now comes the key observation.

In Section 5.1.1, we discovered that the time-evolution of *any* phase space function $F(q, p)$ is given by Hamilton's general equation (Eq. 5.36)

$$\frac{d}{dt}F = \{F, H\}. \qquad (10.18)$$

Therefore, the time-evolution of a generator $G(q, p)$ is also given by

$$\frac{d}{dt}G = \{G, H\}. \qquad (10.19)$$

If we now recall that in Eq. 10.17, we've learned that $\{H, G\} = 0$ if G generates a symmetry, we can learn something important about G. But first, take note that the Poisson bracket is antisymmetric:

$$\{A, B\} = -\{B, A\} \qquad (10.20)$$

which follows directly from its definition (Eq. 5.34):

$$\begin{aligned}\{A,B\} &= \frac{\partial A}{\partial q}\frac{\partial B}{\partial p} - \frac{\partial A}{\partial p}\frac{\partial B}{\partial q} \\ &= -\left(\frac{\partial A}{\partial p}\frac{\partial B}{\partial q} - \frac{\partial A}{\partial q}\frac{\partial B}{\partial p}\right) \\ &= -\left(\frac{\partial B}{\partial q}\frac{\partial A}{\partial p} - \frac{\partial B}{\partial p}\frac{\partial A}{\partial q}\right) \\ &= -\{B,A\}. \end{aligned} \quad (10.21)$$

Therefore, we can conclude that $\{H,G\} = 0$ implies

$$\{G,H\} = -\{H,G\} \stackrel{(10.17)}{=} -0 = 0. \quad (10.22)$$

And putting this into Eq. 10.19 tells us:

$$\boxed{\frac{d}{dt}G = \{G,H\} \stackrel{(10.22)}{=} 0.} \quad (10.23)$$

We therefore learn that if G generates a symmetry, it automatically describes a conserved quantity.[17] This is **Noether's theorem**.

[17] In general, a conserved quantity is defined as something with a vanishing rate of change, i.e.,

$$\frac{d}{dt}F = 0.$$

To summarize:

$\boxed{\{H,G\} = 0 \text{ (symmetry)}} \longrightarrow \boxed{\frac{d}{dt}G = \{G,H\} = 0 \text{ (conserved quantity)}}$

As a concrete example, let's consider the simplest system there is: a single free object. The Hamiltonian for this system reads[18]

[18] Here the Hamiltonian is simply equal to the kinetic energy because there is no potential.

$$H = \frac{1}{2}\frac{p^2}{m}. \quad (10.24)$$

Our first task is to find a symmetry of this system. Above, we learned that mathematically a symmetry is a canonical transformation generated by some phase space function $G = G(q,p)$

which has a vanishing Poisson bracket with the Hamiltonian (Eq. 10.17):
$$\{H, G\} = 0. \tag{10.25}$$

Let's try what is arguably the simplest phase space function
$$G(q, p) = p. \tag{10.26}$$

Using the explicit definition of the Poisson bracket (Eq. 5.34), we find

$$\{H, p\} \stackrel{(5.34)}{=} \frac{\partial H}{\partial q}\frac{\partial p}{\partial p} - \frac{\partial H}{\partial p}\frac{\partial p}{\partial q}$$

$$= \frac{\partial \left(\frac{1}{2}\frac{p^2}{m}\right)}{\partial q}\frac{\partial p}{\partial p} - \frac{\partial \left(\frac{1}{2}\frac{p^2}{m}\right)}{\partial p}\frac{\partial p}{\partial q} \quad \circlearrowleft \text{ Eq. 10.24}$$

$$= 0. \quad \circlearrowleft \frac{\partial p}{\partial q} = 0 \tag{10.27}$$

Therefore, $G(q, p) = p$ indeed generates a symmetry. But what kind of transformation is generated by this generator?

To understand this, we need to recall that a generator acts on the phase space coordinates via the Poisson bracket (Eq. 7.120)

$$q \to Q = q + \epsilon\{q, F\} = q + \epsilon\frac{\partial G}{\partial p}$$
$$p \to P = p + \epsilon\{p, G\} = p - \epsilon\frac{\partial G}{\partial q}. \tag{10.28}$$

Therefore, we calculate

$$\frac{\partial G}{\partial p} \stackrel{(10.26)}{=} \frac{\partial p}{\partial p} = 1$$
$$\frac{\partial G}{\partial q} \stackrel{(10.26)}{=} \frac{\partial p}{\partial q} = 0. \tag{10.29}$$

And, using Eq. 10.28, this tells us that the phase space coordinates transform as follows:

$$q \to Q = q + \epsilon\frac{\partial G}{\partial p} \stackrel{(10.29)}{=} q + \epsilon$$
$$p \to P = p - \epsilon\frac{\partial G}{\partial q} \stackrel{(10.29)}{=} p + \epsilon 0 = p. \tag{10.30}$$

This means in words that the location coordinate q is shifted by a constant amount, while the momentum coordinate p remains unchanged. We can therefore conclude that p generates spatial translations.

Finally, it's time to recall the punchline of Noether's theorem: $\{H, G\} = 0$ implies directly $\frac{d}{dt}G = 0$ (Eq. 10.23). For our concrete example here this implies that the momentum p is conserved:

$$\frac{dp}{dt} = 0. \tag{10.31}$$

To summarize: momentum p is conserved whenever the system is invariant under spatial translations because spatial translations are *generated* by p.

As a second example, let's consider the canonical transformation generated by the phase space function H (i.e., the Hamiltonian itself)[19]

$$G(q, p) = H. \tag{10.32}$$

This is an interesting choice because H certainly generates a symmetry since[20]

$$\{H, H\} = \frac{\partial H}{\partial q}\frac{\partial H}{\partial p} - \frac{\partial H}{\partial p}\frac{\partial H}{\partial q}$$

$$= 0. \tag{10.33}$$

Now what kind of transformation is generated by H?

As before, we can understand this by recalling that our generators act on the phase space coordinates via the Poisson bracket (Eq. 7.120). We find

$$\{q, H\} \stackrel{(5.37)}{=} \frac{\partial H}{\partial p}$$

$$\{p, H\} \stackrel{(5.38)}{=} -\frac{\partial H}{\partial q}. \tag{10.34}$$

[19] We discovered in Section 7.3.3 that we can use *any* phase space function to generate an infinitesimal canonical transformation.

[20] Reminder: the criterion that a specific phase space function G generates a symmetry is $\{G, H\} = 0$ (Eq. 10.17).

Therefore, after the infinitesimal transformation generated by H, our coordinates read (Eq. 10.28)

$$q \to Q = q + \epsilon\{q, H\} = q + \epsilon \frac{\partial H}{\partial p}$$

$$p \to P = p + \epsilon\{p, H\} = p - \epsilon \frac{\partial H}{\partial q}. \quad (10.35)$$

And this implies

$$\Delta q \equiv Q - q = \epsilon\{q, H\} = \epsilon \frac{\partial H}{\partial p} \stackrel{(5.15)}{=} \epsilon \frac{dq}{dt}$$

$$\Delta p \equiv P - p = \epsilon\{p, H\} = -\epsilon \frac{\partial H}{\partial q} \stackrel{(5.15)}{=} \epsilon \frac{dp}{dt} \quad (10.36)$$

where in the last step we used Hamilton's equations (Eq. 5.15).[21] On the left-hand side, we have the total change in the coordinates $\Delta q, \Delta p$. On the right-hand side, we have their rates of change $\frac{dq}{dt}, \frac{dp}{dt}$ multiplied by some interval ϵ. This means that the change we get by using the transformation generated by H is exactly the same as the change that we get by waiting for ϵ seconds. We can therefore conclude that H generates temporal translations, i.e., shifts to a later point in time!

And once more, we can recall the punchline of Noether's theorem: $\{H, G\} = 0$ implies that $\frac{d}{dt}G = 0$ (Eq. 10.23). For our concrete Hamiltonian this implies

$$\frac{dH}{dt} = 0. \quad (10.37)$$

To summarize: the Hamiltonian H represents a conserved quantity whenever the system is invariant under temporal translations because temporal translations are *generated* by the Hamiltonian. For many systems, the conserved quantity represented by H is exactly the total energy.

Before we move on and discuss how we can derive Noether's theorem in the Lagrangian formulation, it makes sense to talk about two slight modifications of Noether's theorem in the Hamiltonian formulation.

[21] For your convenience: Hamilton's equations (Eq. 5.15) read

$$\frac{dq}{dt} = \frac{\partial H}{\partial p}$$

$$\frac{dp}{dt} = -\frac{\partial H}{\partial q}.$$

10.3.1 Noether's Extended Theorem

In the previous sections, we've used the condition (Eq. 10.11)

$$\tilde{H}(Q,P) \stackrel{!}{=} H(Q,P) \qquad (10.38)$$

to define which transformations are symmetries. However, this condition is a bit too strict. In Section 7.3.2, we learned that after a gauge transformation we have (Eq. 7.100)

$$H'(Q,P) = H(q,p) - \frac{\partial F}{\partial t} \qquad (10.39)$$

since

$$H'(Q,P) \stackrel{(7.101)}{=} P\dot{Q} - L'(Q,P) \qquad (10.40)$$

and

$$L'(Q,P) \stackrel{(7.94)}{=} L(Q,P) + \frac{dF}{dt}. \qquad (10.41)$$

Moreover, we have

$$H(Q,P) \stackrel{(5.12)}{=} P\dot{Q} - L(Q,P). \qquad (10.42)$$

By using these equations, we can calculate

$$H'(Q,P) - H(Q,P) = P\dot{Q} - L'(Q,P) - \left(P\dot{Q} - L(Q,P)\right)$$

$$= L(Q,P) - L'(Q,P)$$

$$= L(Q,P) - \left(L(Q,P) + \frac{dF}{dt}\right) \qquad \text{Eq. 10.41}$$

$$= -\frac{dF}{dt}. \qquad (10.43)$$

Additionally, we've learned in Section 7.3.2 that gauge transformations have no influence on Hamilton's equations and are therefore always symmetries.[22]

---

[22] Recall that a gauge transformation is a shift in the abstract "action space", i.e., a constant shift of the action functional. But the only thing we care about is the path of least action and if we shift the action associated with all paths equally, the path of least action will remain the path of least action. We discussed this at the beginning of Section 7.2.3.

[23] It is conventional to call transformations which fulfill the stricter condition in Eq. 10.11 **symmetries**, and transformations which only fulfill the more generous condition (Eq. 10.44) **quasi-symmetries**. Moreover, take note that the minus sign is not important because we can simply absorb it into a redefinition of the arbitrary function F.

Therefore, it seems reasonable to relax our symmetry condition (Eq. 10.11) a little bit:[23]

$$\tilde{H}(Q,P) \stackrel{!}{=} H(Q,P) - \frac{dF}{dt}. \tag{10.44}$$

Equipped with this new condition, we can once more follow the exact same steps as in Eq. 10.17:

$$\tilde{H}(Q,P) \stackrel{!}{=} H(Q,P) - \frac{dF}{dt}$$

$$\vdots$$

$$0 \stackrel{!}{=} \{H,G\} - \frac{dF}{dt}$$

$$\circlearrowright \ \{H,G\} = -\{G,H\}, \text{ Eq. 10.21}$$

$$0 \stackrel{!}{=} -\{G,H\} - \frac{dF}{dt}. \tag{10.45}$$

This means that even if $\{H,G\}$ is non-zero, the transformation generated by G can be a symmetry as long as $\{H,G\}$ is equal to the total derivative of an arbitrary function $F = F(Q,t)$.

Now let's assume that we have found a function G for which Eq. 10.45 holds:

$$0 = \{G,H\} + \frac{dF}{dt}. \tag{10.46}$$

We can then use, analogous to what we did in Section 10.3, that the time-evolution of the function $G(q,p)$ is given by Hamilton's general equation of motion (Eq. 10.19)

$$\frac{d}{dt}G = \{G,H\}. \tag{10.47}$$

Substituting this into Eq. 10.46 yields

$$0 = \{G,H\} + \frac{dF}{dt}$$

$$\circlearrowright \ \text{Eq. 10.47}$$

$$0 = \frac{d}{dt}G + \frac{dF}{dt}$$

$$\circlearrowright \ \text{rearranging terms}$$

$$0 = \frac{d}{dt}(G+F). \tag{10.48}$$

This means that we find once more a conserved quantity

$$\boxed{\frac{d}{dt}Q = \frac{d}{dt}(G+F) \stackrel{(10.48)}{=} 0.} \tag{10.49}$$

This is the **extended Noether theorem**. We can see that if we are dealing with a transformation that does not fulfill the strict symmetry condition (Eq. 7.98) but instead only the extended condition (Eq. 10.44), the corresponding conserved quantity is the sum of the generator of the point transformation G and the additional function F. From a slightly different perspective, this means that the corresponding conserved quantity is described solely by the function F if we are dealing with a pure gauge transformation.

There is one final interesting aspect of Noether's theorem that we can discuss using the Hamiltonian formalism.

10.3.2 Noether's Converse Theorem

In the previous sections, we derived that we find a conserved quantity for each symmetry. But now we want to investigate whether the converse statement is true too.[24]

To that end, we start with a specific phase space function G which describes a conserved quantity:[25]

$$\frac{dG}{dt} = 0. \qquad (10.50)$$

Using Hamilton's equation of motion (Eq. 5.36), this tells us:

$$\frac{d}{dt}G = \{G, H\} \stackrel{(10.50)}{=} 0. \qquad (10.51)$$

The key idea is that we discovered in Section 7.3.3 that we can use any phase space function to generate an infinitesimal canonical transformation. Moreover, we discovered in Section 10.3.1 that the transformation generated by a specific function G is a symmetry if (Eq. 10.46)[26]

$$\{G, H\} \stackrel{!}{=} 0. \qquad (10.52)$$

[24] Take note that the inverse statement would be: G does not generate a symmetry → G is not conserved. In contrast, the converse statement is: G is conserved → G generates a symmetry.

[25] For simplicity, we assume that G is not explicitly time dependent. However, the proof for the time-dependent case works completely analogously.

[26] Take note that the following reasoning also works if we use the relaxed condition in Eq. 10.45:

$$\{G, H\} \stackrel{!}{=} -\frac{dF}{dt}.$$

This follows because for $F = $ const., we have $\frac{dF}{dt} = 0$. Formulated differently, the condition is fulfilled because we can write zero as the total derivative of a function.

Therefore, our goal is to check if this condition is fulfilled for any function that describes a conserved quantity. Using Eq. 10.51 we can check this explicitly:

$$\{G, H\} \stackrel{!}{=} 0$$

$$\quad\quad\quad\quad\quad\quad\quad\quad\quad\quad\quad \text{Eq. 10.51}$$

$$0 \stackrel{!}{=} 0 \ \checkmark \quad\quad\quad\quad\quad\quad\quad\quad (10.53)$$

Therefore, each phase space function which describes a conserved quantity indeed generates a symmetry. This is the **converse Noether theorem**.

10.4 Noether's Theorem in the Lagrangian Formalism

In the previous sections, we discussed Noether's theorem in the context of the Hamiltonian formalism. We discovered that Noether's theorem is an extremely natural consequence of the general structure of the formalism. Now we want to understand how Noether's theorem manifests itself in the Lagrangian formalism. While the derivation is not particularly difficult, be warned that the whole discussion will be a bit more abstract because, well, the Lagrangian formalism itself is a bit abstract.

First of all, we need to clarify what we mean by a symmetry in the context of the Lagrangian formulation of mechanics.

In Section 7.2.2, we introduced the following notation

$$\tilde{L}(q', \dot{q}') = L\Big(q(q'), \dot{q}(q', \dot{q}')\Big), \quad\quad (10.54)$$

for the Lagrangian after a point transformation $q, \dot{q} \to q', \dot{q}'$. So $\tilde{L}$ is the Lagrangian that we find if we replace all coordinates in the old Lagrangian with our new coordinates *using the explicit transformation rules*.[27]

The special kind of transformations for which[28]

$$\tilde{L}(q', \dot{q}') = L(q', \dot{q}') \quad\quad\quad\quad (10.55)$$

[27] This distinction is discussed in more detail in Appendix D.

[28] Take note that this condition is completely equivalent to our condition in the Hamiltonian formalism, i.e., Eq. 10.11.

holds are called **invariance transformations** and describe symmetries of the system. We can understand this definition by noting that under arbitrary point transformations $q \to q'$, the Euler-Lagrange equation only keeps its form but is not necessarily invariant.[29] If Eq. 10.55 holds, the resulting equation of motion is actually invariant, i.e., completely unchanged. As discussed in detail in Section 10.1, this is exactly our criterion for a symmetry.

[29] We discussed this explicitly in Section 7.2.2.

Next, we consider again an infinitesimal transformation

$$q \to q' = q + \epsilon g$$
$$\Rightarrow \quad \dot{q} \to \dot{q}' = \dot{q} + \epsilon \dot{g}, \qquad (10.56)$$

where $g = g(q)$ is a configuration space function that describes the infinitesimal transformation in question. By using these transformation rules, we can rewrite the symmetry condition in Eq. 10.55 as follows:

$$\tilde{L}(q', \dot{q}') = L(q', \dot{q}')$$
  ⟳ Eq. 10.54 and Eq. 10.56

$$L\big(q(q'), \dot{q}(q', \dot{q}')\big) = L(q + \epsilon g, \dot{q} + \epsilon \dot{g})$$
  ⟳ Taylor expansion, c.f. Eq. 4.20

$$L(q, \dot{q}) \stackrel{!}{=} L(q, \dot{q}) + \epsilon \frac{\partial L}{\partial q} g + \epsilon \frac{\partial L}{\partial \dot{q}} \dot{g}$$
  ⟳ ~~$L(q,\dot{q})$~~

$$0 = \epsilon \frac{\partial L}{\partial q} g + \epsilon \frac{\partial L}{\partial \dot{q}} \dot{g}$$
  ⟳ Euler-Lagrange equation $\frac{\partial L}{\partial q} = \frac{d}{dt}\left(\frac{\partial L}{\partial \dot{q}}\right)$, Eq. 4.25

$$0 = \epsilon \frac{d}{dt}\left(\frac{\partial L}{\partial \dot{q}}\right) g + \epsilon \frac{\partial L}{\partial \dot{q}} \dot{g}$$
  ⟳ product rule and ~~ϵ~~

$$0 = \frac{d}{dt}\left(\frac{\partial L}{\partial \dot{q}} g\right) \qquad (10.57)$$

This tells us that if the transformation in Eq. 10.56 is indeed a symmetry, the function

$$\boxed{Q_0 \equiv \frac{\partial L}{\partial \dot{q}} g} \qquad (10.58)$$

describes a conserved quantity since Eq. 10.57 tells us that $\frac{d}{dt} Q_0 = 0$. This is **Noether's theorem in the Lagrangian for-**

malism. It is conventional to call the conserved quantities Q_0 that we find this way **Noether charges**.

As an example, let's consider the simplest configuration space transformation, a spatial translation:[30]

$$q \to q' = q + \epsilon$$
$$\Rightarrow \quad \dot{q} \to \dot{q}' = \dot{q}. \tag{10.59}$$

[30] This is Eq. 10.56 with $g = 1$. Therefore, $\dot{g} = \frac{dg}{dt} = 0$.

For concreteness, let's use the Lagrangian that describes a free object

$$L = \frac{1}{2}m\dot{q}^2, \tag{10.60}$$

which is invariant under translations:

$$\tilde{L}(q', \dot{q}') \stackrel{(10.60)}{=} \frac{1}{2}m\dot{q}'^2$$

↪ Eq. 10.59

$$= \frac{1}{2}m\dot{q}^2$$

↪

$$= L(q, \dot{q}). \tag{10.61}$$

Therefore, Noether's theorem tells us that (Eq. 10.58)

$$Q_0 \equiv \frac{\partial L}{\partial \dot{q}} g$$

↪ $g = 1$ here

$$= \frac{\partial L}{\partial \dot{q}}$$

↪ Eq. 10.60

$$= \frac{\partial \left(\frac{1}{2}m\dot{q}^2\right)}{\partial \dot{q}}$$

↪

$$= m\dot{q} \tag{10.62}$$

is a conserved quantity. We can see that the conserved quantity we get from the invariance of the Lagrangian under translations is exactly the momentum once more.

Before we summarize what we've learned in this chapter, there are two important generalization of Noether's theorem that we need to talk about.

10.4.1 Noether's Extended Theorem

In the previous section, we used the condition (Eq. 10.55)

$$\tilde{L}(q',\dot{q}') \stackrel{!}{=} L(q',\dot{q}'), \tag{10.63}$$

to define a symmetry. This definition is motivated by the observation that if this equation holds, the Euler-Lagrange equation remains unchanged. However, this condition is actually a bit too strict. In Section 7.2.3, we learned that the Euler-Lagrange equation also remains unchanged when the Lagrangian changes by a total derivative (Eq. 7.39):

$$L \to L' = L - \frac{dF(q,t)}{dt}, \tag{10.64}$$

where F is an arbitrary function which only depends on the locations q and t, and we introduced a conventional minus sign.[31]

This means that we can relax our symmetry condition a little bit:

$$\tilde{L}(q',\dot{q}') \stackrel{!}{=} L(q',\dot{q}') - \frac{dF(q',t')}{dt}. \tag{10.65}$$

[31] This minus sign is not important because we could simply absorb it into the definition of the function F, i.e., $\tilde{F} = -F$ and then

$$L \to L' = L + \frac{d\tilde{F}(q,t)}{dt}.$$

Using this modified condition, we can repeat the steps that we performed in Eq. 10.57:

$$\tilde{L}(q',\dot{q}') = L(q',\dot{q}') - \frac{dF(q',t')}{dt}$$

$$\vdots$$

$$0 = \epsilon \frac{d}{dt}\left(\frac{\partial L}{\partial \dot{q}}g\right) - \frac{dF}{dt}$$

$$\circlearrowright \text{ rearranging terms}$$

$$0 = \frac{d}{dt}\left(\frac{\partial L}{\partial \dot{q}}g - \frac{1}{\epsilon}F\right). \tag{10.66}$$

This is the **generalized Noether theorem**. In words, it tells us that if our Lagrangian changes at most by a total derivative of some function $F = F(q,t)$, we get a conserved quantity

$$\boxed{Q \equiv \frac{\partial L}{\partial \dot{q}}g - \frac{1}{\epsilon}F \stackrel{(10.58)}{=} Q_0 - \frac{1}{\epsilon}F.} \tag{10.67}$$

In this context it is conventional to call Q the **full Noether charge** and Q_0 the **bare Noether charge**.

A great example where we need the extended Noether theorem is infinitesimal time translation symmetry:[32]

$$t \to t' = t + \epsilon. \tag{10.68}$$

Our first task is to calculate how such a shift influences the configuration space coordinates:

$$q(t) \to q'(t) = q(t + \epsilon) = q(t) + \epsilon \frac{dq}{dt}, \tag{10.69}$$

where we used the Taylor expansion and that all higher order terms vanish since ϵ is infinitesimal. By comparing this result with Eq. 10.16, we can conclude that the function which describes the transformation is[33]

$$g_t = \frac{dq}{dt}. \tag{10.70}$$

For the velocity coordinates we find analogously

$$\dot{q}(t) \to \dot{q}'(t) = \dot{q}(t + \epsilon) = \dot{q}(t) + \epsilon \frac{d\dot{q}}{dt}. \tag{10.71}$$

So in words, our new coordinates are the original ones at an infinitesimally later point in time.

For the Lagrangian, an infinitesimal time translation therefore implies:

$$
\begin{aligned}
L(q', \dot{q}', t') &\stackrel{(10.69)}{=} L\left(q + \epsilon \frac{dq}{dt}, \dot{q} + \epsilon \frac{d\dot{q}}{dt}, t + \epsilon\right) \\
&\quad \curvearrowright \text{Taylor expansion and } \epsilon^2 = 0, \text{ etc.} \\
&= L(q, \dot{q}, t) + \epsilon \frac{\partial L}{\partial q} \frac{\partial q}{\partial t} + \epsilon \frac{\partial L}{\partial \dot{q}} \frac{\partial \dot{q}}{\partial t} + \epsilon \frac{\partial L}{\partial t} \\
&\quad \curvearrowright \text{factoring out } \epsilon \\
&= L(q, \dot{q}, t) + \epsilon \left(\frac{\partial L}{\partial q} \frac{\partial q}{\partial t} + \frac{\partial L}{\partial \dot{q}} \frac{\partial \dot{q}}{\partial t} + \frac{\partial L}{\partial t}\right) \\
&\quad \curvearrowright \text{terms are exactly the total derivative} \\
&= L(q, \dot{q}, t) + \epsilon \left(\frac{dL}{dt}\right)
\end{aligned}
\tag{10.72}
$$

---

[32] We need the extended Noether theorem here because, in general, the Lagrangian is not invariant under time translation. (In contrast to the Hamiltonian, the Lagrangian does not describe a conserved quantity). But we can still find a conserved quantity associated with time-translation symmetry which, in fact, we already discovered at the end of Section 10.3 using the Hamiltonian version of Noether's theorem.

[33] For your convenience: Eq. 10.56 reads

$$q \to q' = q + \epsilon g$$
$$\Rightarrow \dot{q} \to \dot{q}' = \dot{q} + \epsilon \dot{g}.$$

Below, we need g to calculate the resulting conserved quantity using Noether's theorem.

Substituting this result into our relaxed symmetry condition (Eq. 10.65) yields

$$\tilde{L}(q',\dot{q}',t') \stackrel{!}{=} L(q',\dot{q}',t') - \frac{dF}{dt}$$

↪ Eq. 10.72

$$\tilde{L}(q',\dot{q}',t') \stackrel{!}{=} L(q,\dot{q},t) + \epsilon\left(\frac{dL}{dt}\right) - \frac{dF}{dt}$$

↪ $\tilde{L}(q',\dot{q}',t') \equiv L(q(q',t'),\dot{q}(q',\dot{q}',t'),t(t'))$

$$L(q,\dot{q},t) \stackrel{!}{=} L(q,\dot{q},t) + \epsilon\left(\frac{dL}{dt}\right) - \frac{dF}{dt}.$$

(10.73)

We can see that the condition is indeed fulfilled for

$$F = \epsilon L. \tag{10.74}$$

In other words, since the Lagrangian changes under an infinitesimal time translation by at most a total derivative term $\frac{dL}{dt}$, we have time translation symmetry.[34] We therefore don't need to consider a specific Lagrangian because the conserved quantity we will find in the following is conserved for all Lagrangians.

[34] Recall that in the Hamiltonian formalism we reached the same conclusion because $\{H, H\} = 0$, for any Hamiltonian H.

The extended Noether theorem (Eq. 10.66) tells us that the associated conserved quantity is

$$Q \equiv \frac{\partial L}{\partial \dot{q}} g - \frac{1}{\epsilon} F$$

↪ Eq. 10.70 and Eq. 10.74

$$= \frac{\partial L}{\partial \dot{q}} \frac{dq}{dt} - L$$

↪ $p \equiv \frac{\partial L}{\partial \dot{q}}$ and $\dot{q} = \frac{dq}{dt}$

$$= p\dot{q} - L$$

↪ $H \equiv p\dot{q} - L$, (Eq. 5.12)

$$= H. \tag{10.75}$$

In the final step we used that $p\dot{q} - L$ is exactly the definition of the Hamiltonian. So the conserved quantity that we find by considering time translation symmetry is exactly the Hamiltonian.[35]

[35] We discovered this already in Eq. 10.37 using the Hamiltonian version of Noether's theorem.

10.5 Summary

In this chapter we discovered that whenever a system possesses a specific symmetry, there is a corresponding conserved quantity. This beautiful insight is commonly known as Noether's theorem.[36]

We started by talking about symmetries. In general, a symmetry is a transformation that leaves a given object unchanged.

The symmetries in which we are interested in physics are transformations that leave the equations of motion or the Lagrangian unchanged.[37]

We then discussed that in the Hamiltonian formalism, we can understand Noether's theorem beautifully by noting that many important phase space functions play a double role.

On the one hand, they represent conserved quantities. On the other hand, they generate symmetry transformations.[38] In general, a phase space function G generates a symmetry if its Poisson bracket with the Hamiltonian can be written as the total derivative of a function $F = F(q,t)$ (Eq. 10.46):

$$\boxed{\{G, H\} \stackrel{!}{=} -\frac{dF}{dt}.} \qquad (10.76)$$

In particular, this implies that we are dealing with a symmetry when the Poisson bracket $\{G, H\}$ yields zero since for $F = $ const. we have $\frac{dF}{dt} = 0$. Moreover, sometimes transformations which fulfill the more strict condition

$$\{G, H\} \stackrel{!}{=} 0$$

are called real symmetries, while transformations which only fulfill

$$\{G, H\} \stackrel{!}{=} -\frac{dF}{dt}$$

but not the more strict condition, are called quasi-symmetries.

---

[36] Just for the record: what we talked about in this chapter is known as Noether's first theorem. Her second theorem deals with infinite-dimensional "symmetries" which are commonly discussed in the context of certain field theories. The theorem tells us that these infinite-dimensional "symmetries" do not give rise to conserved quantities but to constraints. However, these infinite-dimensional "symmetries" are not really symmetries but rather redundancies which can't be observed and are only part of the formalism. Therefore it shouldn't be too surprising that they don't lead to new conserved quantities.

[37] Most importantly, while any point transformation leaves the *form* of the equation of motion unchanged, only a few specific ones leave the equation itself completely unchanged.

[38] In my humble opinion, this beautiful insight alone justifies the introduction of the Hamilton formalism. Moreover, if we try to apply the same idea to a new mathematical arena (Hilbert space) we end up with quantum mechanics. In this sense, Noether's theorem can act as a bridge between classical and quantum mechanics. For a discussion of this perspective, we see .

Jakob Schwichtenberg. *No-Nonsense Quantum Mechanics*. No-Nonsense Books, Karlsruhe, Germany, 2018c. ISBN 978-1719838719

As soon as we've found a function which fulfills this condition, Noether's theorem tells us that there is a conserved quantity (Eq. 10.49):[39]

$$Q = G + F. \quad (10.77)$$

[39] If $\{G, H\} = 0$ the conserved quantity is simply G.

Concrete examples are the momentum p which (in some systems) is a conserved quantity and, at the same time, generates spatial translations.

A second example is the Hamiltonian H itself which represents a conserved quantity and, at the same time, generates temporal translations.[40]

[40] As mentioned above, the conserved quantity described by the Hamiltonian is often the total energy.

In the Lagrangian formalism, we were able to derive similar conclusions although the interpretation of Noether's theorem is a bit less transparent.

We defined a symmetry as a transformation which at most changes the Lagrangian by a total derivative (Eq. 10.65):[41]

$$\tilde{L}(q', \dot{q}') \stackrel{!}{=} L(q', \dot{q}') - \frac{dF(q', t')}{dt}. \quad (10.78)$$

[41] Transformations which fulfill the more strict condition

$$\tilde{L}(q', \dot{q}') \stackrel{!}{=} L(q', \dot{q}')$$

are sometimes called a real symmetry while transformations which only fulfill

$$\tilde{L}(q', \dot{q}') \stackrel{!}{=} L(q', \dot{q}') - \frac{dF(q', t')}{dt}$$

but not the more strict condition are called quasi-symmetries.

Again, as soon as we've found a transformation which fulfills this condition, Noether's theorem tells us there is a conserved quantity (Eq. 10.67):[42]

$$Q = \frac{\partial L}{\partial \dot{q}} g - \frac{1}{\epsilon} F. \quad (10.79)$$

[42] Here ϵ is the infinitesimal parameter which parameterizes the transformation.

Concrete examples are again the conservation of momentum, which follows if the Lagrangian is invariant under spatial translations, and the conservation of the Hamiltonian, which follows if the Lagrangian only changes by a total derivative under temporal translations.[43]

[43] Take note that it's also possible to derive the conservation of electric charge using Noether's theorem. This is usually discussed in the context of quantum field theory.

11

Additional Formulations of Classical Mechanics

So far we've talked about the three most famous formulations of classical mechanics.[1] Additionally there are further not-so-famous formulations and in this chapter we will talk briefly about three of them. However, strictly speaking, only one of them is a completely new formulation. The other two are better described as *methods* which can be useful to simplify problems.[2]

To understand this remark, recall that the defining feature of the three formulations we have talked about so far is that each of them describes classical mechanics using a different mathematical arena:

▷ The Newtonian formulation describes classical mechanics using trajectories in physical space.

▷ The Lagrangian formulation describes classical mechanics using paths in configuration space.

▷ The Hamiltonian formulation describes classical mechanics using paths in phase space.

[1] Newtonian, Lagrangian, Hamiltonian.

[2] There are further alternative "formulations" like Routhian mechanics, which is a hybrid of the Lagrangian and Hamiltonian formalism. However, since it doesn't reveal any new facets of classical mechanics and is only useful in engineering applications, we will not discuss it here.

In this sense, a completely new formulation of classical mechanics requires a description in terms of some new mathematical arena. There is exactly one such arena which is used all the time in modern physics: Hilbert space. To understand how we end up with Hilbert space, recall that the "trick" which we used as we progressed from physical space via configuration space to phase space was to make our mathematical arena "bigger".[3] From this perspective we can say that Hilbert space is the next logical step in this direction.[4] The formulation of classical mechanics in Hilbert space is known as the **Koopman-von Neumann formulation**.

[3] For example, for 2 free particles physical space is, as always, 3-dimensional, configuration space is $2 \times 3 = 6$-dimensional and phase space is $2 \times 2 \times 3 = 12$-dimensional.

[4] The Hilbert space for a free single particle is infinite-dimensional! To quote Carlton Caves: *"Hilbert space is a big place!"* Nevertheless, Hilbert spaces are not the end of the ladder. For example, in quantum field theory we use so-called Fock spaces to describe various quantum particles at once. In some sense, a Fock space is a product of the Hilbert spaces of the individual particles and therefore even larger.

One of the other remaining "formulations", known as **Hamilton-Jacobi mechanics**, does not require any new mathematical arena and is therefore really just an alternative perspective on the existing formulations. Similarly, the second alternative "formulation" we will talk about, known as statistical mechanics, is more accurately described as a tool that allows us to describe uncertainty in the context of the Hamiltonian formalism.

Be warned that we will not discuss any details. We will focus on the fundamental ideas, and one of the main points you should take away is simply that these alternative formulations exist. Moreover, while these alternatives are often not really useful for practical applications, they allow us to develop a deeper understanding of classical mechanics. This will be especially important when we want to understand the origin of classical mechanics.[5]

[5] We will discuss the origin of classical mechanics in Chapter 12

Let's start by talking about Hamilton-Jacobi mechanics which, in some sense, is a clever strategy to solve Hamilton's equations.

11.1 Hamilton-Jacobi Mechanics

While Hamilton-Jacobi mechanics is a useful tool for some applications (especially chaotic systems and geometrical optics), from a more fundamental perspective we are mainly interested in it because it provides a bridge to quantum mechanics.[6]

To put it in a nutshell, a German mathematician called Carl Gustav Jacob Jacobi came up with a clever method to simplify classical mechanics problems using canonical transformations.[7] And although it's a method to simplify problems and not a real new formulation of mechanics like the Newtonian or Lagrangian formulations, it's conventional to call this method the Hamilton-Jacobi formulation of classical mechanics or simply Hamilton-Jacobi mechanics.

The key idea is to simplify the Hamiltonian by using the fact that it isn't invariant under canonical transformations. If we use a canonical transformation generated by some function $F = F(q, Q, t)$, the Hamiltonian becomes (Eq. 7.141)[8]

$$H \to H' = H + \frac{\partial}{\partial t} F. \quad (11.1)$$

But this means that we can try to find a specific function $W = W(q, Q, t)$ for which the Hamiltonian becomes zero:

$$H \to H' = 0 = H + \frac{\partial}{\partial t} W. \quad (11.2)$$

It is conventional to call this special function W **Hamilton's principal function**. If we can find such a function, Hamilton's equations become trivial to solve

$$\frac{dp}{dt} = -\frac{\partial H}{\partial q} \to \frac{dP}{dt} = -\frac{\partial H'}{\partial Q} \stackrel{(11.2)}{=} 0$$
$$\frac{dq}{dt} = \frac{\partial H}{\partial p} \to \frac{dQ}{dt} = \frac{\partial H'}{\partial P} \stackrel{(11.2)}{=} 0. \quad (11.3)$$

This implies that P and Q are constant! Because these constants will be important, we give them concrete names:

$$Q = \alpha \quad P = \beta. \quad (11.4)$$

Therefore, finding the generating function $W(q, Q, t)$ for which Eq. 11.2 holds is really our only job because once we've found it, we can write down the new coordinates directly by using[9]

---

[6] We will see in Chapter 12 that Hamilton-Jacobi mechanics is invaluable when we want to understand the connection between classical and quantum mechanics.

[7] We discussed canonical transformation in Section 7.3.1.

[8] In Eq. 7.141, we derived

$$H \to H' = H - \frac{\partial}{\partial t} F.$$

Therefore, we introduce an additional minus sign here. We do this to derive the conventional form of the Hamilton-Jacobi equation. However, this additional minus sign has no deeper significance since it simply corresponds to a redefinition of the function $F \to \tilde{F} = -F$.

[9] This is how, in general, we can calculate the new coordinates using a given generating function, i.e., Eq. 7.124. However, since we now work with a new generating function which is related to the original one by $\tilde{F} = -F$, we have switched signs in the equations here.

$$P = -\frac{\partial W}{\partial Q}$$
$$p = \frac{\partial W}{\partial q}. \tag{11.5}$$

To find the differential equation which allows us to find W, we substitute Eq. 11.5 into Eq. 11.2:

$$\boxed{0 = H\left(q, \frac{\partial W}{\partial q}, t\right) + \frac{\partial W}{\partial t}.} \tag{11.6}$$

This is the famous **Hamilton-Jacobi equation**.

If we are dealing with a Hamiltonian of the usual form

$$H(q, p, t) = T(p) + V(q) = \frac{p^2}{2m} + V(q), \tag{11.7}$$

the Hamilton-Jacobi equation reads in more explicit terms:

$$0 = H\left(q, \frac{\partial W}{\partial q}, t\right) + \frac{\partial W}{\partial t}$$

$$\circlearrowright \text{ Eq. 11.7 with } p = \frac{\partial W}{\partial q}$$

$$0 = \frac{1}{2m}\left(\frac{\partial W}{\partial q}\right)^2 + V(q) + \frac{\partial W}{\partial t}. \tag{11.8}$$

So to write down the Hamilton-Jacobi equation, we take the given Hamiltonian H and use Eq. 11.5 to substitute the momentum coordinates by derivatives of the function W. This yields a differential equation that we need to solve for W. Moreover, since our new coordinates Q are constant, the generating function W is really only a function of the old coordinates q:[10]

[10] Take note that the α and β are integration constants which we need to determine by using specific initial conditions.

$$W = W(q, Q, t) \stackrel{(11.4)}{=} W(q, \alpha, t).$$

This is a clever approach because finding the solution of the Hamilton-Jacobi equation for a specific Hamiltonian is equivalent to finding the solution of Hamilton's original equations. In other words, we've replaced the problem of finding a solution of Hamilton's equations with the problem of finding a solution of the Hamilton-Jacobi equation. A solution of the Hamilton-Jacobi equation yields a generating function that allows us to "trivialize" Hamilton's equations (see Eq. 11.3).

ADDITIONAL FORMULATIONS OF CLASSICAL MECHANICS 255

As soon as we've found the solution $W(q, \alpha, t)$ of the Hamilton-Jacobi equation, we can calculate the momenta directly using the second line in Eq. 11.5. Moreover, we can calculate the constant values of our new momenta P by using the first line in Eq. 11.5:

$$P \stackrel{(11.4)}{=} \beta \stackrel{(11.5)}{=} -\frac{\partial W}{\partial Q} \stackrel{(11.4)}{=} -\frac{\partial W(q, \alpha, t)}{\partial \alpha} \qquad (11.9)$$

This, in turn, allows us to determine our original coordinates q, p in terms of the constants α, β:

$$q = q(\alpha, \beta, t), \quad p = p(\alpha, \beta, t). \qquad (11.10)$$

This is the solution of Hamilton's equations we've been looking for.

Here we are primarily interested in the Hamilton-Jacobi formalism as a theoretical framework and not so much as a practical tool.[11] Therefore, let's try to understand the principal function W a little better.

11.1.1 Meaning of Hamilton's Principal Function

First of all, let's calculate the total time derivative of $W = W(q, \alpha, t)$:[12]

$$\frac{dW}{dt} = \frac{\partial W}{\partial q}\frac{dq}{dt} + \frac{\partial W}{\partial t}$$

$$\curvearrowright \ H + \frac{\partial W}{\partial t} = 0, \text{ Eq. 11.6}$$

$$= \frac{\partial W}{\partial q}\frac{dq}{dt} - H$$

$$\curvearrowright \ p = \frac{\partial W}{\partial q}, \text{ Eq. 11.5}$$

$$= p\frac{dq}{dt} - H. \qquad (11.11)$$

But this is exactly what the Lagrangian L looks like in terms of the Hamiltonian (Eq. 5.12)! We can therefore conclude:

$$\frac{dW}{dt} = L. \qquad (11.12)$$

[11] Using the Hamilton-Jacobi formalism in practice is quite complicated and we will not discuss any details. Moreover, for the majority of examples, the Hamilton-Jacobi approach is not particularly useful. Instead, we are mainly interested in it because it reveals a deep connection between classical mechanics and quantum mechanics. But if you're interested in how the Hamilton-Jacobi formalism is used in practice, you can find numerous examples and applications in the books listed in Chapter 11.

[12] Here it is crucial to recall that α is constant in time (Eq. 11.4).

And integrating this equation yields

$$W = \int L dt. \qquad (11.13)$$

This tells us that there is a direct connection between the principal function W and the action, which is defined as the integral over the Lagrangian, too:

$$S = \int_{t_i}^{t_f} L dt. \qquad (11.14)$$

But there is an important difference between the two. The action functional $S[q(t)]$ assigns a number to each path between *fixed* initial and final points. Using specific initial conditions, the least action principle allows us to calculate the correct path.

In contrast, for Hamilton's principal function W the boundary conditions are not fixed, but the path $\big(q(t), p(t)\big)$ is. Formulated differently, Hamilton's principal function tells us the amount of action required by the *correct* path $\big(q(t), p(t)\big)$ between *arbitrary* initial and final times.

In particular, this means that we can't use Eq. 11.13 to calculate Hamilton's principal function W before we've solved the problem. To evaluate Eq. 11.13, we need to know the correct path which is, well, the solution of the problem. Therefore, Eq. 11.13

only helps us to understand the meaning of Hamilton's principal functions.

All this probably makes little sense to you at this point. But this will change as soon as we talk about the origin of classical mechanics and, in particular, its relationship to quantum mechanics in Chapter 12.

To understand how the Hamilton-Jacobi method works in practice, let's once more consider the harmonic oscillator.

11.1.2 Harmonic Oscillator

In Section 8.3, we already derived the correct harmonic oscillator Hamiltonian (Eq. 8.14):

$$H = \frac{1}{2}\frac{p^2}{m} + \frac{1}{2}kx^2. \tag{11.15}$$

To simplify the following calculations, we rewrite the Hamiltonian a little bit

$$H = \frac{1}{2m}\left(p^2 + m^2\omega^2 q^2\right), \tag{11.16}$$

where we introduced $\omega \equiv \sqrt{k/m}$.

Substituting this Hamiltonian into the Hamilton-Jacobi equation (Eq. 11.6) yields

$$0 = H\left(q, \frac{\partial W}{\partial q}, t\right) + \frac{\partial W}{\partial t}$$

↻ Eq. 11.16 with $p = \frac{\partial W}{\partial q}$ (Eq. 11.5)

$$= \frac{1}{2m}\left(\left(\frac{\partial W}{\partial q}\right)^2 + m^2\omega^2 q^2\right) + \frac{\partial W}{\partial t}. \tag{11.17}$$

This is a differential equation that we need to solve for W.

This is possible by making the ansatz[13]

$$W(q,t) = A(q) + B(t). \tag{11.18}$$

[13] We will see in a moment why this ansatz is clever. As mentioned before, solving differential equations is somewhat of an art. The key observation here is that such a separation ansatz will be successful because one term only contains derivatives with respect to t, and the remaining terms only involve q. But for our purposes here, it is only important to take note that the equation *can* be solved.

If we rearrange the terms in Eq. 11.17 and then use this ansatz, we find

$$\frac{1}{2m}\left(\left(\frac{\partial W}{\partial q}\right)^2 + m^2\omega^2 q^2\right) = -\frac{\partial W}{\partial t}$$

↷ Eq. 11.18

$$\frac{1}{2m}\left(\left(\frac{\partial(A(q)+B(t))}{\partial q}\right)^2 + m^2\omega^2 q^2\right) = -\frac{\partial(A(q)+B(t))}{\partial t}$$

↷ $\frac{\partial B(t)}{\partial q} = 0, \frac{\partial A(q)}{\partial t} = 0$

$$\frac{1}{2m}\left(\left(\frac{\partial A(q)}{\partial q}\right)^2 + m^2\omega^2 q^2\right) = -\frac{\partial B(t)}{\partial t}.$$

(11.19)

Now we can see why the ansatz in Eq. 11.18 is clever. The expression on the left-hand side only depends on q and the expression on the right-hand side only on t. But q and t are completely independent.

Still, both sides must be equal. This is only possible if both sides are constant. If, for example, the right-hand side is not constant, this would mean that we could change its value by varying t. But since the left-hand side does not depend on t at all, there is then no way that both sides are still equal. This observation allows us to convert our quite complicated differential equation into two simpler ones.

Let's call the constant that both sides of the equation are equal to α because we know that our principal function depends not only on q and t but also on Q, and the whole point of the Hamilton-Jacobi method is that this new location coordinate is actually a constant that we conventionally call α. With this in mind we find:[14]

[14] Take note that what we have on the left-hand side is really just the Hamiltonian. Therefore, the constant α is actually simply the energy of the harmonic oscillator.

$$\frac{1}{2m}\left(\left(\frac{\partial A(q)}{\partial q}\right)^2 + m^2\omega^2 q^2\right) \stackrel{!}{=} \alpha \stackrel{!}{=} -\frac{\partial B(t)}{\partial t}$$

⇒ 1.) $\frac{1}{2m}\left(\left(\frac{\partial A(q)}{\partial q}\right)^2 + m^2\omega^2 q^2\right) = \alpha$

⇒ 2.) $-\frac{\partial B(t)}{\partial t} = \alpha.$

(11.20)

We can integrate the equation for $B(t)$ directly:

$$-\frac{\partial B(t)}{\partial t} = \alpha$$

$$-\int \frac{\partial B(t)}{\partial t} dt = \int \alpha \, dt \qquad \circlearrowright \int dt$$

$$-B(t) = \alpha t. \qquad (11.21)$$

The equation for $A(q)$ is solved by

$$A(q) = \int \sqrt{2m\alpha - m^2\omega^2 q^2} \, dq \qquad (11.22)$$

as we can check explicitly:

$$\frac{1}{2m}\left(\left(\frac{\partial A(q)}{\partial q}\right)^2 + m^2\omega^2 q^2\right) = \alpha$$

$\circlearrowright$ Eq. 11.22

$$\frac{1}{2m}\left(\left(\frac{\partial \left(\int \sqrt{2m\alpha - m^2\omega^2 q^2} \, dq\right)}{\partial q}\right)^2 + m^2\omega^2 q^2\right) = \alpha$$

$$\frac{1}{2m}\left(\left(\sqrt{2m\alpha - m^2\omega^2 q^2}\right)^2 + m^2\omega^2 q^2\right) = \alpha$$

$$\frac{1}{2m}\left(2m\alpha - m^2\omega^2 q^2 + m^2\omega^2 q^2\right) = \alpha$$

$\circlearrowright \; \cancel{m^2\omega^2 q^2}$

$$\alpha = \alpha \; \checkmark$$

$$(11.23)$$

Therefore, our full solution of the Hamilton-Jacobi equation reads

$$W(q,t) = A(q) + B(t) \qquad \text{this was our ansatz, Eq. 11.18}$$

$\circlearrowright$ Eq. 11.22 and Eq. 11.21

$$= \int \sqrt{2m\alpha - m^2\omega^2 q^2} \, dq - \alpha t. \qquad (11.24)$$

Next, we can use this solution to determine the solutions we are really interested in, i.e., $q(t)$ and $p(t)$. Specifically, we now use Eq. 11.5 and recall that our new coordinates Q, P are actually constants (α, β).[15] Then, the first line in Eq. 11.5 tells us

[15] Recall that these are constants that we need to determine using specific initial conditions.

$$\beta = -\frac{\partial W}{\partial \alpha}$$

$$= -\frac{\partial \left(\int \sqrt{2m\alpha - m^2\omega^2 q^2}\, dq - \alpha t \right)}{\partial \alpha}$$ ⟩ Eq. 11.24

⟩

$$= -\int \frac{m}{\sqrt{2m\alpha - m^2\omega^2 q^2}}\, dq + t$$

⟩ $\frac{d}{dx}\sin^{-1}(cx) = \frac{c}{\sqrt{1-c^2x^2}}$

$$= -\frac{1}{\omega}\sin^{-1}\left(q\sqrt{\frac{m\omega^2}{2\alpha}} \right) + t . \qquad (11.25)$$

And finally, we can solve this equation for q:

$$q = -\sqrt{\frac{2\alpha}{m\omega^2}}\sin\left(\omega t + \omega\beta \right). \qquad (11.26)$$

This is the solution of Hamilton's equation we've been looking for and exactly the same result that we obtained in Chapter 8.[16]

———————————————

Next, let's talk about statistical mechanics.

———

[16] Specifically, in Eq. 8.36 we found

$$x = \sqrt{\frac{2E}{k}}\sin\left(\sqrt{\frac{k}{m}}t - \phi_0 \right).$$

Using the definition $\omega \equiv \sqrt{k/m}$ this is equal to

$$x = \sqrt{\frac{2E}{m\omega^2}}\sin(\omega t - \phi_0).$$

By comparing this to Eq. 11.26, we can conclude that α is indeed equal to the total energy. Moreover, β is directly connected to the initial configuration as specified by ϕ_0.

11.2 Statistical Mechanics

In this section, we'll discuss how we can describe situations in classical mechanics in which we don't know all the details. This may seem like an additional complication which we can safely ignore if we are only interested in fundamental aspects, analogous to how we usually ignore friction. But this couldn't be further from the truth.

Many of the most important discoveries in modern physics are about how we can describe situations without absolute knowledge. For example, in quantum mechanics and quantum field theory we are exclusively calculating probabilities because there is a fundamental uncertainty we seemingly can't get rid of.

But statistical mechanics is not only a great way to get familiar with probabilistic concepts in a familiar context. Statistical mechanics is also essential to understand the fundamental difference between quantum mechanics and classical mechanics.

So far, we've always assumed that we are dealing with perfectly known initial conditions. But whenever we perform a real experiment, there is at least *some* uncertainty.

We never know the location or momentum of a given object with 100% accuracy because there are always technical limits to how precisely we can measure them. Of course, for many applications this uncertainty is so small that we can safely ignore it. For example, if we want to describe a big ball, it usually doesn't really matter whether its center is at $x = 1.0$ m or at $x = 1.0000001$ m.

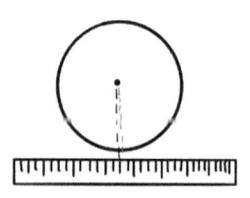

[17] For example, think about a gas which consists of so many individual molecules that it's impossible to specify all the individual locations and momenta exactly.

Especially when we describe small objects or many objects at once, such uncertainties can become extremely significant.[17] Luckily, we already have everything that we need to understand how we can take uncertainty into account.

First of all, let's recall that a point in phase space corresponds to one *specific* state of the system. Formulated differently: a point in phase space corresponds to perfectly known locations and momenta for all objects in the system.

As time passes by, this point moves through phase space since, in general, the locations and momenta of the various objects change. Therefore, a trajectory describes the time evolution of the system in phase space.

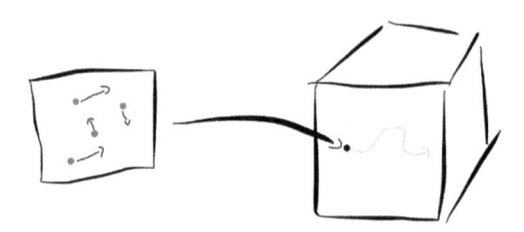

Therefore, whenever we are not 100% certain about the state of the system, we need more than a single point in phase space to describe it. Each initial state we think is possible corresponds to one particular phase space point.

Formulated differently, our limited accuracy means that we don't know the exact locations of our objects but only that they are in a certain spatial region. The same is true for the momenta which have to be within some range. Now, if we take uncertainty into account, our time evolution is no longer just a trajectory in phase space but a collection of trajectories.

Imagine you put a pencil down on each possible initial state of the system. Then, as time passes by, each of these pencils traces out the path in phase space that describes the time evolution if the system was in the corresponding initial state. When taken together, all of these pencils trace out what we call a **flow** in phase space. Formulated differently, while in the previous chapters we always considered the trajectories of individual phase space points, we are now interested in the trajectories of a swarm of points in phase space.

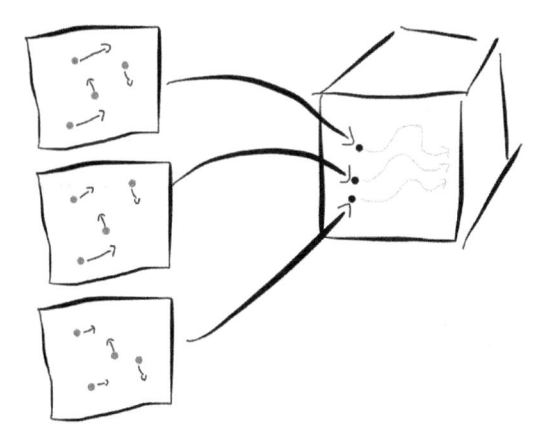

We will usually not deal with a finite set of points in practice. Instead we consider a whole *region* in phase space as our initial state. Each point within this region corresponds to a specific possible initial state of the system. In other words, this means that we assume that a whole range of locations and momenta are possible and not just a finite set.

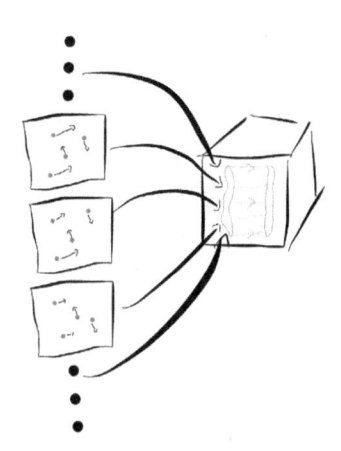

Figure 11.1: In general, for a continuous set of possible initial states, we get a region in phase space.

To understand this, let's imagine that we want to describe a single object but are somewhat uncertain about its location. We can then split up the possible locations into discrete boxes. Each such box corresponds to a range of possible locations. And in phase space, each box corresponds to a specific region.

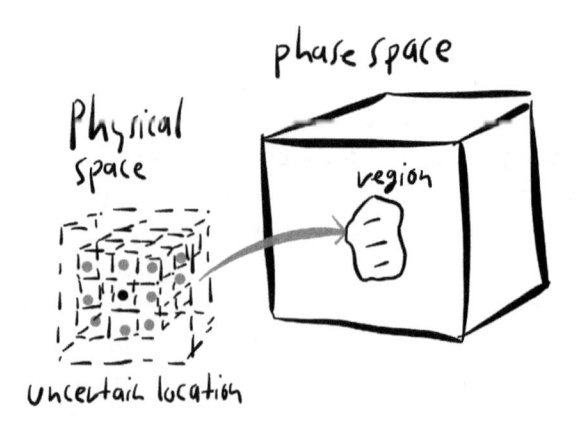

Of course, typically all points in such an initial region are not equally likely. For example, often we are somewhat sure that one particular initial state is much more likely than the other ones. But even if we are 99% certain, it can make sense to consider all of the less likely states too.[18]

[18] We will see why below.

Mathematically, we can take this into account by using a so-called probability density.[19]

[19] An alternative name for a probability density is **probability distribution**. For example, the probability distribution for a regular die is

$$\chi(1) = 1/6$$
$$\chi(2) = 1/6$$
$$\chi(3) = 1/6$$
$$\chi(4) = 1/6$$
$$\chi(5) = 1/6$$
$$\chi(6) = 1/6.$$

But we can also imagine that we are dealing with an oddly shaped die. Then, our probability distribution could look as follows:

$$\chi(1) = 1/2$$
$$\chi(2) = 1/6$$
$$\chi(3) = 1/12$$
$$\chi(4) = 1/12$$
$$\chi(5) = 1/12$$
$$\chi(6) = 1/12.$$

In words this means that for this die it is much more likely to get a 1 than, say, a 6.

11.2.1 Probability Density

A probability density $\rho(x, p, t)$ is a function which eats a phase space point (x, p) plus a specific point in time t and spits out the probability that this point corresponds to the correct state of the system.

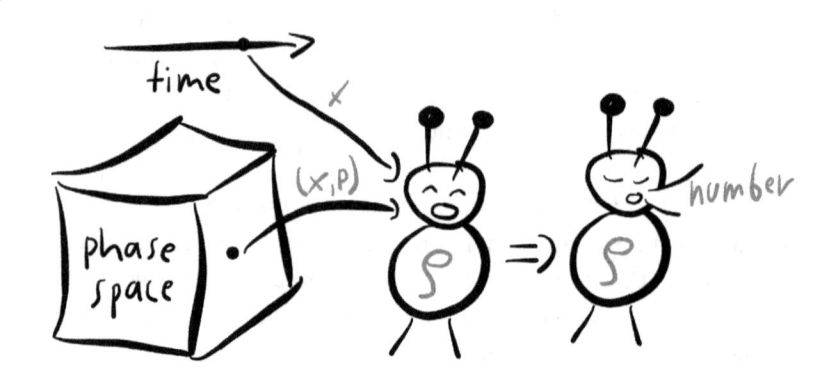

Before we discuss this in more technical terms, maybe the following idea helps: If we have definite knowledge about the state of the system, we have just one point. If we are not exactly sure, this one point splits up into many points. We can imagine that now, instead of one perfectly black point (probability = 100%), we have lots of points which are all a little bit transparent (probability < 100%). In other words, the level of transparency indicates how confident we are to find the system in this state. More transparency corresponds to a smaller probability to find the system in the corresponding state. The crucial point is then that if we add up all the points we get a completely non-transparent point which is identical to the definite knowledge point. In physical terms, this means that probability is conserved. In the definite knowledge case, we have just one point since we are 100% certain that the system is in this state. If we are not sure, the 100% gets distributed among all possible states.[20]

[20] We will discuss this using an explicit example in a moment.

To understand the concept of "probability density" a bit better, let's imagine a one-dimensional system with two objects and that (for some unspecified reason) only very specific positions and momentum values are possible. This means that only a few initial states are viable and not a continuous set. Further, let's say we are pretty certain that our system is in the state A where $q_1 = 2$ m, $p_1 = 3$ kg·m/s and $q_2 = 3$ m, $p_2 = 4$ kg·m/s. However, we can't exclude the state B where $q_1 = 3$ m, $p_1 = 4$ kg·m/s and $q_2 = 4$ m, $p_2 = 5$ kg·m/s or state C where $q_1 = 1$ m, $p_1 = 2$ kg·m/s and $q_2 = 2$ m, $p_2 = 3$ kg·m/s. Our (now discrete) initial probability density is then[21]

[21] To unclutter the notation we dropped all units.

$$\rho(t=0, A) = 0.7$$
$$\rho(t=0, B) = 0.2$$
$$\rho(t=0, C) = 0.1 \qquad (11.27)$$

In words this means that we are 70% likely to find the system in state A, 20% likely to find it in state B, and 10% likely to find it in state C.[22] An important property is that the probabilities add to 1 = 100 %. This is necessarily the case because one of the outcomes must happen and a total probability of more than 100% doesn't make sense.

[22] From a slightly different perspective we can imagine that when we prepare a given experiment multiple times, we expect to find it at $t = 0$ in state A in 70% of all runs, in state B in 20% of the runs and in state C only in 10% of the runs.

[23] Take note that the state C is not in our region and therefore does not appear in the sum.

At $t = 0$, to get the probability of finding our two objects in the regions $(q_1, q_2) \in \{2\ldots3, 3\ldots4\}$ and with momenta in the range $(p_1, p_2) \in \{3\ldots4, 4\ldots5\}$, we have to "integrate" over the corresponding volume. Since we are dealing with a discrete set of possible states, our "integration" is simply a sum[23]

$$P\Big(t=0, (q_1,q_2) \in \{2\ldots3, 3\ldots4\}, (p_1,p_2) \in \{3\ldots4, 4\ldots5\}\Big) = \sum_{q_1=2, q_2=3, p_1=3, p_2=4}^{q_1=3, q_2=4, p_1=4, p_2=5} \rho(t=0, p_1, q_1, p_2, q_2)$$
$$= \rho(t=0, A) + \rho(t=0, B)$$
$$= 0.7 + 0.2 = 0.9 \, .$$

If we are dealing with a continuous set of possible states (which is usually the case), the probability of finding our objects with positions within the region R_p and momenta within the range R_m at a specific moment in time t is[24]

[24] As usual, x is our shorthand notation for all phase space location coordinates $x_1, x_2, \ldots$ and p represents all momentum coordinates $p_1, p_2, \ldots$. In particular, this means that $dx = dx_1 dx_2 \ldots$ and $dp = dp_1 dp_2 \ldots$.

$$P(R_m, R_p, t) = \int_{R_p} dx \int_{R_m} dp\, \rho(x, p, t) \, . \tag{11.28}$$

Usually, we will use a shorthand notation for this kind of integral and simply talk about a region R in phase space and a phase space volume $dV = dqdp$:

$$\boxed{P(R, t) = \int_R \rho(x, p, t) dV \, .} \tag{11.29}$$

Now, what we are really interested in is, given an initial probability density describing our system at $t = 0$, what does the system look like after some time (i.e., at $t \neq 0$)? In other words, our main goal is, as usual, to calculate the time evolution of some given initial data.

But first, we need to talk about an important general property of phase space in classical mechanics.

11.2.2 Conservation of Phase Space Paths

In the example we discussed above, we started with three possible initial phase space points (Eq. 11.27). Each of them corresponds to a specific state our system can be in and the probability density tells us how likely they are. Moreover, given any specific phase space point, we can use Hamilton's equations to calculate how it evolves in time.

The key observation is that this implies that if we start with three possible initial states, we will always end up with exactly three possible final states. Formulated in more general terms:[25]

> The number of final points is always equal to the number of initial points.

Each initial point follows a specific path in phase space as specified by Hamilton's equations. And for each initial point there is always exactly one phase space path. So a different way of putting our observation is that:

> The number of phase space paths we have to consider is constant in time.

To understand this a bit better, recall that we can imagine that we trace out the phase space flow by putting down a pencil on each point and then drawing the trajectories in phase space for each point as a line. The statement here is that you never suddenly need more pencils than at the beginning since the trajectories you are drawing are completely specified by Hamilton's equations.[26]

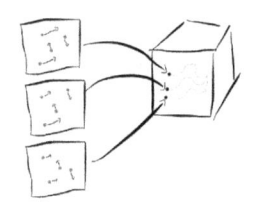

[25] This is one key difference between classical and quantum mechanics. In quantum mechanics one initial state can lead to many different final states because the fundamental equation of motion (Schrödinger's equation) is intrinsically probabilistic.

[26] It is conventional to reformulate this observation by saying that in classical mechanics the phase space flow is incompressible.

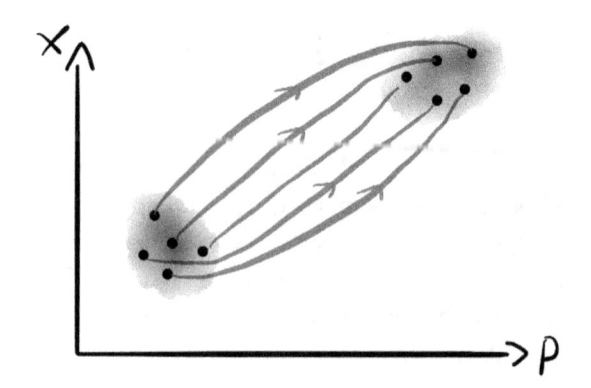

Let's try to understand what this implies in mathematical terms.[27]

[27] We will see that by putting the simple observation discussed above in mathematical terms, we can derive one of the most important statistical mechanics equations, which is known as the Liouville equation.

[28] A microstate corresponds to a specific choice of locations and momenta for all individual gas molecules. In contrast, a macrostate is defined by macroscopic observables like temperature and pressure. In general, there are lots of microstates which correspond to the same macrostate.

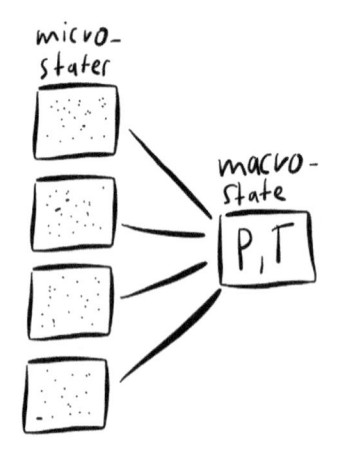

Let's assume for simplicity that, in total, N different states of our system are possible and we think each of them occurs with the same probability. This kind of situation is very common in statistical mechanics. For example, when we try to describe a gas there is no way we can determine the locations and momenta of all gas molecules at the same time. The best we can do is to determine macroscopic properties like the temperature and pressure of the gas and then collect all microstates of the system which are compatible with these macroscopic properties.[28]

Since we want to think about the behavior of individual phase space points, it is helpful to introduce a new density $\tilde{\rho}(q, p, t)$ which tells us how many viable phase space points are within some given region.[29] So when we integrate $\tilde{\rho}$ over all phase space, we get the total number of microstates which are compatible with our specific macrostate at a particular moment in time t:

$$N(t) = \int_{\text{all}} \tilde{\rho}(q, p, t) \, dV. \tag{11.30}$$

And in general, if we integrate $\tilde{\rho}$ over some phase space region R, we find out how many viable phase space points there are

[29] We will clarify how this density is related to the probability density we discussed above.

within this region:

$$N_R(t) = \int_R \tilde{\rho}(q,p,t)\, dV. \tag{11.31}$$

Since we assume that each microstate occurs with equal probability, the corresponding probability to find our system in a state described by a point in the phase space region R is[30]

$$P(R,t) = \frac{N_R(t)}{N(t)}. \tag{11.32}$$

[30] Take note that the probability to find our system in *any* state described by a phase space point is
$$P(\text{all},t) = \frac{N(t)}{N(t)} = 1 = 100\%.$$

This allows us to relate the original *probability* density $\rho(q,p,t)$ to our new *number* density $\tilde{\rho}(q,p,t)$:

$$\rho(q,p,t) = \frac{\tilde{\rho}(q,p,t)}{N(t)}. \tag{11.33}$$

But using this definition we find the correct probability for any region R by integrating over $\rho(q,p,t)$:

$$\int_R \rho(q,p,t) dV \stackrel{(11.33)}{=} \int_R \frac{\tilde{\rho}(q,p,t)}{N(t)} dV$$

$$\circlearrowright \quad N(t) \text{ does not depend on } q \text{ and } p$$

$$= \frac{\int_R \tilde{\rho}(q,p,t) dV}{N(t)}$$

$$\stackrel{(11.31)}{=} \frac{N_R(t)}{N(t)}$$

$$\stackrel{(11.32)}{=} P(R,t) \checkmark. \tag{11.34}$$

Equipped with these new tools, we can finally discuss what it really means to say that each possible initial phase space point evolves into exactly one final phase space point.

First of all, we can conclude that the *total* number of viable microstates $N(t)$ will always be the same because each initial state evolves into exactly one final state:

$$N(t) = \text{const.} \equiv N. \tag{11.35}$$

But what about the number of microstates within an arbitrary given phase space region R?

We need to be very careful when we answer this question because there are multiple aspects which we can easily miss. For concreteness, let's say that at some specific moment in time $t = t_0$, we have $N_R(t_0)$ microstates within the region R

$$N_R(t_0) = \int_R \tilde{\rho}(q, p, t_0) \, dV. \tag{11.36}$$

Now, we can immediately conclude that after some short time interval $t_0 \to t_0 + \epsilon$, the number of viable points within the region R is

$$N_R(t_0 + \epsilon) = \int_R \tilde{\rho}(q, p, t_0 + \epsilon) \, dV. \tag{11.37}$$

In general, this number will not be equal to $N_R(t_0)$ because phase space points move around as time passes:

$$N_R(t_0) \neq N_R(t_0 + \epsilon). \tag{11.38}$$

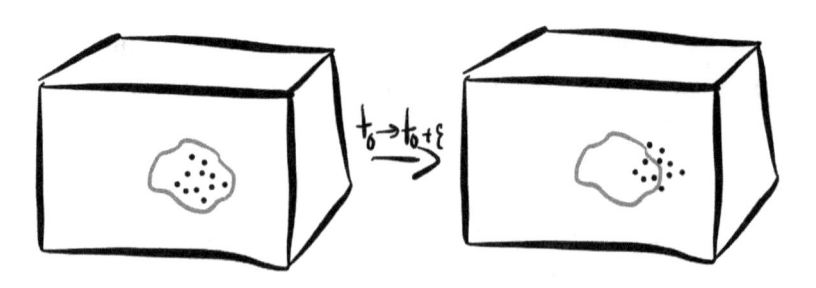

But for our purposes this is not a particularly useful insight.[31] However, we can modify the situation a little bit and derive one of the most important equations in statistical mechanics.

We can see in the image above that while the number of points did not change in principle, the number of points *within the*

---

[31] Recall that our goal is to encode the observation that each initial point evolves into exactly one final point in mathematical terms. And we discovered that the number of viable points within a fixed region R changes over time. So this is certainly not the kind of statement we are looking for.

region R got smaller because some of the points crossed the boundary of the region.[32]. This observation hints at how we can remedy this situation.

[32] No point representing a viable microstate was created or destroyed since each phase space point evolves according to Hamilton's equations.

First of all, we need to recall that the phase space of a specific system is not simply a static set of points (q, p). Instead, these points are equipped with a concrete Hamiltonian which tells us how each phase space point moves around.

Therefore, we can imagine that all phase space points are constantly moving around as described by Hamilton's equations. This implies that our region R is moved around, too:[33]

$$R \to \tilde{R}. \quad (11.39)$$

[33] A region is a set of points and since each point is moving around as described by Hamilton's equations, the whole region is moving around.

If we take this into account, we can calculate the number of viable phase space points within the *moved* region $\tilde{R}$

$$N_{\tilde{R}}(t_0 + \epsilon) = \int_{\tilde{R}} \tilde{\rho}(q, p, t_0 + \epsilon) \, dV. \quad (11.40)$$

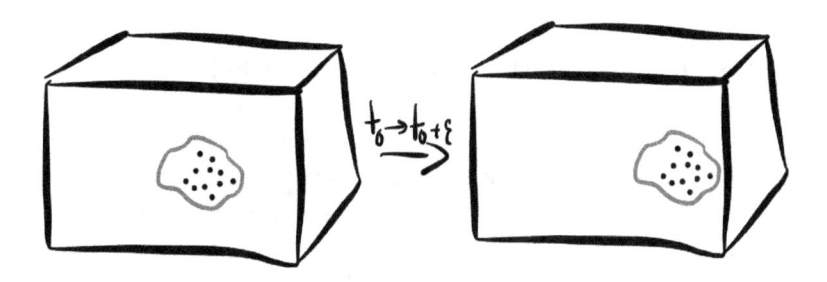

A key observation here is that

$$N_R(t_0) = N_{\tilde{R}}(t_0 + \epsilon). \quad (11.41)$$

This follows because the evolution of each point within the region R is uniquely fixed by Hamilton's equations and the viable phase space points we are interested in are simply phase space points in the region R too.

As time passes, all these points get "dragged" around by the Hamiltonian. Analogously, all points outside the region R get dragged around. But no new point will ever enter our moving region.

This is necessarily the case because:

> The paths of different phase space points never intersect each other.

Otherwise we would have a situation in which the time-evolution of at least one specific state of the system (corresponding to the phase space point where the paths intersect) wouldn't be fixed. Instead, there would be two possible trajectories that our state could take as time evolves. Each possibility would be described by one of the paths.

In other words, if paths in phase space could intersect, the time evolution of at least one initial state wouldn't be unique.

But we can put any phase space point (q, p) into Hamilton's equations and get a specific phase space path as a result. Therefore, we never run into such a situation in which the further time evolution is not uniquely fixed. Translated into phase space language this means that paths cannot intersect each other.

Therefore, the paths of all phase space points at the boundary of our region are like a shield that prevents new phase space points from entering our moving region.

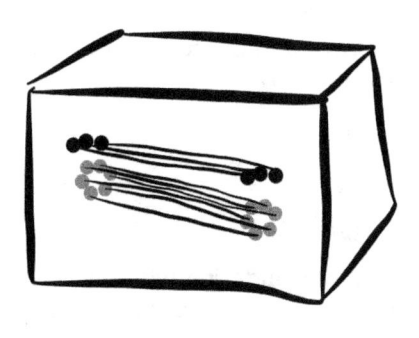

A key observation is now that Eq. 11.41 encodes exactly the kind of statement made at the beginning of this section.[34] To turn this equation into a more useful form, there are two things that we need to take into account. First of all, the region R only changes because our phase space points q and p evolve in time. So instead of using a transformed R, we can simply modify the arguments in

$$\tilde{\rho}(q, p, t_0 + \epsilon) \to \tilde{\rho}(Q, P, t_0 + \epsilon), \qquad (11.42)$$

[34] Each initial state evolves uniquely into one final state.

where (Eq. 10.35)

$$\begin{aligned} Q &= q + \epsilon\{q, H\} \\ P &= p + \epsilon\{p, H\}. \end{aligned} \qquad (11.43)$$

But then, there is one additional thing that we need to take into account: the infinitesimal volume elements dV we are integrating over can be different because $dV = dqdp$ and q and p change in time:[35]

$$dV = dqdp \to d\tilde{V} = JdQdP, \qquad (11.44)$$

[35] We will discuss this in much more detail in the following section.

where J is a possible scale factor that describes how the old and new infinitesimal volume elements differ.

Therefore, what we end up with is

$$\begin{aligned} N_{\tilde{R}}(t_0 + \epsilon) &= \int_{\tilde{R}} \tilde{\rho}(q, p, t_0 + \epsilon)\, dV \\ &\quad \circlearrowright \ \text{Eq. 11.42 and Eq. 11.44} \\ &= \int_{R} \tilde{\rho}(Q, P, t_0 + \epsilon)\, d\tilde{V}. \end{aligned} \qquad (11.45)$$

Using this formula, we can rewrite our key observation from above (Eq. 11.41) as follows:

$$N_R(t_0) = N_{\tilde{R}}(t_0 + \epsilon)$$

$$\int_R \tilde{\rho}(q, p, t_0) \, dV = \int_{\tilde{R}} \tilde{\rho}(Q, P, t_0 + \epsilon) \, d\tilde{V}. \quad (11.46)$$

↷ Eq. 11.31 and Eq. 11.45

Now, you are probably wondering why this is useful at all. So to avoid you getting too bored, let's discuss the punchline first and the details afterwards.

11.2.3 Liouville's Equation

In the previous section, we've simply put the statement that the number of viable initial and final states is always equal into a mathematical form. In the following section, we will derive that $dV = d\tilde{V}$. This is known as Liouville's theorem and tells us that the infinitesimal phase space volume elements we are integrating over remain unaffected by the time-evolution. This result allows us to rewrite the main result of the previous section (Eq. 11.46) as:

$$\int_R \tilde{\rho}(q, p, t_0) \, dV = \int_{\tilde{R}} \tilde{\rho}(Q, P, t_0 + \epsilon) \, d\tilde{V}$$

↷ $dV = d\tilde{V}$, see next section

$$\int_R \tilde{\rho}(q, p, t_0) \, dV = \int_{\tilde{R}} \tilde{\rho}(Q, P, t_0 + \epsilon) \, dV. \quad (11.47)$$

[36] Eq. 10.35:
$$Q = q + \epsilon\{q, H\}$$
$$P = p + \epsilon\{p, H\}.$$

By comparing the left-hand and right-hand side we can conclude:

$$\tilde{\rho}(q, p, t_0) = \tilde{\rho}(Q, P, t_0 + \epsilon). \quad (11.48)$$

Now this is *really* interesting because $\tilde{\rho}(Q, P, t_0 + \epsilon)$ is our number density $\tilde{\rho}$ at an infinitesimally later point in time since Q and P are the phase space coordinates at an infinitesimally later point in time.[36] But when our phase space density does not change during an infinitesimal interval, it never does. After all, we made no assumptions about the initial moment in time t_0 and therefore the equality holds in general. So what we learn here is that the *total* derivative of $\tilde{\rho}$ vanishes:[37]

[37] Take note that the partial derivative $\frac{\partial \tilde{\rho}}{\partial t}$ would vanish if instead of Eq. 11.48, we would have

$$\tilde{\rho}(q, p, t_0) = \tilde{\rho}(q, p, t_0 + \epsilon),$$

which, in general, is not the case as we've seen above (Eq. 11.38). The difference between the partial and total derivative is discussed in Appendix A.3. Moreover, we will discuss this from a different perspective in Section 11.2.5.

$$\frac{d\tilde{\rho}(q, p, t)}{dt} = 0. \quad (11.49)$$

Moreover, we can use the relationship between the number density $\tilde{\rho}$ and the probability density ρ (Eq. 11.33) to conclude:[38]

$$\frac{d\rho(q,p,t)}{dt} = 0. \quad (11.51)$$

Next, by recalling that Hamilton's general equation (Eq. 5.42) holds for any phase space function, we can rewrite Eq. 11.51 as follows:

$$\frac{d\rho}{dt} = 0$$

$$\downarrow \text{Eq. 5.42}$$

$$\{\rho, H\} + \frac{\partial \rho}{\partial t} = 0. \quad (11.52)$$

By rearranging the terms, we find:

$$\boxed{\frac{\partial \rho}{\partial t} = -\{\rho, H\}} \quad (11.53)$$

This is the famous **Liouville equation** which describes how the probability density ρ evolves in time.

[38] For your convenience Eq. 11.33 reads:

$$\rho(q,p,t) = \frac{\tilde{\rho}(q,p,t)}{N}$$

where N is constant because the total number of viable states does not change since each initial state evolves into exactly one final state. Therefore:

$$\frac{d\tilde{\rho}(q,p,t)}{dt} = 0$$

$$\therefore \frac{d\left(N\rho(q,p,t)\right)}{dt} = 0$$

$$\therefore N\frac{d\rho(q,p,t)}{dt} = 0$$

$$\therefore \frac{d\rho(q,p,t)}{dt} = 0. \quad (11.50)$$

In the following section we will discuss why infinitesimal phase space volume elements remain unaffected as we move forward in time.

11.2.4 Liouville's Theorem

We now consider one of the smallest possible regions in phase space: an infinitesimal rectangle spanning from[39]

$$A \equiv (A_q, A_p) \equiv (q, p) \quad (11.54)$$

via

$$B \equiv (B_q, B_p) \equiv (q + dq, p) \quad (11.55)$$

to

$$C \equiv (C_q, C_p) \equiv (q, p + dp) \quad (11.56)$$

[39] We are simply giving convenient names to the coordinates of the various points.

and
$$D \equiv (D_q, D_p) \equiv (q+dq, p+dp). \tag{11.57}$$

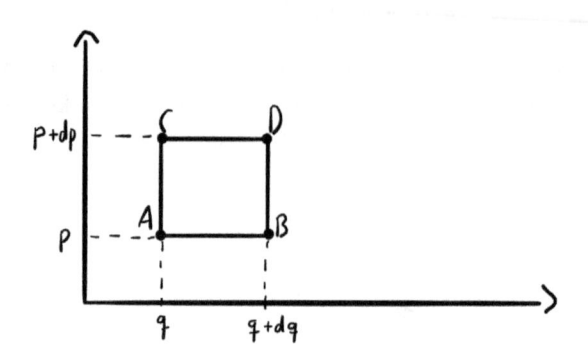

Take note that the area of this region is $a = dqdp$.[40]

[40] This will be important in a moment.

As time passes, our region moves through phase space. What does our region look like after an infinitesimal time interval dt, i.e., at $t+dt$?

To understand this, we need to calculate explicitly how each of the individual points moves. We can do this by using Hamilton's equations, but there are subtleties that we need to take into account.

First of all, Hamilton's equations give us the velocity $\dot{q} \equiv \frac{dq}{dt}$ in the q-direction and the velocity $\dot{p} \equiv \frac{dp}{dt}$ in the p-direction. Therefore, when we want to know how a specific phase space point $A = (q, p)$ moves, we need to take into account that, in general, its q coordinate and its p coordinate will change.[41]

[41] We will use Hamilton's equations below, but first we express everything in terms of the general rates of change.

Therefore, after the infinitesimal time interval dt, the q coordinate of our point A will have moved to

$$A'_q = A_q + \dot{q}(q, p, t)dt \stackrel{(11.54)}{=} q + \dot{q}(q, p, t)dt \tag{11.58}$$

and the p coordinate will have moved to

$$A'_p = A_p + \dot{p}(q, p, t)dt \stackrel{(11.54)}{=} p + \dot{p}(q, p, t)dt. \tag{11.59}$$

This means that our point A is now located at

$$A' = (A'_q, A'_p) = \left(q + \dot{q}dt, p + \dot{p}dt\right). \quad (11.60)$$

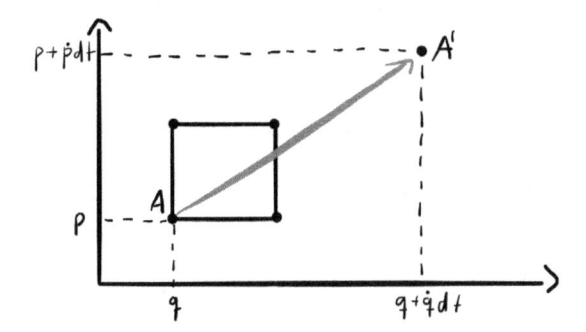

For the remaining points, we need to talk about a second subtlety. As before, we need to take the rates of change in the q direction and the p direction into account. But we need to be careful because these rates are, in general, different for different phase space points.[42] For example, the rate of change of the point B in the q direction is $\dot{q}(q + dq, p, t)$ and its rate of change in the p direction is $\dot{p}(q + dq, p, t)$.

[42] Recall that different phase space points correspond to distinct states of the system. Therefore, there is no general reason why the way these different points evolve in time should be equal.

Therefore, we will find that the q coordinate of the phase space point B is now

$$B'_q = B_q + \dot{q}(q + dq, p, t)dt$$

$$\quad \curvearrowright \text{Taylor expansion}$$

$$= B_q + \left(\dot{q}(q, p, t) + \frac{\partial \dot{q}(q, p, t)}{\partial q}dq\right)dt, \quad (11.61)$$

and its p coordinate reads

$$B'_p = B_p + \dot{p}(q + dq, p, t)dt$$

$$\quad \curvearrowright \text{Taylor expansion}$$

$$= B_p + \left(\dot{p}(q, p, t) + \frac{\partial \dot{p}(q, p, t)}{\partial q}dq\right)dt, \quad (11.62)$$

where we used, as usual, that for infinitesimal intervals, we can neglect higher order terms in the Taylor expansion. So our point

B will therefore be located at

$$B' = (B'_q, B'_p) = \left(B_q + \left(\dot{q} + \frac{\partial \dot{q}}{\partial q} dq \right) dt, B_p + \left(\dot{p} + \frac{\partial \dot{p}}{\partial q} dq \right) dt \right)$$

$$\stackrel{(11.55)}{=} \left(q + dq + \left(\dot{q} + \frac{\partial \dot{q}}{\partial q} dq \right) dt, p + \left(\dot{p} + \frac{\partial \dot{p}}{\partial q} dq \right) dt \right).$$

(11.63)

Analogously, we can calculate the new coordinates of C:

$$C'_q = C_q + \dot{q}(q, p + dp, t) dt$$

$$\quad \curvearrowright \text{Taylor expansion}$$

$$= C_q + \left(\dot{q}(q, p, t) + \frac{\partial \dot{q}(q, p, t)}{\partial p} dp \right) dt$$

$$C'_p = C_p + \dot{p}(q, p + dp, t) dt$$

$$\quad \curvearrowright \text{Taylor expansion}$$

$$= C_p + \left(\dot{p}(q, p, t) + \frac{\partial \dot{p}(q, p, t)}{\partial p} dp \right) dt.$$

(11.64)

Therefore[43]

$$C' = (C'_q, C'_p) = \left(C_q + \left(\dot{q} + \frac{\partial \dot{q}}{\partial p} dp \right) dt, C_p + \left(\dot{p} + \frac{\partial \dot{p}}{\partial p} dp \right) dt \right)$$

$$\stackrel{(11.56)}{=} \left(q + \left(\dot{q} + \frac{\partial \dot{q}}{\partial p} dp \right) dt, p + dp + \left(\dot{p} + \frac{\partial \dot{p}}{\partial p} dp \right) dt \right).$$

(11.65)

[43] Completely analogously, we can calculate the coordinates of D'. We will skip this step since we don't need it for the following discussion.

Now, a first important observation is that our region is no longer a rectangle because each point moves with a different velocity.

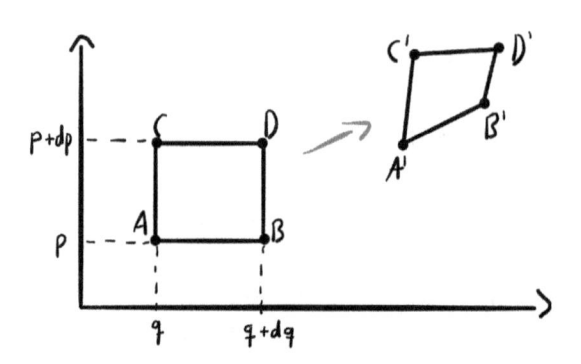

But what about the area of the region?[44]

In general, the area of the parallelogram spanned by two 2-dimensional vectors $\vec{v} = (v_1, v_2)^T$ and $\vec{w} = (w_1, w_2)^T$ is given by[45]

$$\text{area}(v \times w) = v_1 w_2 - v_2 w_1. \tag{11.66}$$

The vectors spanning our region at t are

$$\vec{AB} = \vec{B} - \vec{A} = \begin{pmatrix} q + dq \\ p \end{pmatrix} - \begin{pmatrix} q \\ p \end{pmatrix} = \begin{pmatrix} dq \\ 0 \end{pmatrix}$$

$$\vec{AC} = \vec{C} - \vec{A} = \begin{pmatrix} q \\ p + dp \end{pmatrix} - \begin{pmatrix} q \\ p \end{pmatrix} = \begin{pmatrix} 0 \\ dp \end{pmatrix}. \tag{11.67}$$

Therefore the original area of the region, as already mentioned above, is

$$\text{area}(\vec{AB} \times \vec{AC}) \stackrel{(11.66)}{=} (\vec{AB})_1 (\vec{AC})_2 - (\vec{AB})_2 (\vec{AC})_1$$
$$\stackrel{(11.67)}{=} dq\,dp - 0 = dq\,dp. \tag{11.68}$$

The vectors spanning our region at $t + dt$ are

$$\vec{A'B'} = \vec{B'} - \vec{A'}$$

⟩ Eq. 11.60 and Eq. 11.63

$$= \begin{pmatrix} q + dq + \left(\dot{q} + \frac{\partial \dot{q}}{\partial q} dq\right) dt \\ p + \left(\dot{p} + \frac{\partial \dot{p}}{\partial q} dq\right) dt \end{pmatrix} - \begin{pmatrix} q + \dot{q}dt \\ p + \dot{p}dt \end{pmatrix}$$

$$= \begin{pmatrix} dq + \frac{\partial \dot{q}}{\partial q} dq\, dt \\ \frac{\partial \dot{p}}{\partial q} dq\, dt \end{pmatrix}$$

$$\vec{A'C'} = \vec{C'} - \vec{A'}$$

⟩ Eq. 11.60 and Eq. 11.65

$$= \begin{pmatrix} q + \left(\dot{q} + \frac{\partial \dot{q}}{\partial p} dp\right) dt \\ p + dp + \left(\dot{p} + \frac{\partial \dot{p}}{\partial p} dp\right) dt \end{pmatrix} - \begin{pmatrix} q + \dot{q}dt \\ p + \dot{p}dt \end{pmatrix}$$

$$= \begin{pmatrix} \frac{\partial \dot{q}}{\partial p} dp\, dt \\ dp + \frac{\partial \dot{p}}{\partial p} dp\, dt \end{pmatrix}. \tag{11.69}$$

Therefore, the area of our region at $t + dt$ reads

---

[44] What we will discover in the following is one of the most important insights in all of physics. So while the following calculation is quite cumbersome, the final result is worth the hassle.

[45] We can understand this by noting that a parallelogram is basically a rectangle minus some extra stuff.

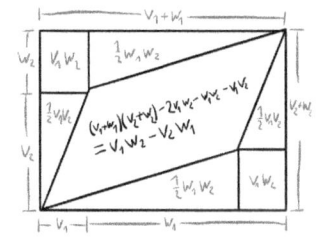

A more abstract way to understand this is by noting that the area of the parallelogram spanned by two 2-dimensional vectors $\vec{v} = (v_1, v_2)^T$ and $\vec{w} = (w_1, w_2)^T$ is exactly the determinant of the matrix $\begin{pmatrix} v_1 & w_1 \\ v_2 & w_2 \end{pmatrix}$.

$$\text{area}(\vec{A'B'} \times \vec{A'C'}) \stackrel{(11.66)}{=} (\vec{A'B'})_1(\vec{A'C'})_2 - (\vec{A'B'})_2(\vec{A'C'})_1$$

$$\stackrel{(11.69)}{=} \left(dq + \frac{\partial \dot{q}}{\partial q}dqdt\right)\left(dp + \frac{\partial \dot{p}}{\partial p}dpdt\right) - \left(\frac{\partial \dot{p}}{\partial q}dqdt\right)\left(\frac{\partial \dot{q}}{\partial p}dpdt\right)$$

⟩ $dt^2 = 0$ (infinitesimal)

$$= dqdp + dq\frac{\partial \dot{p}}{\partial p}dpdt + dp\frac{\partial \dot{q}}{\partial q}dqdt$$

⟩ Hamilton's equations, Eq. 5.15

$$= dqdp + dq\frac{\partial}{\partial p}\left(-\frac{\partial H}{\partial q}\right)dpdt + dp\frac{\partial}{\partial q}\left(\frac{\partial H}{\partial p}\right)dqdt$$

⟩ $\frac{\partial}{\partial q}\frac{\partial}{\partial p} = \frac{\partial}{\partial p}\frac{\partial}{\partial q}$

$$= dqdp - dq\frac{\partial}{\partial q}\frac{\partial H}{\partial p}dpdt + dp\frac{\partial}{\partial q}\frac{\partial H}{\partial p}dqdt$$

⟩ $dq\frac{\partial}{\partial q}\frac{\partial H}{\partial p}dpdt$

$$= dqdp.$$
(11.70)

This is exactly equal to the area of the region at t (Eq. 11.68)![46] What we've therefore discovered is that while the shape of our phase space region has changed wildly, its area remains unchanged. Since any phase space region consists of such infinitesimal regions and any finite time interval consists of many infinitesimal ones, we can conclude that this result holds in general. Moreover, since phase space is usually not 2-dimensional but higher-dimensional, we are dealing with volumes instead of surfaces. Therefore, we can conclude:

> Phase space volume remains constant in time.

This is famously known as **Liouville's theorem**. And, as shown in the previous section, it is essential in the derivation of the Liouville equation (Eq. 11.53) which governs the behavior of the probability density in classical mechanics.

11.2.5 Summary and Comments

In the previous section, we've started talking about how we can describe situations in which we don't know all details with 100% accuracy.

We learned that we can still make predictions by describing sys-

[46] If you're unfamiliar with the transformation rules of an integral under coordinate changes, ignore this sidenote. Otherwise, take note that from a more technical perspective, we can understand what is going on here by noting that time translation $t \to t + \epsilon$ is a canonical transformation generated by the Hamiltonian H. (We discovered this in Section 10.3.) But a canonical transformation is, in some sense, simply a change in coordinates. Thus, what we do here is switch to new coordinates which correspond to the old ones at a later point in time. From this perspective, we can use the usual transformation rules for an integral under a coordinate transformation. In particular, we need to keep in mind that after the coordinate transformation the Jacobian determinant appears under the integral. So what Liouville's theorem really says is that the Jacobi determinant is 1 for the transformation corresponding to time translations. For a short proof, see page 90 here: http://www.damtp.cam.ac.uk/user/tong/dynamics/clas.pdf

tems in terms of probability densities. In general, if we integrate a probability density over a phase space region R, we get the probability of finding our system in a state described by a phase space point within the region (Eq. 11.29):

$$\boxed{P(R,t) = \int_R \rho(x,p,t)dV.} \qquad (11.71)$$

We then used the key observations that each possible initial state evolves into exactly one final state, that phase space paths never intersect each other, and that phase space volumes are constant in time (**Liouville's theorem**) to derive the **Liouville equation** (Eq. 11.53)[47]

[47] We discussed Liouville's theorem in Section 11.2.4.

$$\boxed{\frac{\partial \rho}{\partial t} = -\{\rho, H\}.} \qquad (11.72)$$

Of course, we've barely scratched the surface. Statistical mechanics is a vast and beautiful branch of physics. But for the modest goals of this book we have everything that we need to move forward.

Nevertheless, there is one additional perspective on Liouville's equation that I would like to talk about because so far it may appear a bit nebulous. In particular, you might find it puzzling that we first discovered that, in general, the total time derivative of the probability density vanishes (Eq. 11.51)

$$\frac{d\rho}{dt} = 0 \qquad (11.73)$$

but then, using Hamilton's general equation of motion (Eq. 5.36), we derived that the corresponding partial time derivative $\frac{\partial \rho}{\partial t}$ does not vanish in general (Eq. 11.53).

The observation which allows us to understand this is that the key fact encoded in the Liouville equation is that in classical mechanics there is nothing which really influences and modifies probabilities.

In particular, this implies that if we start with an initial state $A = (q_A, p_A)$ which occurs with a probability of 70% at $t = t_0$, at some later point in time $t = t_1$ the *corresponding* final state $\tilde{A} = (Q_A, P_A)$ will occur with a probability of 70% too.[48]

[48] Or, to come full circle, imagine that we are 100% certain that a particular initial state describes our system. In classical mechanics we can then simply use Hamilton's equations to calculate the corresponding final state at a later point in time. And most importantly, since there is no intrinsically probabilistic aspect in classical mechanics, this final state will occur with 100% certainty too. So there is just one phase space path and the probability of 100% stays attached to each point along the way. (The situation is very different in quantum mechanics where one fixed initial state can lead to many different final states which each occur with a certain probability.)

In mathematical terms, when we are 100% certain that $(q = 0, p = 0)$ is the correct initial point, we use the probability density

$$\rho(q, p, t = 0) = \delta(q, p),$$

where $\delta(x)$ denotes the delta distribution. (Don't worry if you don't know yet what the delta distribution is. Speaking colloquially, it's simply a mathematical tool which we invent to describe situations in which something is localized exactly at one point.) The Liouville equation then tells us that as time passes, our probability density is described by

$$\rho(q, p, t) = \delta(q - q(t), p - p(t)),$$

where $q(t)$ and $p(t)$ describe the correct phase space path as given by Hamilton's equations. In words, this means that our 100% probability simply is dragged along the path defined by $q(t)$ and $p(t)$. And in particular, we don't get a 50% probability of finding the final state X and a 50% probability of finding the final state Y.

So if we start with a fixed set of possible initial states and specific probabilities for each of them, the corresponding final states will each occur with exactly the same probabilities.

Mathematically we can imagine that we start with

$$\begin{aligned} P(A, t_0) &= 70\% \\ P(B, t_0) &= 10\% \\ P(C, t_0) &= 5\% \\ &\vdots \end{aligned} \quad (11.74)$$

And at a later point in time we find

$$\begin{aligned} P(\tilde{A}, t_1) &= 70\% \\ P(\tilde{B}, t_1) &= 10\% \\ P(\tilde{C}, t_1) &= 5\% \\ &\vdots \end{aligned} \quad (11.75)$$

where $\tilde{A}, \tilde{B}, \tilde{C}$ are the phase space points that we get by evolving our initial points A, B, C using Hamilton's equations.

In contrast, we have

$$\begin{aligned} P(A, t_1) &\neq 70\% \\ P(B, t_1) &\neq 10\% \\ P(C, t_1) &\neq 5\% \\ &\vdots \end{aligned} \quad (11.76)$$

because, in general, our phase space points move around. Therefore, the probability that the original phase space point

ADDITIONAL FORMULATIONS OF CLASSICAL MECHANICS 283

A still describes the state of the system at a later point in time is not necessarily equal to the probability at the initial moment in time.

This is exactly what $\frac{d\rho}{dt} = 0$ but $\frac{\partial \rho}{\partial t} \neq 0$ means in the context of probability densities.

In terms of the phase space probability density, the statements above read:[49]

$$\text{Eq. 11.74} \neq \text{Eq. 11.76} \quad \leftrightarrow \quad \rho(q, p, t_0) \neq \rho(q, p, t_1) \quad (11.77)$$

because, for example, state $A = (q_A, p_A)$ will have moved to a new location $\tilde{A} = (Q_A, P_A)$. In terms of derivatives, this means that $\frac{\partial \rho}{\partial t} \neq 0$.

But in contrast, we have[50]

$$\text{Eq. 11.74} = \text{Eq. 11.75} \quad \leftrightarrow \quad \rho(q, p, t_0) = \rho(Q, P, t_1) \quad (11.78)$$

because the probabilities remain attached to their points as they move through phase space.

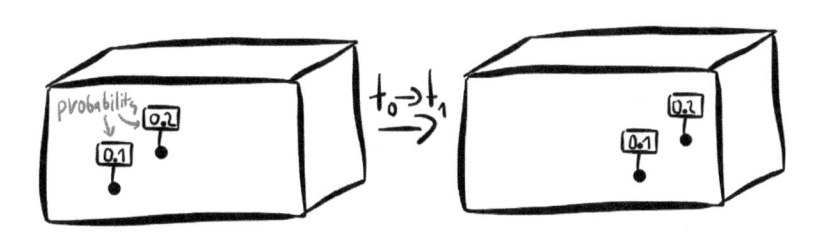

[49] Recall the definition of the partial derivative:

$$\frac{\partial \rho}{\partial t} = \lim_{\Delta t \to 0} \left(\frac{\rho(t + \Delta t, x(t), p(t))}{\Delta t} - \frac{\rho(t, x(t), p(t))}{\Delta t} \right).$$

(See Appendix A.3.)

[50] Recall that here

$$Q = q + \epsilon\{q, H\}$$
$$P = p + \epsilon\{p, H\}.$$

In words this means that Q and P are our old coordinates at an infinitesimally later point in time.

The state described by $A = (q_A, p_A)$ has now evolved into the state described by $\tilde{A} = (Q_A, P_A)$. Therefore, the probability of observing this state at t_1 is exactly the same as the probability of observing the state corresponding to A at t_0. And in terms of derivatives, this means that $\frac{d\rho}{dt} = 0$.[51]

To summarize: the pattern of probabilities is frozen ($\frac{d\rho}{dt} = 0$) since each probability simply remains attached to its corresponding phase space point as it moves around. The Liouville

[51] Recall the definition of the total derivative:

$$\frac{d\rho}{dt} = \lim_{\Delta t \to 0} \left(\frac{\rho(t + \Delta t, x(t + \Delta t), p(t + \Delta t))}{\Delta t} - \frac{\rho(t, x(t), p(t))}{\Delta t} \right).$$

(See Appendix A.3.)

[52] The pattern of probabilities is fixed by the initial probabilities.

[53] Take note that the zero probabilities which are now found at the original locations, moved there from some other location. To understand this, take note that there is really a probability assigned to all points. But for most points, the probability is zero. Nevertheless, as time moves on, all these probabilities (zero or not) move around as described by Hamilton's equations.

equation allows us to keep track of how the *fixed* probabilities are dragged around as time passes.[52]

On the other hand, this also implies that the probabilities at a *fixed* phase space point change all the time (i.e., $\frac{\partial \rho}{\partial t} \neq 0$) because all points drag their probabilities with them as time moves on.[53]

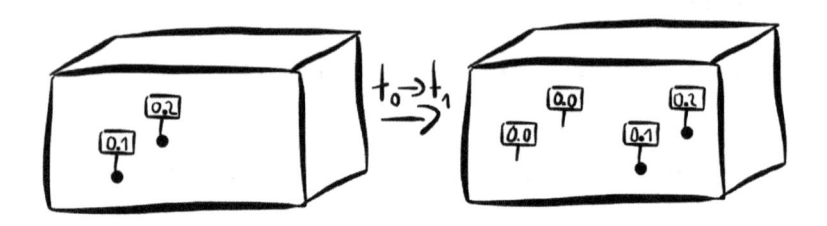

Maybe it helps to imagine that we place a little boat at each phase space point which has a real chance to describe the system in question accurately. Inside each such boat we put the probability that the corresponding phase space point describes our system. All points without a boat have zero probability of describing our system.

Then, as time passes, our little boats get dragged around phase space as a result of the Hamiltonian flow.[54]

[54] In this analogy, the time evolution as given by Hamilton's equations yields a flow which drags our boats in certain directions.

So when we wait some time and then check the original locations of the boats, we will most likely not find any boats there. And if we don't find any boat at a given phase space point at a specific moment in time, the probability that this point describes our system is zero. Usually, our boats have moved on and therefore the probabilities are now found elsewhere. This is what $\rho(q, p, t_0) \neq \rho(q, p, t_1)$ means.

But on the other hand, we know exactly where we can find our boats since their movement is determined completely by Hamilton's equations. We will find the boat which was located at the specific phase space point (q, p) after an infinitesimal time interval at

$$Q = q + \epsilon\{q, H\}$$
$$P = p + \epsilon\{p, H\}.$$

This is what we mean by $\rho(q, p, t_0) = \rho(Q, P, t_1)$.

The key observation here is that the probabilities that we put inside the boats at $t = 0$ do not change at all and will be the same at any later point in time. It's just that we then find these boats at different locations. The probabilities inside them remain unchanged.[55]

[55] Once more, this is only the case in classical mechanics. In quantum mechanics, there are multiple possible outcomes for a specific initial state and therefore one boat splits up into many even smaller boats which each carry a fraction of the original probability.

As a final comment, take note that what we discussed above does not imply that the *uncertainties* are constant. To understand this imagine that we've measured the location and momentum of a single object quite accurately. In phase space this means that we describe our initial configuration using a small phase space region.

Now, as time passes, the various individual points inside the region start moving around. However, it is by no means guaranteed that they stay close to each other. Instead, our region can split up, since even small differences in the initial condition can lead to wildly different final states.

This is known as **phase space filamentation.**

Just think about a billiard ball. A small difference in its initial location and momentum can lead to a completely different final location.

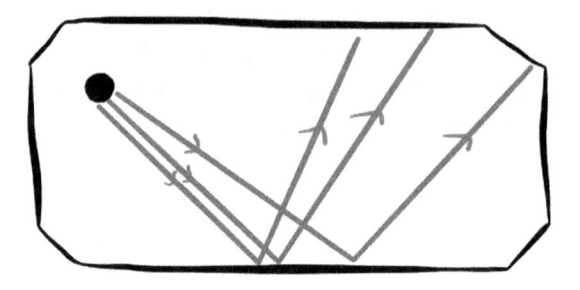

However, as we've learned above, the probabilities remain attached to the viable phase space points as they move around. Therefore, we find a much larger range of possible locations and momenta for our object in the final state. And this implies that we can end up with huge uncertainties in the final state although we started with a fairly well known initial state.

To quote Roger Penrose[56]

[56] Roger Penrose. *The Emperor's New Mind : Concerning Computers, Minds and the Laws of Physics.* Oxford University Press, Oxford, 2016. ISBN 9780198784920

For a somewhat analogous situation, think of a small drop of ink placed in a large container of water. Whereas the actual volume of material in the ink remains unchanged, it eventually becomes thinly spread over the entire contents of the container. [...] We may ask, in view of this spreading throughout phase space, how is it possible at all to make predictions in classical mechanics? That is, indeed, a good question. What this spreading tells us is that, no matter how accurately we know the initial state of a system (within some reasonable limits), the uncertainties will tend to grow in time and our initial information may become almost useless. Classical mechanics is, in this kind of sense, essentially unpredictable.

But, of course, we can often make quite good predictions in classical mechanics because our uncertainties are so small compared to the size of the objects that we want to describe so that a moderate growth in the uncertainties doesn't make a huge difference.

Now, let's move on to a final formulation of classical mechanics which is as close to quantum mechanics as it gets.

11.3 Koopman-von Neumann Mechanics

At the beginning of this chapter, I've already mentioned that Koopman-von Neumann mechanics, in some sense, is the logical next step after the Newtonian, Lagrangian and Hamiltonian formulations of classical mechanics. We climbed this ladder by successively increasing the number of dimensions of the mathematical arena that we use. In Newtonian mechanics, we have physical space with just 3 dimensions. In configuration space we add 3 dimensions for each object moving around freely. And in phase space, we add 6 dimensions for each object because we not only introduce new axes to keep track of the location of each individual object, but also new axes for the momenta of the various objects.

So how can we construct a space that is even larger than phase space?

11.3.1 Hilbert Space

Well, we can add one *axis* for each possible state a given object can be in. This certainly yields a huge space, but why would anyone want to do that? And what does an "axis for each possible state" mean?

To answer these questions, we need to talk again about uncertainty. Let's imagine that for some reason, we don't know the position and momentum of a particular object with 100% certainty.[57] In such a scenario it's not too stupid to introduce an

[57] Possible reasons for such an uncertainty are that there is always a limit to how precisely we can measure the location of objects, and when it comes to elementary particles, there is a fundamental uncertainty we seemingly can't get rid of. This, in fact, is the defining feature of quantum mechanics and we will talk more about it in the next chapter.

[58] We will discuss how this works below.

[59] In general, an object can be in infinitely many states.

individual axis for each possible state. We can then describe how likely a given state is by using a particular point on the corresponding axis.[58]

For simplicity, let's assume that our object can only be in four different states (q_1, p_1), (q_2, p_2), (q_3, p_3), and (q_4, p_4).[59] Mathematically, we describe these four possible states using four different basis vectors $\hat{e}_1, \hat{e}_2, \hat{e}_3, \hat{e}_4$ because we now introduce an individual axis for each possible state.

If we are 100% certain to find our object in the first state, we would describe this situation by the vector

$$\Psi = \hat{e}_1. \tag{11.79}$$

Analogously, if we are 100% certain to find it in the second state, we would use

$$\Psi = \hat{e}_2. \tag{11.80}$$

But this kind of framework is only useful when we are dealing with uncertainties. So let's assume that each of the four states is equally likely. We can then describe the situation by the vector

$$\Psi = \frac{1}{2}\hat{e}_1 + \frac{1}{2}\hat{e}_2 + \frac{1}{2}\hat{e}_3 + \frac{1}{2}\hat{e}_4. \tag{11.81}$$

Analogously, if we think there is a 50% chance that our object is in the first state and a 50% that we find it in the second state, we use the vector

$$\Psi = \frac{1}{\sqrt{2}}\hat{e}_1 + \frac{1}{\sqrt{2}}\hat{e}_2. \tag{11.82}$$

So the numbers in front of the basis vectors encode the corresponding probabilities. And in this sense, the projection of a state onto a specific axis tells us how likely the corresponding state is.

But how exactly are the numbers in front of the basis vectors related to probabilities?

The basic rule is that the absolute value squared of the number in front of a basis vector yields the corresponding probability.[60] So for our vector in Eq. 11.81, the probability of each of the four states is $\left(\frac{1}{2}\right)^2 = \frac{1}{4} = 0.25 = 25\%$. And for the vector in Eq. 11.82, we have a $\left(\frac{1}{\sqrt{2}}\right)^2 = \frac{1}{2} = 0.5 = 50\%$ probability for each of the two states and a probability of zero for the remaining two states because we can rewrite our vector as

$$\Psi = \frac{1}{\sqrt{2}}\hat{e}_1 + \frac{1}{\sqrt{2}}\hat{e}_2 + 0\hat{e}_3 + 0\hat{e}_4. \quad (11.83)$$

[60] We will see later how this comes about.

So in general, if there are N possible states, we describe a specific state of the system using the vector

$$\Psi = \sum_{i=1}^{N} c_i \hat{e}_i = c_1 \hat{e}_1 + c_2 \hat{e}_2 + c_3 \hat{e}_3 + \ldots, \quad (11.84)$$

where c_i are the coefficients in front of the basis vectors $\hat{e}_i$ which encode the corresponding probabilities.

Usually, there are infinitely many possible states and we then have to replace our sum with an integral

$$\Psi(x, p, t) = \int c(x, p, t) \hat{e}_{x,p} \, dx dp, \quad (11.85)$$

where our coefficients depend on the time: $c = c(x, p, t)$, because the probability to find an object in a specific state changes over time.

In general, we call the space spanned by the basis vectors $\hat{e}_{x,p}$, which we introduce for each possible state, a **Hilbert space**. Moreover, we call the vectors which describe a specific physical situation Ψ a **state vector**.

A key feature of basis vectors is that they are orthogonal and normalized. So, for example,

$$\hat{e}_1 \cdot \hat{e}_2 = 0$$
$$\hat{e}_1 \cdot \hat{e}_1 = 1 \quad (11.86)$$

and we have, in general,

$$\hat{e}_i \cdot \hat{e}_j = 0 \quad \text{for} \quad i \neq j$$
$$\hat{e}_i \cdot \hat{e}_j = 1 \quad \text{for} \quad i = j. \tag{11.87}$$

This allows us to extract the concrete probability that a specific state, say state 3, is the right one by multiplying the corresponding state vector by the corresponding basis vector:

$$P(3) = |\hat{e}_3 \cdot \Psi|^2$$
$$\stackrel{(11.84)}{=} \left| \hat{e}_3 \cdot \left(\sum_{i=1}^{N} c_i \hat{e}_i \right) \right|^2$$
$$= \left| \sum_{i=1}^{N} c_i \hat{e}_3 \cdot \hat{e}_i \right|^2$$
$$= |c_3|^2 . \qquad \curvearrowright \hat{e}_3 \cdot \hat{e}_i = 0 \text{ unless } i = 3, \text{ Eq. 11.87} \tag{11.88}$$

As mentioned above, $|c_3|^2$ is exactly the probability that we find our system in the state 3.

So far, we've simply constructed a new mathematical arena which you can find useful or awkward depending on your personal preferences.[61] In any case, the crucial next step is to find a way to describe classical mechanics in this new mathematical arena.

[61] Take note that Hilbert space is an essential tool to describe things in quantum mechanics.

11.3.2 Koopman-von Neumann Equation

There are two key ideas which allow us to describe classical mechanics in Hilbert space:

1. The probability density $\rho(x, p, t)$ is related to the state vectors Ψ by[62]

$$\rho(x, p, t) = |\Psi(x, p, t)|^2 . \tag{11.89}$$

This follows directly if we recall what we've learned in the previous section. By evaluating Eq. 11.89 explicitly using the

[62] We discussed probability densities in Section 11.2.1.

general form of our state vector (Eq. 11.85) and that our basis vectors are orthogonal, we find:[63]

$$\rho(x,p,t) = |\Psi(x,p,t)|^2 = |c(x,p,t)|^2. \quad (11.90)$$

Moreover, as mentioned in the previous section, the absolute value squared of the coefficients $c(x,p,t)$ yields exactly the probability to measure a specific state (x,p).

Therefore if we integrate $c(x,p,t)$ over a specific phase space region R, we get the probability to find our system in a state described by a phase space point within this region at a specific moment in time t. This is exactly how we defined the probability density in the first place (Eq. 11.29).[64]

2. The time-evolution of our state vectors is described by the equation

$$\boxed{i\frac{\partial}{\partial t}\Psi(x,p,t) = \hat{L}\Psi(x,p,t),} \quad (11.91)$$

where $\hat{L}$ is the so-called **Liouville operator**:[65]

$$\hat{L} \equiv i\left(-\frac{\partial H}{\partial p}\frac{\partial}{\partial x} + \frac{\partial H}{\partial x}\frac{\partial}{\partial p}\right). \quad (11.92)$$

We call Eq. 11.91 the **Koopman-von Neumann equation**.

We can understand why the Koopman-von Neumann equation is the right choice by noting that if it holds, the corresponding probability density $\rho(x,p,t) = |\Psi(x,p,t)|^2$ automatically fulfills the Liouville equation (Eq. 11.53). Let's check this explicitly.

We need two puzzle pieces here. The first one is that we rewrite the Koopman-von Neumann equation a little bit and then multiply it by $\Psi^\star$:[66]

---

[63] Reminder: Eq. 11.85 reads

$$\Psi(x,p,t) = \int c(x,p,t)\hat{e}_{x,p}\,dxdp.$$

The integral vanishes because basis vectors are orthogonal. This is analogous to what we discussed in Eq. 11.88.

[64] Reminder: Eq. 11.29 reads

$$P(R,t) = \int_R \rho(x,p,t)dV.$$

[65] An operator is a mathematical object which eats a particular kind of object and spits out another object of the same kind. A simple example is a differential operator like $\hat{\partial}_x \equiv \frac{\partial}{\partial x}$ which eats a function $f(x)$ and spits out a new function

$$\hat{\partial}_x f(x) = \frac{\partial f(x)}{\partial x},$$

, i.e., its derivative. Another example would be a matrix M, which acts on vectors $\vec{v}$ and yields a new vector

$$\vec{v}' = M\vec{v}.$$

Moreover, take note that we introduce an additional factor i because it makes our operator Hermitian. This is a requirement on all physical operators in Hilbert space because it guarantees that the corresponding eigenvalues are real. You'll understand this once you learn more about quantum mechanics.

[66] This is clever because want an equation for $|\Psi|^2 = \Psi^\star\Psi$.

$$i\frac{\partial}{\partial t}\Psi = \hat{L}\Psi$$

⟩ Eq. 11.92

$$i\frac{\partial}{\partial t}\Psi = i\left(-\frac{\partial H}{\partial p}\frac{\partial \Psi}{\partial x} + \frac{\partial H}{\partial x}\frac{\partial \Psi}{\partial p}\right)$$

⟩ cancel i

$$\frac{\partial}{\partial t}\Psi = -\frac{\partial H}{\partial p}\frac{\partial \Psi}{\partial x} + \frac{\partial H}{\partial x}\frac{\partial \Psi}{\partial p}$$

⟩ multiplying by $\Psi^\star$

$$\Psi^\star \frac{\partial}{\partial t}\Psi = -\Psi^\star \frac{\partial H}{\partial p}\frac{\partial \Psi}{\partial x} + \Psi^\star \frac{\partial H}{\partial x}\frac{\partial \Psi}{\partial p}. \tag{11.93}$$

The second puzzle piece is that we take the complex conjugate of the Koopman-von Neumann equation and multiply it by Ψ:

$$\frac{\partial}{\partial t}\Psi = -\frac{\partial H}{\partial p}\frac{\partial \Psi}{\partial x} + \frac{\partial H}{\partial x}\frac{\partial \Psi}{\partial p}$$

⟩ complex conjguate

$$\left(\frac{\partial}{\partial t}\Psi\right)^\star = \left(-\frac{\partial H}{\partial p}\frac{\partial \Psi}{\partial x} + \frac{\partial H}{\partial x}\frac{\partial \Psi}{\partial p}\right)^\star$$

⟩ only Ψ is complex

$$\frac{\partial}{\partial t}\Psi^\star = -\frac{\partial H}{\partial p}\frac{\partial \Psi^\star}{\partial x} + \frac{\partial H}{\partial x}\frac{\partial \Psi^\star}{\partial p}$$

⟩ multiplying by Ψ

$$\Psi \frac{\partial}{\partial t}\Psi^\star = -\Psi \frac{\partial H}{\partial p}\frac{\partial \Psi^\star}{\partial x} + \Psi \frac{\partial H}{\partial x}\frac{\partial \Psi^\star}{\partial p}. \tag{11.94}$$

If we calculate the sum of Eq. 11.93 and Eq. 11.94, we find

$$\Psi^\star \frac{\partial}{\partial t}\Psi + \Psi \frac{\partial}{\partial t}\Psi^\star = -\Psi^\star \frac{\partial H}{\partial p}\frac{\partial \Psi}{\partial x} + \Psi^\star \frac{\partial H}{\partial x}\frac{\partial \Psi}{\partial p} - \Psi \frac{\partial H}{\partial p}\frac{\partial \Psi^\star}{\partial x} + \Psi \frac{\partial H}{\partial x}\frac{\partial \Psi^\star}{\partial p}$$

⟩ product rule

$$\frac{\partial}{\partial t}(\Psi^\star \Psi) = -\frac{\partial H}{\partial p}\frac{\partial(\Psi^\star \Psi)}{\partial x} + \frac{\partial H}{\partial x}\frac{\partial(\Psi^\star \Psi)}{\partial p}$$

⟩ $\rho \equiv |\Psi|^2 = \Psi^\star \Psi$, Eq. 11.89

$$\frac{\partial}{\partial t}(\rho) = -\frac{\partial H}{\partial p}\frac{\partial(\rho)}{\partial x} + \frac{\partial H}{\partial x}\frac{\partial(\rho)}{\partial p}$$

⟩ definition of $\{,\}$, Eq. 5.33

$$\frac{\partial}{\partial t}\rho = \{H, \rho\}$$

⟩ antisymmetry: $\{A, B\} = -\{B, A\}$, Eq. 10.21

$$\frac{\partial}{\partial t}\rho = -\{\rho, H\}. \tag{11.95}$$

This is exactly the Liouville equation (Eq. 11.53)! Therefore, we can conclude that the Koopman-von Neumann equation is indeed the correct choice because the corresponding probability density will fulfill the correct equation of motion.

Now unfortunately, I have to tell you that our journey into the wonderful world of Koopman-von Neumann mechanics will already end here before it really started.

I don't think it makes much sense to study Koopman-von Neumann mechanics (= classical mechanics in Hilbert space) before you study Schrödinger mechanics (= quantum mechanics in Hilbert space). Hilbert space, in some sense, is really the natural arena of quantum theories and while we can study a non-quantum theory like classical mechanics in Hilbert space, the resulting formalism is quite awkward.[67] It's much easier to learn how physics in Hilbert space really works in the context of quantum mechanics.[68]

Nevertheless, I think it's really important that at this stage you take note that, in principle, we can formulate classical mechanics in Hilbert space. The resulting formalism is not very pretty but the main point to take away is that it is possible at all.[69]

To quote Nobel laureate Frank Wilczek in his "Notes on Koopman von Neumann Mechanics, and a Step Beyond":

Classical mechanics, as usually formulated, seems to inhabit a different conceptual universe from quantum mechanics. Koopman and von Neumann developed a mathematical formulation of classical mechanics using concepts usually associated with quantum theory. [...] It deserves to be better known among physicists, because it gives a new perspective on the conceptual foundations of quantum theory, and it may suggest new kinds of approximations and even new kinds of theories.

So in summary, it makes a lot more sense to study Koopman-von Neumann mechanics in more detail once you've learned a thing or two about quantum mechanics. Great resources to learn more about Koopman-von Neumann mechanics are:

[67] Similarly, we can study quantum mechanics in physical space but we end up with something quite awkward. The formulation of quantum mechanics in physical space is known as **pilot wave mechanics**. Moreover, take note that we can also formulate quantum mechanics in configuration space, which is known as the **path integral formulation**, and also in phase space, which is known as the **phase space formulation of quantum mechanics**.

[68] In a sense, we automatically end up with quantum mechanics if we apply to Hilbert space the lessons we've learned in Section 10.3 regarding the double role certain objects play as conserved quantities plus generators of symmetries. This is discussed, for example, in my book

Jakob Schwichtenberg. *No-Nonsense Quantum Mechanics*. No-Nonsense Books, Karlsruhe, Germany, 2018c. ISBN 978-1719838719

[69] As a final side remark, take note that the Koopman-von Neumann equation (Eq. 11.91) is strikingly similar to the Schrödinger equation

$$i\hbar \frac{\partial}{\partial t} \Psi(x,t) = \hat{H}\Psi(x,t), \quad (11.96)$$

which describes the time-evolution of state vectors Ψ in quantum mechanics. The main difference is how the Hamiltonian acts on the state vector. In quantum mechanics it acts as an ordinary "product" ($\hat{H}\Psi(x,t)$) while in classical mechanics it acts via the Poisson bracket. (This is what the Liouville operator describes, c.f. Eq. 11.92.)

Additionally, in quantum mechanics our state vectors only depend on either the location or momentum but not both. The reason for these differences is that in quantum mechanics we can't determine the momentum and location of a given object with arbitrary precision at the same time. Instead, there is always a fundamental uncertainty we seemingly can't get rid of.

▷ "Topics in Koopman-von Neumann Theory" by Danilo Mauro[70]

▷ "Notes on Koopman von Neumann Mechanics, and a Step Beyond" by Frank Wilczek[71]

[70] https://arxiv.org/abs/quant-ph/0301172

[71] http://frankwilczek.com/2015/koopmanVonNeumann02.pdf

12

The Origins of Classical Mechanics

Let me paraphrase a famous quote by the Danish philosopher Søren Kierkegaard:[1] *"Physics can only be understood backwards; but it must be discovered forwards."*

[1] The original quote is *"Life can only be understood backwards; but it must be lived forwards."*

And in fact, I'm not really sure how anybody could *understand* classical mechanics before quantum mechanics, special relativity and general relativity were discovered. There are just so many mysteries which we cannot explain within the framework of classical mechanics. For example:

▷ Why is the Lagrangian exactly the difference between kinetic and potential energy ($L = T - V$)?

▷ How does an object know which path is the path of least action? Why does the principle of least action work so remarkably well?[2]

[2] Recall that the least action principle states that a given system follows the path in configuration space which minimizes the time integral of the difference between the kinetic and potential energies, i.e., the action.

Luckily, nowadays we can beautifully answer these questions. This is what this chapter is about.

[3] Take note that here we always start with a more fundamental theory (quantum mechanics, etc.) and then derive how classical mechanics emerges in a specific limit. In contrast, in many textbooks you'll find attempts to derive, for example, quantum mechanics by starting with classical mechanics. But this is not a very fruitful endeavor because usually it's simply impossible to derive a more fundamental theory by starting from a less fundamental one ("zooming in"). In fact, there is still research going on about how we can properly "quantize" a given classical theory. But the reverse procedure ("zooming out") is usually possible without problems. This shouldn't be too surprising because if we zoom out we lose information, e.g., about the quantum behavior of individual atoms. But if we try to zoom in purely by using theoretical (not experimental) tools there is no unique way how we can recover this information. If there were a method that allowed us to systematically derive more fundamental theories, we would use it immediately to derive the correct theory of quantum gravity or the theory of everything. The fact that no one has successfully done this so far shows that we can't zoom in purely by using theoretical tools.

[4] $L = T - V = T$ for $V = 0$.

But these answers require some knowledge of quantum mechanics, special relativity and general relativity.[3] And, of course, we can't discuss these topics properly here. So instead my goal is to give you a rough overview so that you know what to look out for as you continue your journey in physics.

We will start by talking about the relationship between classical mechanics and these more fundamental theories. Then we will learn how quantum mechanics allows us to understand why the least action principle works at all. Afterwards, using special relativity we will discuss why the Lagrangian for a *free* object is equal to the kinetic energy.[4] And finally, we understand how general relativity helps us to understand why the potential energy V enters the Lagrangian with a relative minus sign.

As usual here's a diagram which summarizes our plan for this chapter:

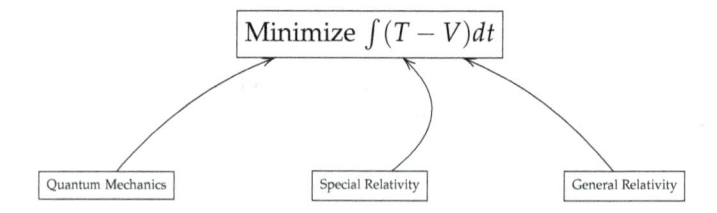

Minimize $\int (T - V) dt$

Quantum Mechanics   Special Relativity   General Relativity

12.1 The Cube of Physics

The place of classical mechanics in the hierarchy of physical theories can be understood beautifully by considering the so-called cube of physics.

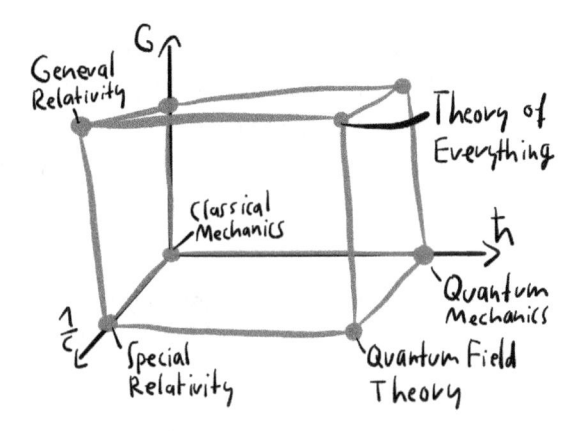

This cube is a map which helps us to navigate the landscape of fundamental physics. The main idea is to make sense of this landscape by talking about the following fundamental constants:[5]

▷ The speed of light $c = 2.9979 \times 10^8$ m/s, which encodes an upper speed limit for all physical processes.
▷ The gravitational constant $G = 6.6741 \times 10^{-11} \frac{m^3}{kg \cdot s^2}$, which encodes the strength of gravitational interactions.
▷ The Planck constant $\hbar = 1.05457 \times 10^{-34} \frac{m^2 \cdot kg}{s}$, which encodes the magnitude of quantum effects.[6]

While these are, well, constants, we now imagine what happens when we vary them. This is motivated by the observation that, for example, when every object in a given system moves extremely slowly compared to the speed of light $v \ll c$, we can act as if $c \to \infty$ to simplify our equations.[7] Similarly, by considering the limit $G \to 0$, we end up with theories in which there is no

---

[5] Take note that the numerical values of these constants are not really important because they depend on which units we use. If we use inches instead of meters, we get different numerical values. It is even possible to choose so-called natural units in which the numerical value of these constants is exactly 1.

[6] Take note that $\hbar$ is the **reduced Planck constant** which is related to the original Planck constant by

$$\hbar = \frac{h}{2\pi}.$$

Since Planck's constant usually appears together with a factor 2π, it is conventional to use the reduced Planck constant everywhere.

[7] The speed of light is an upper speed limit. No object can move with a velocity faster than c. So taking $c \to \infty$ corresponds to a situation in which there is no such speed limit at all. While there is *always* this speed limit in physics, we can act as if there were none if we only consider slowly moving objects. Just imagine there was a highway with an upper speed limit of $v_{max} = 100000000 \frac{km}{h}$ while no car can drive faster than $v \approx 300 \frac{km}{h}$. So while technically there is a speed limit it doesn't matter at all and we can act as if there was none.

gravity at all.[8] And by considering the limit $\hbar \to 0$, we end up with a theory in which quantum effects play no role.[9]

So the most accurate theory of physics takes the upper speed limit ($c \neq \infty$), gravitational interactions ($G \neq 0$) and quantum effects ($\hbar \neq 0$) into account. This would be a **theory of everything** and, so far, no one has succeeded in writing it down.

While this is certainly a big problem, it doesn't stop us from making astonishingly accurate predictions. Depending on the system at hand, we can act as if certain effects don't exist at all. And this is how we end up with the various theories which live at the corner points of the cube of physics:[10]

▷ Whenever it is reasonable to ignore gravitational interactions $G \to 0$ (e.g., for elementary particles), we can use **quantum field theory**.

▷ For systems in which it is reasonable to ignore quantum effects $\hbar \to 0$ (e.g., planets), we can use **general relativity**.

▷ If we can ignore quantum effects *and* gravitational interactions ($\hbar \to 0$ and $G \to 0$), we can use **special relativity**.

▷ Moreover, when it is reasonable to ignore gravitational interactions and that there is an upper speed limit ($G \to 0$ and $\frac{1}{c} \to 0$), we can use **quantum mechanics**.

▷ For systems in which we can ignore quantum effects *and* that there is an upper speed limit ($\hbar \to 0$ and $\frac{1}{c} \to 0$), we can use classical mechanics with Newton's laws to describe gravity. (The resulting model is often called Newtonian gravity.)

▷ And finally, if we can ignore quantum effects, the upper speed limit, *and* gravitational interactions ($\hbar \to 0$, $\frac{1}{c} \to 0$ and $G \to 0$), we can use non-relativistic classical mechanics without gravity.

If you still find the cube of physics confusing, here's an alternative perspective:[11]

---

[8] Again, while there are always gravitational interactions even between elementary particles, we can often ignore them because gravity is much weaker than all other kinds of interactions. So by taking the limit $G \to 0$ we consider systems which only contain objects which are so light that gravity plays no important role.

[9] While quantum effects are extremely important for elementary particles, they have no notable effects on everyday objects. Such everyday objects consist of so many elementary particles that the quantum effects average out and we end up with no quantum effects at all, i.e., $\hbar \to 0$.

[10] The remaining corner point is at $\frac{1}{c} \to 0$ and corresponds to Non-Relativistic Quantum Gravity which, so far, is a speculative topic that we will not discuss here.

[11] Take note that we can, of course, take gravitational effects in classical mechanics into account as long as these are not too wild (black holes, etc.) So the distinction between classical mechanics and Newtonian mechanics is really just semantics. Here we define Newtonian mechanics as classical mechanics including gravity.

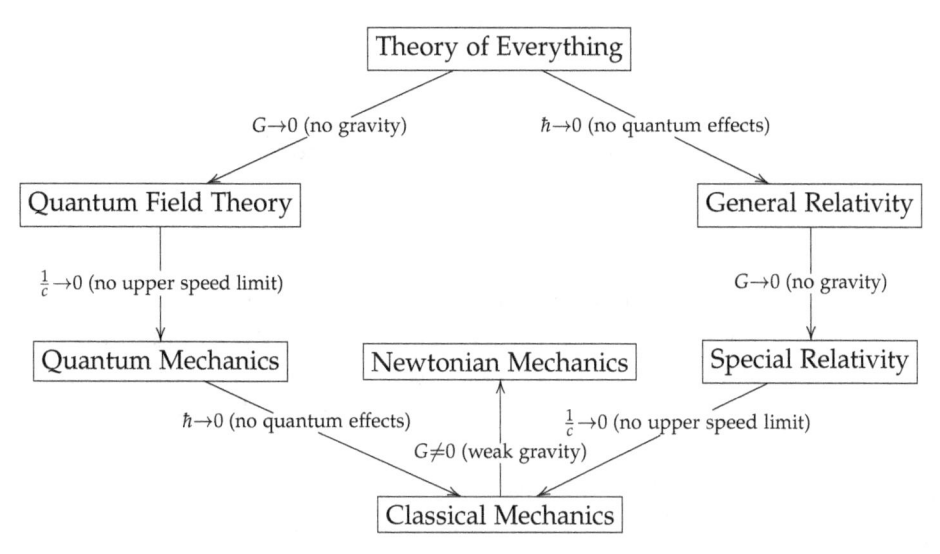

```
                    ┌─────────────────────┐
                    │ Theory of Everything│
                    └─────────────────────┘
              G→0 (no gravity)    ℏ→0 (no quantum effects)
             ↙                              ↘
┌─────────────────────┐              ┌─────────────────────┐
│ Quantum Field Theory│              │  General Relativity │
└─────────────────────┘              └─────────────────────┘
  1/c→0 (no upper speed limit)         G→0 (no gravity)
         ↓                                      ↓
┌──────────────────┐  ┌────────────────────┐  ┌──────────────────┐
│ Quantum Mechanics│  │ Newtonian Mechanics│  │ Special Relativity│
└──────────────────┘  └────────────────────┘  └──────────────────┘
     ℏ→0 (no quantum effects)  ↑   1/c→0 (no upper speed limit)
                      G≠0 (weak gravity)
                    ┌─────────────────────┐
                    │  Classical Mechanics│
                    └─────────────────────┘
```

Take note that although it might seem as if classical mechanics only plays a niche role, its range of applicability is actually extremely large.

▷ Only at the scale of atoms ($\sim 10^{-8}$ m) and below do quantum effects become important and classical mechanics fails to give accurate predictions. In particular, this means that for any macroscopic object (~ 1 m), classical mechanics makes perfectly accurate predictions.

▷ Only for objects which move at a substantial fraction of the speed of light ($c \approx 10^9$ km/h) does special relativity become important. To put this into perspective: the fastest man-made objects ever produced were the solar probes Helios-A and Helios-B. But even those only reached a speed of $v \approx 10^5$ km/h which is not really that fast compared to the speed of light: $v/c \approx 0.0001$. So the predictions of classical mechanics without corrections from special relativity are perfectly fine, as long as we don't want *extremely* fine measurements.

▷ And finally, only for cosmological objects like a complete solar system does general relativity lead to significant effects. Near the surface of the Earth, corrections from general relativity are of order $\sim 10^{-8}$ and can therefore be neglected for most measurements.

Now after this big picture overview, let's talk about the origin of classical mechanics in more detail.

12.2 The Origin of the Least Action Principle

For a long time, the principle of least action was something magical. How exactly does nature "know" which path is the path of least action? Unsurprisingly, many shared the skepticism of people like Cureau de la Chambre who argued:[12]

[12] Cureau de la Chambre was a friend of Fermat and talked about Fermat's principle which states that light always takes the shortest path. (We discussed Fermat's principle at the beginning of Chapter 4.) But exactly the same philosophical objection can be and was raised against the least action principle.

"The principle which you take as the basis for your proof, namely that Nature always acts by using the simplest and shortest paths, is merely a moral, and not a physical one. It is not, and cannot be, the cause of any effect in Nature."

But this changed quickly when in 1948 Richard Feynman developed the configuration space formulation of quantum mechanics. The key idea is that *every* path between a fixed initial and final point in configuration space is possible in principle.

[13] As an aside, take note that we can see here that Planck's constant $\hbar$ has the same units as our action $S[q(t)]$. This is necessarily the case because the argument of the exponential function has to be dimensionless. This follows if we recall that we can define the exponential function in terms of a series:

$$e^x = \sum_{n=0}^{\infty} \frac{x^n}{n!}.$$

(This is discussed in Appendix F.) So if the argument of the exponential function X wasn't dimensionless, we would have a sum over objects with different dimensions which doesn't make sense.

However, the **probability amplitude** of each path $q(t)$ is a different one:

$$\Psi = e^{iS[q(t)]/\hbar}, \qquad (12.1)$$

where $S[q(t)]$ is the action associated with the path, and $\hbar$ is a fundamental constant known as **Planck's constant** which encodes the strength of quantum effects.[13]

To calculate the total probability that our system evolves from a specific initial point A to a specific final point B we sum over the probability amplitudes for all possible paths:

$$\psi_{A \to B} = \sum_{i=1}^{N} e^{iS[q_i(t)]/\hbar} \qquad (12.2)$$

and then take the absolute square:[14]

$$P(A \to B) = |\psi_{A \to B}|^2. \qquad (12.3)$$

But usually there are infinitely many paths possible, so we need to replace our sum with an integral:

$$\boxed{\psi_{A \to B} = \int Dq(t) e^{iS[q(t)]/\hbar}.} \qquad (12.4)$$

The expression on the right-hand side is known as the **path integral** and $\int Dq(t)$ indicates that we consider a "sum" over all possible paths. The total probability is given by the absolute square of this total probability amplitude.

[14] In this sense, the probability amplitudes are analogous to the coefficients in front of the basis vectors we discussed in Section 11.3. To get the corresponding probability, we need to calculate the absolute value squared of these coefficients too.

Let's try to understand all this a little better.[15]

First of all, recall that the action is just a number for any given path. Some paths require a lot of action (i.e., $S[q(t)]$ is large for these paths between A and B) while others require only a little action. The action here appears as the argument of the complex exponential function: $e^{iS[q(t)]}$. In general, since the action for a specific path ($S[q(t)]$) is an ordinary number, $e^{iS[q(t)]}$ is a complex number with absolute value 1. In the complex plane, these numbers lie on the unit circle.[16]

[15] The following visual interpretation of the path integral was mainly popularized by Richard Feynman, for example, in his excellent book

Richard Feynman. *QED : the Strange Theory of Light and Matter*. Princeton University Press, Princeton, NJ, 2014. ISBN 978-0691164090

[16] We can understand this using Euler's formula

$$z = e^{i\phi}$$
$$= \cos(\phi) + i\sin(\phi)$$
$$= \text{Re}(z) + i\text{Im}(z).$$

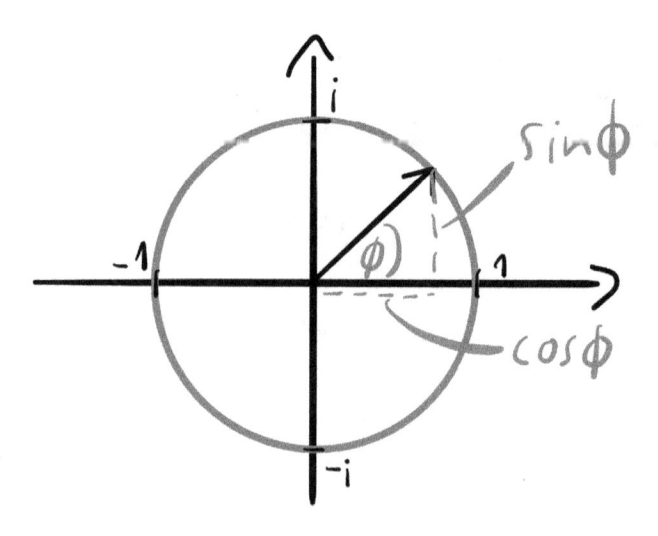

The contribution of each path to the total path integral is therefore simply a unit complex number. The total path integral is a sum over infinitely many unit complex numbers. Therefore, it is useful to imagine that there is a little stopwatch attached to the particle as it travels a given path. At the beginning of each path, the dial points directly to the right.[17] which in our complex plane corresponds to $z = 1 = e^{i0}$. The clocks move as our configuration space point traces out a specific path. At the end of each particular path, the dial points to one specific number on the clock. For example, for one path the situation may look like this:

[17] On a real clock it would point to the 3.

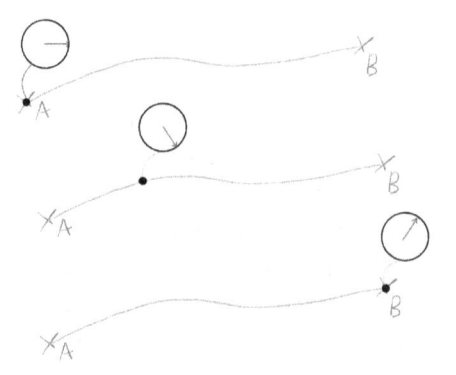

While for another path we have:

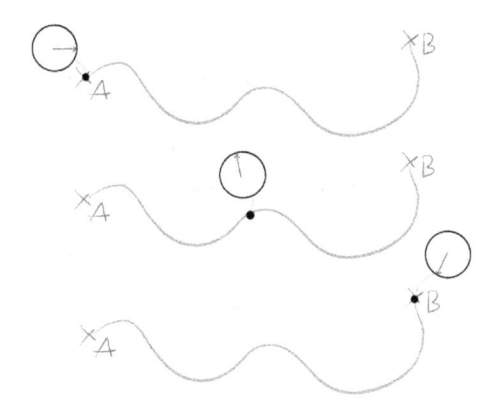

To calculate the path integral, we have to add the little arrows for each path like we would add vectors. The total value of the path integral is then the resulting arrow.

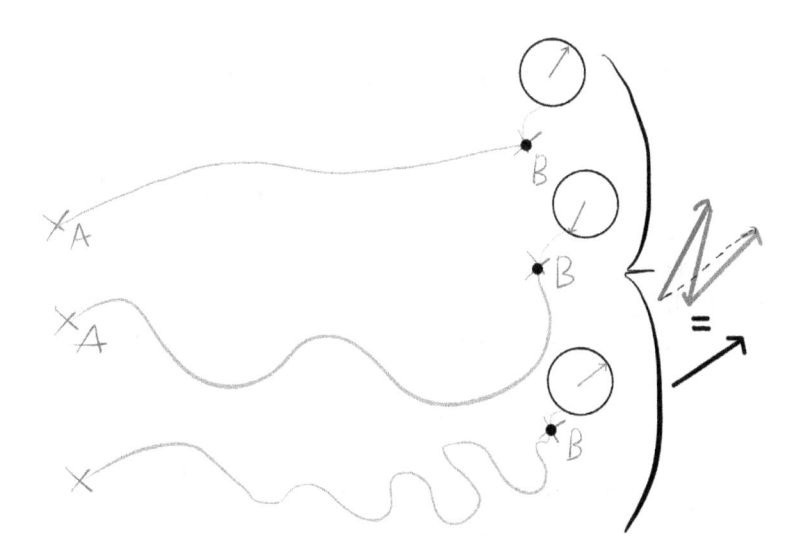

Since the resulting arrows do not necessarily all point in the same direction, the resulting arrow can be quite small. Here, we have three paths, but to get the final result we have to include all possible paths, not just three. The final result depends on the starting locations A and B. For some final point B' most of the arrows cancel each other. The resulting arrow is tiny. In physical terms, this means that the probability to find the particle here is tiny. In addition, we can imagine that for another final point B'', lots of arrows point in the same direction and the resulting arrow is large. This means that it is quite probable to find the particle at B'' at the end of our time interval.

The main point to take away, however, is that each path between a fixed initial and final point is possible. Each path contributes exactly one little arrow (one probability amplitude) to the total path integral.

But why then is the path of least action so important in classical mechanics?

12.2.1 The Origin of the Classical Path

What we have learned above is that the probability of a given final position depends crucially on the relative positions of the final arrows. If the arrows point mostly in the same direction, we get a long final arrow. In such a situation we say that we have **constructive interference**. If the arrows point wildly in different directions, they mostly average out and we end up with a short total arrow. This is known as **destructive interference**.

This observation allows us to understand why the path of least action is so important in classical mechanics.

In our quantum context, the classical path is just one path out of many. But we can understand why the classical path is so important in classical mechanics by exploring the contributions of neighboring paths. For concreteness, let's consider two neighboring paths $q(t)$ and $q'(t)$ where the second path is a variation of the first one $q'(t) = q(t) + \eta(t)$, and where $\eta(t)$ denotes a small variation.

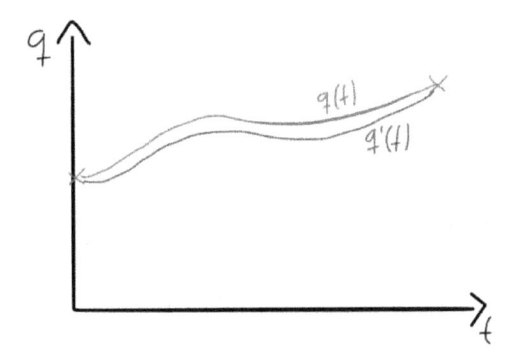

[Figure: coordinate axes with q vertical and t horizontal, showing two paths labeled $q(t)$ and $q'(t)$ between two endpoints]

The first path contributes $e^{iS[q(t)]/\hbar}$ while the second path contributes $e^{iS[q'(t)]/\hbar}$. We can expand the action of the second path around the first one

$$S[q'] = S[q+\eta] = S[q] + \int dt\, \eta(t) \frac{\delta S[q]}{\delta q(t)} + O(\eta^2).$$

Now, if $q(t)$ is the path with minimal action $q_{cl}(t)$, the first order variation vanishes[18]

$$S[q'] = S[q_{cl}+\eta] = S[q_{cl}] + \underbrace{\int dt\, \eta(t) \frac{\delta S[q]}{\delta q(t)}}_{=0 \text{ for } q(t)=q_{cl}(t)} + O(\eta^2)$$

$$= S[q_{cl}] + O(\eta^2).$$

[18] Reminder: the minimum of the action functional is characterized by a vanishing first order variation. We discussed this in Section 4.2.

The physical implication for our path integral is that paths in the neighborhood of the path with minimal action $q_{cl}(t)$ yield arrows that point in approximately the same direction since $S[q'] \approx S[q_{cl}]$. In other words, paths around the classical path interfere constructively.

This is why the classical path is important. In contrast, for an arbitrary path far away from the classical path, the resulting arrows of neighboring paths vary wildly, and we get destructive interference.

This effect becomes even more dominant if we consider systems in which the action of each path is much larger than Planck's constant $S[q(t)] \gg \hbar$.[19] Since the probability amplitude associated with each path is

$$\Psi = e^{iS[q(t)]/\hbar}, \qquad (12.5)$$

we can see that for $S[q(t)] \gg \hbar$ even tiny differences in the action of neighboring paths lead to vastly different probability amplitudes.[20] And paths with vastly different probability amplitudes interfere destructively.

Therefore, for systems for which $S[q(t)] \gg \hbar$ the only paths which interfere constructively are those surrounding the classical paths of least action. The limit $S[q(t)] \gg \hbar$ is known as the **classical limit** because for macroscopic objects, the energies involved are much higher than for elementary particles and therefore lead to much larger values for the action functional $S[q(t)]$.

There is an interesting alternative perspective which allows us to understand why classical mechanics works the way it does. While the main idea is quite similar to what we discussed above, there are a few additional things that we can understand

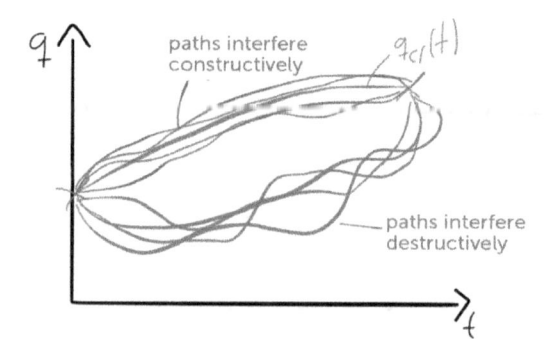

[19] This is the case for macroscopic objects which we describe in classical mechanics.

[20] In some sense, by dividing the term in the exponent by a, in comparison, tiny number $\hbar$ differences become especially significant. (Dividing by a tiny number is equal to multiplying by a huge number.)

this way. In particular, we can understand why the Hamilton-Jacobi equation (Eq. 11.6) plays an important role in classical mechanics.

12.2.2 The Origin of the Hamilton-Jacobi Equation

The fundamental equation of motion in quantum mechanics is the Schrödinger equation[21]

$$i\hbar \frac{\partial}{\partial t} \Psi(q, t) = \hat{H} \Psi(q, t), \qquad (12.6)$$

where Ψ is a complex function which we interpret as a probability amplitude and

$$\hat{H} \equiv -\frac{\hbar^2}{2m} \frac{\partial^2}{\partial q^2} + V(q) \qquad (12.7)$$

is the Hamiltonian operator.[22] Like in classical mechanics, the Hamiltonian represents the total energy. We get the operator form by using the classical Hamiltonian

$$H = \frac{p^2}{2m} + V(q) = T + V$$

and then replacing the momentum p with the momentum operator $\hat{p} = i\hbar \frac{\partial}{\partial q}$. This identification can be motivated by considering Noether's theorem in Hilbert space. The generator of translations in Hilbert space is $i\frac{\partial}{\partial q}$, and in Chapter 10, we learned that the generator of translations is the momentum.[23] Moreover, as mentioned above, $\hbar$ is a fundamental constant which encodes the magnitude of quantum effects.

In general, we can write a complex function in the polar form[24]

$$\Psi(x, t) = R(x, t) e^{iW(x,t)/\hbar}, \qquad (12.8)$$

where R and W are real functions. In physical terms, R^2 is the probability density and W is known as the phase.[25]

If we now put the polar form of the probability amplitude (Eq. 12.8) into the Schrödinger equation (Eq. 12.6), we discover something remarkable:

---

[21] The Schrödinger equation was already mentioned in Section 11.3 where we noted its similarity to the Koopman-von Neumann equation (Eq. 11.91):

$$i\frac{\partial}{\partial t} \Psi = \hat{L} \Psi.$$

[22] Take note that in this context Ψ is usually called the **wave function**.

[23] This is discussed in more detail in my book *No-Nonsense Quantum Mechanics*.

[24] It is conventional in the context of quantum mechanics to factor out $1/\hbar$. It would be equally possible to absorb it into the definition of the function in the exponent

$$\tilde{W} \equiv W/\hbar.$$

The polar form follows if we use Euler's formula:

$$e^{i\phi} = \cos(\phi) + i\sin(\phi).$$

[25] We have

$$\begin{aligned}|\Psi|^2 &= \Psi^* \Psi \\ &= \left(Re^{iW/\hbar}\right)^* \left(Re^{iW/\hbar}\right) \\ &= Re^{-iW/\hbar} Re^{iW/\hbar} \\ &= R^2.\end{aligned}$$

To get to the last line, we used that $e^{-iW/\hbar} e^{iW/\hbar} = e^{-iW/\hbar + iW/\hbar} = e^0 = 1$.

THE ORIGINS OF CLASSICAL MECHANICS

$$i\hbar \frac{\partial}{\partial t}\Psi = \hat{H}\Psi$$

⟩ Eq. 12.7

$$i\hbar \frac{\partial}{\partial t}\Psi = -\frac{\hbar^2}{2m}\frac{\partial^2 \Psi}{\partial q^2} + V(q)\Psi$$

⟩ Eq. 12.8

$$i\hbar \frac{\partial}{\partial t}\left(Re^{iW/\hbar}\right) = -\frac{\hbar^2}{2m}\frac{\partial^2 \left(Re^{iW/\hbar}\right)}{\partial q^2} + V(q)\left(Re^{iW/\hbar}\right)$$

⟩ product rule

$$i\hbar \left(\frac{\partial}{\partial t}R\right)e^{iW/\hbar} + i\hbar R\left(\frac{\partial}{\partial t}e^{iW/\hbar}\right) = -\frac{\hbar^2}{2m}\left(e^{iW/\hbar}\frac{\partial^2 R}{\partial q^2} + 2\frac{\partial R}{\partial q}\frac{\partial e^{iW/\hbar}}{\partial q} + R\frac{\partial^2 e^{iW/\hbar}}{\partial q^2}\right) + V(q)Re^{iW/\hbar}$$

⟩ chain rule

$$i\hbar e^{iW/\hbar}\frac{\partial}{\partial t}R + i\hbar e^{iW/\hbar}R\frac{\partial(iW/\hbar)}{\partial t} = -\frac{\hbar^2}{2m}\left(e^{iW/\hbar}\frac{\partial^2 R}{\partial q^2} + 2e^{iW/\hbar}\frac{\partial R}{\partial q}\frac{\partial(iW/\hbar)}{\partial q} + R\frac{\partial\left(\frac{\partial(iW/\hbar)}{\partial q}e^{iW/\hbar}\right)}{\partial q}\right) + V(q)Re^{iW/\hbar}$$

⟩ product rule

$$i\hbar e^{iW/\hbar}\frac{\partial}{\partial t}R + i\hbar e^{iW/\hbar}R\frac{\partial(iW/\hbar)}{\partial t} = -\frac{\hbar^2}{2m}\left(e^{iW/\hbar}\frac{\partial^2 R}{\partial q^2} + 2e^{iW/\hbar}\frac{\partial R}{\partial q}\frac{\partial(iW/\hbar)}{\partial q} + Re^{iW/\hbar}\left(\frac{\partial(iW/\hbar)}{\partial q}\right)^2 + Re^{iW/\hbar}\frac{\partial^2(iW/\hbar)}{\partial q^2}\right) + V(q)Re^{iW/\hbar}$$

⟩ rearranging

$$i\hbar e^{iW/\hbar}\frac{\partial R}{\partial t} - e^{iW/\hbar}R\frac{\partial W}{\partial t} = -\frac{\hbar^2}{2m}e^{iW/\hbar}\frac{\partial^2 R}{\partial q^2} - i\frac{\hbar}{m}e^{iW/\hbar}\frac{\partial R}{\partial q}\frac{\partial W}{\partial q} + \frac{1}{2m}Re^{iW/\hbar}\left(\frac{\partial W}{\partial q}\right)^2 - i\frac{\hbar}{2m}Re^{iW/\hbar}\frac{\partial^2 W}{\partial q^2} + V(q)Re^{iW/\hbar}$$

⟩ $e^{iW/\hbar}$

$$i\hbar \frac{\partial R}{\partial t} - R\frac{\partial W}{\partial t} = -\frac{\hbar^2}{2m}\frac{\partial^2 R}{\partial q^2} - i\frac{\hbar}{m}\frac{\partial R}{\partial q}\frac{\partial W}{\partial q} + \frac{1}{2m}R\left(\frac{\partial W}{\partial q}\right)^2 - i\frac{\hbar}{2m}R\frac{\partial^2 W}{\partial q^2} + V(q)R$$

(12.9)

Since W and R are both real functions, we actually have two equations here.[26] One equation is given by the real part (i.e., all terms without an i):

$$-R\frac{\partial W}{\partial t} = -\frac{\hbar^2}{2m}\frac{\partial^2 R}{\partial q^2} + \frac{1}{2m}R\left(\frac{\partial W}{\partial q}\right)^2 + V(q)R$$

⟩ $\times \frac{1}{R}$

$$-\frac{\partial W}{\partial t} = -\frac{\hbar^2}{2mR}\frac{\partial^2 R}{\partial q^2} + \frac{1}{2m}\left(\frac{\partial W}{\partial q}\right)^2 + V(q).$$

(12.10)

The second equation is given by the imaginary part (i.e., all

[26] Since R and W are real, there is no way the real and imaginary parts of the equation can influence each other. Hence, the imaginary part and the real part of the equation must be fulfilled independently.

terms with an i)

$$i\hbar \frac{\partial R}{\partial t} = -i\frac{\hbar}{m}\frac{\partial R}{\partial q}\frac{\partial W}{\partial q} - i\frac{\hbar}{2m}R\frac{\partial^2 W}{\partial q^2}$$

⟩ cancel i

$$\hbar \frac{\partial R}{\partial t} = -\frac{\hbar}{m}\frac{\partial R}{\partial q}\frac{\partial W}{\partial q} - \frac{\hbar}{2m}R\frac{\partial^2 W}{\partial q^2}$$

⟩ $\times R$

$$\hbar R\frac{\partial R}{\partial t} = -\frac{\hbar}{m}R\frac{\partial R}{\partial q}\frac{\partial W}{\partial q} - \frac{\hbar}{2m}R^2\frac{\partial^2 W}{\partial q^2}$$

⟩ product rule

$$\frac{\hbar}{2}\frac{\partial R^2}{\partial t} = -\frac{\hbar}{2m}\frac{\partial}{\partial q}\left(R^2 \frac{\partial W}{\partial q}\right)$$

⟩ $\frac{\hbar}{\hbar}$

$$\frac{\partial R^2}{\partial t} = -\frac{1}{m}\frac{\partial}{\partial q}\left(R^2 \frac{\partial W}{\partial q}\right). \tag{12.11}$$

This is really interesting because $\hbar$ is a constant which encodes the magnitude of quantum effects. Therefore, if we assume that we can ignore quantum effects for a moment ($\hbar \to 0$), our first equation (Eq. 12.10) becomes[27]

$$-\frac{\partial W}{\partial t} = \frac{1}{2m}\left(\frac{\partial W}{\partial q}\right)^2 + V(q). \tag{12.12}$$

But this is *exactly* the Hamilton-Jacobi equation (Eq. 11.8).[28] Therefore, we can conclude that in the **classical limit** $\hbar \to 0$, the phase W of the probability amplitude Ψ obeys the Hamilton-Jacobi equation. If we want to include quantum effects, we only need to modify the Hamilton-Jacobi equation a little bit:

$$-\frac{\partial W}{\partial t} \stackrel{(12.10)}{=} \frac{1}{2m}\left(\frac{\partial W}{\partial q}\right)^2 + V(q) + V_q(q) \tag{12.13}$$

where

$$V_q(q) = -\frac{\hbar^2}{2mR}\frac{\partial^2 R}{\partial q^2} \tag{12.14}$$

is the **quantum potential**.[29]

What we've therefore derived is once more how classical mechanics emerges from quantum mechanics in the limit where we

---

[27] The second equation (Eq. 12.11) is the continuity equation for the probability density $\rho = R^2$. Unfortunately, a proper discussion of how this interpretation comes about is far beyond the scope of this book.

[28] For your convenience: Eq. 11.8 reads
$$0 = \frac{1}{2m}\left(\frac{\partial W}{\partial q}\right)^2 + V(q) + \frac{\partial W}{\partial t}.$$

[29] Like for any other potential, we can calculate the force resulting from this quantum potential. The "quantum force" we find this way allows us to include quantum effects in the physical space formulation of quantum mechanics. Formulated differently, the idea to take the force which results from this quantum potential seriously is the starting point for the pilot wave formulation of quantum mechanics. (The pilot wave formulation of quantum mechanics is quantum mechanics in physical space.)

can ignore quantum effects. But it also allows us to see Hamilton's principal function W in a completely new light.

In Section 11.1.1, we discovered that Hamilton's principal function W tells us the amount of action required by the *correct* path $\big(q(t), p(t)\big)$ between arbitrary initial and final times. Here we learned that W is equal to the phase of the wave function $\Psi = Re^{iW/\hbar}$. Therefore, as in the previous section, we reach the conclusion that the action is really the phase of the probability amplitude Ψ of quantum mechanics.[30]

[30] In the previous section we discussed this in the context of the path integral formulation of quantum mechanics. Each possible path contributes the probability amplitude $\Psi = e^{iS[q(t)]/\hbar}$ (Eq. 12.1).

12.3 The Origin of the Classical Lagrangian

In the previous section, we discussed why the least action principle works at all. In general, the action is defined as the time integral over the Lagrangian L:

$$S = \int L dt. \quad (12.15)$$

Moreover, in classical mechanics we end up with the correct equations of motion if we use the Lagrangian

$$L = T - V, \quad (12.16)$$

where T denotes the kinetic and V the potential energy.

But why is the Lagrangian given by such a strange formula?

To understand this, we need to discuss one crucial observation:[31]

[31] Speaking colloquially, an inertial observer is a non-accelerating observer. Inertial observers are moving with constant velocity relative to each other.

> Physics is the same for all inertial observers.

This is the **principle of relativity**. In words it means that the equations of motion must be the same for all physicists no matter where they are located at (e.g., Tokyo or New York), when a given experiment is performed (e.g., 10 years ago or tomorrow) or from which perspective they look at it.

Moreover, the laws of physics remain unchanged even for physicists who move with a constant velocity relative to other physicists. For example, imagine that you're sitting in a boat and are unable to look outside. In such a situation it's impossible for you to find out if the boat is moving with a constant velocity or is stationary.[32] In contrast, you notice immediately if the boat accelerates because a non-zero acceleration means that a force acts on the boat and thus you can measure this force.

[32] This thought experiment is known as Galileo's boat and we discussed it already in Section 10.1.1.

In more technical terms, we can reformulate the same observation by saying that physics is fundamentally rotational, translational and boost symmetric.[33]

[33] A boost is a transformation into a state which moves with a different constant velocity.

This key fact is essential when we try to write down the correct Lagrangian for a given system:

▷ We use the Lagrangian to derive the equations of motion.

▷ We know that the equations of motion are unchanged by symmetry transformations.

▷ Therefore, we can conclude that our Lagrangian must be constructed in such a way that we get equations of motion which respect the symmetries at hand.

This is a powerful observation because it severely restricts which terms we can use in the Lagrangian.

In general, whenever we want to write down the correct Lagrangian for a given system, our first task is to identify the symmetries of the system. Then we try to construct terms which remain unchanged by these symmetry transformations. These terms are candidates to appear in our Lagrangian. And if then, additionally, we restrict ourselves to the simplest possible, non-trivial terms, we end up with the correct Lagrangian.[34]

[34] Of course, there are many subtleties and technical details missing here. You can find a more complete discussion in

Jakob Schwichtenberg. *Physics from Symmetry*. Springer, Cham, Switzerland, 2018b. ISBN 978-3319666303

This is one of the puzzle pieces that we need to understand why the classical Lagrangian reads $L = T - V$.

12.3.1 Special Relativity

The second puzzle piece that we now need is a curious fact of nature which was discovered by the famous Michelson-Morley experiment:

> The velocity of light has the same value c for all inertial observers.

This is a puzzling statement because, usually, the speed of an object depends on how we move relative to it, i.e., the frame of reference we are using. For example, imagine that an observer *standing* at a train station measures that a train moves at 50 $\frac{km}{h}$.

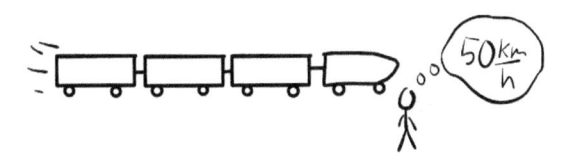

A second observer who runs at 15 $\frac{km}{h}$ parallel to the same train, measures that the train moves at 35 $\frac{km}{h}$.

Curiously, this does not happen for electromagnetic waves. Electromagnetic waves always travel with $c = 1/\sqrt{\epsilon_0 \mu_0} = 2.9979 \times 10^8$ m/s, no matter how you move.[35]

[35] This is only true in free space and not if our wave moves in a medium. The speed of electromagnetic waves in a medium is lower.

The curious fact of nature that the speed of electromagnetic waves has always exactly the same value leads to all kinds of strange consequences. Taking it seriously leads to Einstein's theory of **special relativity**. While there is no way we can discuss special relativity in detail here, we should at least talk about the most famous phenomenon.

12.3.2 Time Dilation

Let's imagine that a person sends a light pulse straight up where it is reflected by a mirror and finally reaches the point from where it was sent.

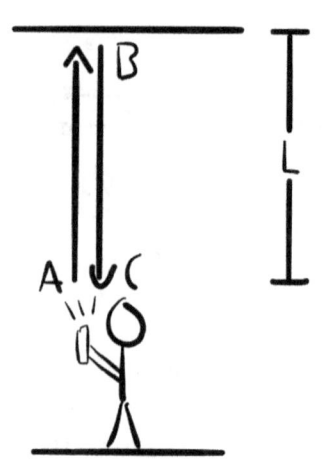

We record three important events:

▷ **A** : the light pulse leaves the starting point
▷ **B** : the light pulse is reflected by the mirror
▷ **C** : the light pulse returns to the starting point.

The time-interval between the two events **A** and **C** is[36]

$$\Delta t = t_C - t_A = \frac{2L}{c}, \quad (12.17)$$

where L denotes the distance between the person and the mirror.

[36] Reminder: for a constant speed v we have $v = \frac{\Delta s}{\Delta t}$, where Δs is the distance and Δt the time interval. Therefore, we have $\Delta t = \frac{\Delta s}{v}$.

So far, nothing interesting has happened. But this changes quickly as soon as we consider how a second person observes exactly the same situation.

We imagine that this second person moves with some constant speed u relative to the first person. For simplicity, we assume that the origins of the two coordinate systems coincide when the light pulse is sent off (t_A). Moreover, we assume that each person stands at the origin of his coordinate system.

A first crucial observation is now that for this second observer, the starting and end points of the light pulse have different coordinates:

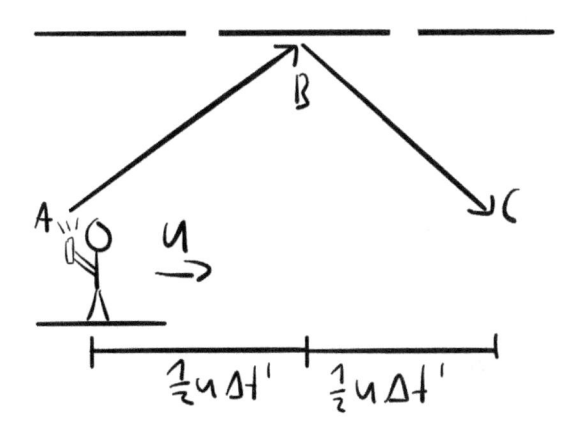

Mathematically, we have

$$x'_A = 0 \neq x'_C = u\Delta t' \quad \rightarrow \quad \Delta x' = u\Delta t', \quad (12.18)$$

where we use primed coordinates for the coordinate system associated with the second person. In words, this means that for this second person the light has apparently also moved in the x-direction. In contrast, for the first person

$$x_A = x_C \quad \rightarrow \quad \Delta x = 0. \quad (12.19)$$

Now, what's the time interval which the second person measures between the event A and the event C?[37]

[37] It will become clear in a moment, why this is an interesting question.

As usual, the time interval $\Delta t' = t'_C - t'_A$ can be calculated as the distance l divided by the speed of the light pulse c.

$$\Delta t' = \frac{l}{c} \quad (12.20)$$

The distance l is for this second observer no longer simply L, but we can calculate it using the Pythagorean theorem[38]

[38] See the triangle in the figure above.

$$l = 2\sqrt{\left(\frac{1}{2}u\Delta t'\right)^2 + L^2}. \quad (12.21)$$

The time interval measured by this second person is therefore

$$\Delta t' \stackrel{(12.20)}{=} \frac{l}{c}$$

$$= \frac{2\sqrt{\left(\frac{1}{2}u\Delta t'\right)^2 + L^2}}{c} \quad \curvearrowright \text{Eq. 12.21}$$

$$= \frac{2\sqrt{\left(\frac{1}{2}u\Delta t'\right)^2 + \left(\frac{c\Delta t}{2}\right)^2}}{c} \quad \curvearrowright L = \frac{c\Delta t}{2}, \text{ Eq. 12.17}$$

$$= \frac{\sqrt{u^2\Delta t'^2 + c^2\Delta t^2}}{c}. \quad \curvearrowright \not{2} \quad (12.22)$$

We can solve this equation for $\Delta t'$ by squaring it:

$$\therefore \Delta t'^2 = \frac{u^2 \Delta t'^2 + c^2 \Delta t^2}{c^2}$$ ⟩ rearranging terms

$$\therefore \Delta t'^2 c^2 - u^2 \Delta t'^2 = c^2 \Delta t^2$$ ⟩ factoring $\Delta t'^2$

$$\therefore \Delta t'^2 (c^2 - u^2) = c^2 \Delta t^2$$ ⟩ dividing by $(c^2 - u^2)$

$$\therefore \Delta t'^2 = \frac{c^2 \Delta t^2}{c^2 - u^2}$$ ⟩ √

$$\therefore \Delta t' = \sqrt{\frac{1}{1 - \frac{u^2}{c^2}}} \Delta t . \qquad (12.23)$$

So for $u \neq 0$ this new time interval $\Delta t'$ is different from the time interval measured by the first observer: $\Delta t' \neq \Delta t$.[39] In words, this means that two observers moving relative to each other do not agree on the time interval between the two events A and C! This phenomenon is usually called **time dilation**. Clocks tick differently for different observers and they count a different number of ticks between two events. However, take note that this effect is usually unnoticeable because the size of the effect depends crucially on $\frac{u^2}{c^2}$. This implies that for all velocities u which are significantly below the speed of light c, we have $\frac{u^2}{c^2} \approx 0$ and therefore, $\Delta t' \approx \Delta t$.

[39] Reminder: $\Delta t = \frac{2L}{c}$, Eq. 12.17.

Analogously, it's possible to derive that different observers do not necessarily agree on the length of objects. This is known as **length contraction** and is another famous consequence of a constant speed of light. A third incredibly important consequence is that the speed of light c is an upper speed limit for everything physical. Unfortunately, discussing all these consequences in detail requires at least another book.

Now we are almost ready to understand the classical Lagrangian. But first, we need to talk about the Lagrangian that we use in special relativity.

12.3.3 The Lagrangian of Special Relativity

For simplicity, let's focus on a single free object. Above we discussed that our action must respect the symmetries of the system at hand. Therefore, our task is now to find an action functional which respects the symmetries of special relativity.[40]

In general, the action is defined as the time integral over the Lagrangian. Therefore, in some sense, we can understand the total action as a time interval multiplied by some constants.[41]

But in the previous section, we've learned that clocks tick differently for different observers. So how can we write down an action or time interval that all observers agree on?

Well, there is exactly one such time interval:[42]

$$\tau = \int \sqrt{1 - \frac{\dot{q}^2}{c^2}}\, dt, \quad (12.24)$$

where c denotes the (constant) speed of light and $\dot{q}$ is the speed of the object that we want to describe as it follows some trajectory $q(t)$. This time interval τ is known as the **proper time**. The proper time is the time measured by a clock which is attached to the object which we want to describe. And all observers agree on the time interval such a clock measures.

Therefore, a reasonable guess for the action functional in special relativity is

$$S[q(t)] = C\tau = C \int \sqrt{1 - \frac{\dot{q}^2}{c^2}}\, dt, \quad (12.25)$$

---

[40] Take note that the constancy of the speed of light is a symmetry requirement too. As discussed in Chapter 10, a symmetry is a transformation which leaves a given object unchanged.

[41] In particular recall that we discussed in Section 4 that light always follows the "least time" path.

[42] For the moment, we only need to know that it's possible to find such an interval. You can find a complete derivation for why all observers agree on this particular time interval in

Jakob Schwichtenberg. *Physics from Symmetry*. Springer, Cham, Switzerland, 2018b. ISBN 978-3319666303

where C is a proportionality constant.

Let's try to guess what this constant is. In general, the action has units "Joule-seconds":[43]

$$[S] = J \cdot s \qquad (12.26)$$

In terms of base units, a Joule can be written as

$$J = \frac{kg \cdot m^2}{s^2}$$

and therefore we have

$$[S] = \frac{kg \cdot m^2}{s}. \qquad (12.27)$$

The integral in Eq. 12.25 yields something with units seconds

$$\left[\int \sqrt{1 - \frac{\dot{q}^2}{c^2}} \, dt \right] = s \qquad (12.28)$$

since the proper time $\tau = \int \sqrt{1 - \frac{\dot{q}^2}{c^2}} \, dt$ is a time interval.

If we combine these observations, we end up with the conclusion that our constant C in Eq. 12.25 must have units:

$$[C] = \frac{kg \cdot m^2}{s^2} \qquad (12.29)$$

because

$$\left[C \int \tau \, dt \right] = [C] \left[\int \tau \, dt \right]$$

$$ = \frac{kg \cdot m^2}{s^2} s \qquad \circlearrowright \text{Eq. 12.28 and Eq. 12.29}$$

$$ = \frac{kg \cdot m^2}{s} = [S] \;\checkmark. \qquad (12.30)$$

Next, we need to recall that we want to describe a single free object. The only property which comes naturally with every free object is its mass m. That's where the kilograms come from in the constant C. The only other ingredient we have is the

[43] We can understand this by recalling that the classical action for a free particle is simply the time integral over the kinetic energy. The kinetic energy is measured in Joules. Therefore the time integral over the kinetic energy yields something with units Joules times seconds. Moreover, we discussed in Section 12.2 that the action appears together with Plank's constant in the exponential function. Since the argument of the exponential function has to be dimensionless, the action has necessarily the same units as Planck's constant which happen to be $[\hbar] = J \cdot s$.

fundamental constant c which describes the speed of light. Since c is a velocity it has units

$$[c] - \frac{m}{s}. \quad (12.31)$$

Therefore, we can construct a constant C with the correct units as follows:

$$C = mc^2 \quad (12.32)$$

because

$$[mc^2] = \frac{kg \cdot m^2}{s^2} \stackrel{(12.29)}{=} [C] \ \checkmark. \quad (12.33)$$

And indeed

$$S[q(t)] \stackrel{(12.25)}{=} C \int \sqrt{1 - \frac{\dot{q}^2}{c^2}} \, dt$$

$$= mc^2 \int \sqrt{1 - \frac{\dot{q}^2}{c^2}} \, dt \quad (12.34)$$

is (almost) the correct action functional for special relativity. The only thing missing is a conventional minus sign:[44]

$$\boxed{S[q(t)] = -mc^2 \int \sqrt{1 - \frac{\dot{q}^2}{c^2}} \, dt.} \quad (12.35)$$

So the main point to take away is that in special relativity, objects follow the minimum proper time path.[45]

Now, how is this Lagrangian related to the classical Lagrangian?

12.3.4 The Free Classical Lagrangian

We can understand classical mechanics as the theory which emerges in the limit where all objects move slowly compared to the speed of light.[46] Mathematically, this implies that for $\dot{q} \ll c$ we should be able to recover the Lagrangian of classical mechanics.

[44] We will see why below. Moreover, recall that we discovered in Section 7.2.3 that a numerical rescaling of the Lagrangian

$$L \to sL,$$

where s is an arbitrary number, makes no difference.

[45] Take note of the similarity to Fermat's principle, which we discussed at the beginning of Chapter 4.

[46] We discussed this in Section 12.1. Moreover, take note that this is analogous to how we can understand classical mechanics as the correct theory in the limit where we can neglect quantum effects, i.e., in which our action is much larger than $\hbar$.

By looking at the special relativity Lagrangian (Eq. 12.35), we can see that in the limit $\dot{q} \ll c$, the fraction which appears under the square root ($\frac{\dot{q}^2}{c^2}$) is a tiny number. Therefore, we can approximate the square root by using the Taylor expansion[47]

$$\sqrt{1-x} = 1 - \frac{x}{2} - \ldots, \quad (12.36)$$

where we can neglect all higher order terms for $x \ll 1$.

[47] The Taylor expansion is discussed in Appendix F.

This means that as long as our object moves slowly compared to the speed of light ($\dot{q} \ll c$), we can rewrite the action functional as follows:

$$S[q(t)] \stackrel{(12.35)}{=} -mc^2 \int \sqrt{1 - \frac{\dot{q}^2}{c^2}} \, dt$$

↷ Eq. 12.36 with $x = \frac{\dot{q}^2}{c^2}$

$$\approx -mc^2 \int \left(1 - \frac{1}{2}\frac{\dot{q}^2}{c^2}\right) dt$$

↷ rearranging terms

$$= -\int mc^2 \, dt + \int m\frac{1}{2}\dot{q}^2 \, dt. \quad (12.37)$$

If we recall that adding a constant to the Lagrangian has no influence on the equations of motion[48], we can conclude that for slowly moving objects, we can use the action functional:[49]

$$S[q(t)] \approx \int \frac{1}{2}m\dot{q}^2 \, dt. \quad (12.39)$$

This is exactly the action functional that we introduced in Chapter 4 for a free object since $\frac{1}{2}m\dot{q}^2$ is the kinetic energy of the object.[50]

We've therefore discovered that in the limit $\dot{q} \ll c$, the special relativity Lagrangian reduces to the classical mechanics Lagrangian. Thus we now understand that objects follow the minimum proper time paths, but for slowly moving objects, the corresponding integral looks like the integral over the kinetic energy.

But so far, we've only derived the correct Lagrangian for a free object. What if there is a non-zero potential V?

[48] We discussed this in detail in Section 7.2.3.

[49] As an aside: the term mc^2 describes a third type of energy known as the **rest energy**:

$$E_{\text{rest}} = mc^2. \quad (12.38)$$

But it yields no significant contribution to the action functional in the limit of slowly moving objects.

[50] Recall that previously we always used

$$S = \int (T - V) \, dt$$

which for a free object ($V = 0$) reads

$$S = \int T \, dt$$

and T denotes the kinetic energy.

12.3.5 Understanding the Minus Sign

One way to understand why we need to add $-V$ to the free Lagrangian $L_{\text{free}} = T$ is by noting that only then do we get, using Noether's theorem, the correct total energy $E = T + V$ as a conserved quantity.[51]

[51] We derived this in Chapter 10.

From a slightly different perspective, we can say that the minus sign in front of mc^2 in Eq. 12.40 is already a hint that different forms of energy appear with a minus sign in the Lagrangian. To understand why, recall that the full free Lagrangian, which we derived in the previous section, reads

$$L \stackrel{(12.37)}{=} -mc^2 + \frac{1}{2}m\dot{q}^2, \qquad (12.40)$$

where $T \equiv \frac{1}{2}m\dot{q}^2$ denotes the usual kinetic energy and $E_{\text{rest}} \equiv mc^2$ another form of energy known as rest energy.[52] Using Noether's theorem, we can derive that there is a conserved quantity which follows from the invariance under temporal shifts $t \to t + \epsilon$ (Eq. 10.75):

[52] In a non-relativistic context, we can neglect this additional form of energy because it only contributes a constant shift to the action functional, and we showed in Section 7.2.3 that such shifts have no influence on the equations of motion. In some sense, this rest energy encodes the intrinsic amount of energy a given object carries in the absence of any movement and external potential. For high-energetic elementary particles this additional form of energy becomes important because these can scatter to produce new particles or decay. In each such process, the total energy needs to be conserved and thus we need to take the rest energy of the original particles and the rest energy of the final particles into account.

$$Q = p\dot{q} - L$$

↓ Eq. 12.40 and $p = \frac{\partial L}{\partial \dot{q}}$

$$= \left(\frac{\partial\left(-mc^2 + \frac{1}{2}m\dot{q}^2\right)}{\partial \dot{q}}\right)\dot{q} - \left(-mc^2 + \frac{1}{2}m\dot{q}^2\right)$$

↓

$$= m\dot{q}^2 + mc^2 - \frac{1}{2}m\dot{q}^2$$

↓

$$= \frac{1}{2}m\dot{q}^2 + mc^2$$

↓

$$= T + E_{\text{rest}}. \qquad (12.41)$$

We can see here that a relative minus sign between the kinetic energy and other forms of energy in the Lagrangian leads to the correct total energy.

But there is also another way in which we can understand the relative minus sign between the kinetic and potential energy in the classical Lagrangian which has to do with the fact that clocks tick differently if there is a non-zero gravitational field. To understand this, we need to talk about general relativity.

12.3.6 General Relativity

We already discussed at the beginning of Section 12.3 that it's impossible to find out from the inside whether a (soundproof) boat (with no windows) is at rest or moves with a constant velocity.

General relativity is based on a similar but more subtle idea.

While no one doubts that there is no difference between frames of reference that move with constant speed relative to each other, accelerating frames are special.

In a soundproof, perfectly smoothly moving train without windows, there is no way to tell if the train moves at all. For example, a glass of water is indistinguishable in a perfectly smoothly moving train from a glass of water in a standing train. However, we notice immediately when the train accelerates. For example, if the train accelerates rapidly, the water in a glass spills over.

But the key insight at the heart of general relativity is that it's impossible to find out whether such an effect is due to an acceleration of the train or due to gravity.

To understand why, imagine a spaceship somewhere in the universe far away from anything. Usually, the astronauts in a spaceship float around if the spaceship isn't accelerating. There is no way to call one of the walls the floor and another one the ceiling.

But what happens if another spaceship starts pulling the original spaceship?

Immediately there is an "up" and a "down". The passengers of the spaceship get pushed toward one of the walls. This wall suddenly becomes the floor of the spaceship. If one of the passengers drops an apple, it falls to the floor.

For an outside observer, this isn't surprising. Through the pulling of the second spaceship, the floor is moving toward the floating apple. This leads to the illusion for the passengers inside the original spaceship that the apple falls to the floor.

The key idea is now that if there is no window in the original spaceship, there is no way for the astronauts to tell if they are stationary on some planet or if they are accelerating. If their spaceship sits on a planet, the apple and the passengers themselves would equally be pulled to the floor. But in this second case this happens because of gravity.[53]

An outside observer would call the force that pushes things to the floor in the accelerating spaceship, a fictitious force. It is merely a result of the floor moving toward the floating objects. However, for the passengers inside the spaceship, the force would be very real. They experience a real floor and a real ceiling and things really fall down if you let them drop. Without getting an outside view, it would be impossible for them to distinguish this fictitious force caused by the acceleration of their spaceship, from the force we call gravity. They can't distinguish between acceleration and being at rest in an appropriate gravitational field.[54]

To summarize:

> Accelerating frames are indistinguishable from resting frames immersed in a gravitational field.

This idea is at the heart of Einstein's theory of general relativity and is commonly known as the **principle of equivalence**.[55]

To understand the principle of equivalence a little better, let's imagine another situation. Instead of a spaceship somewhere in the middle of nowhere, let's consider a spaceship floating 100 kilometers above the Earth. The spaceship is pulled down by the Earth's gravitational field, and for the moment let's imagine the spaceship is stationary. In this situation, the astronauts in

[53] Take note that even if we try to exploit some special property of gravity, there is no way to distinguish these situations. For example, a bowling ball and an apple that are released from the same height would hit the floor at the same moment. This is what Galileo demonstrated for gravity by throwing things down from the Leaning Tower of Pisa. For an observer outside of the original spaceship, this fact would be by no means mysterious. The floor simply moves constantly toward the floating bowling ball and apple. Hence, the floor touches the apple and the bowling ball at exactly the same moment.

[54] Of course, the situation is only indistinguishable if the acceleration has a precise value that mimics the effect of the gravitational field. If you want to mimic the Earth's gravitational field, you need to accelerate quicker than if you want to mimic the weaker gravitational field of the moon.

[55] To clarify the different degrees of relativity: **Galilean relativity** refers to the observation that physics is the same for all inertial observers (e.g., in a moving boat with constant velocity vs. a boat at rest). Special relativity means Galilean relativity *plus* that the speed of light c has the same value for all observers. General relativity is special relativity *plus* the observation that it's impossible to find any difference between a force due to acceleration of the whole system and gravity.

the spaceship are able to distinguish "up" and "down" without problems. An apple falls down, thanks to the Earth's gravitational field.

Then suddenly the spaceship is released from whatever holds it still 100 kilometers above the Earth. What happens now? Of course, the spaceship starts falling down, i.e., moves toward the Earth. At the same time the notions of "up" and "down" start losing their meaning for the astronauts inside the spaceship. Everything inside the spaceship falls down toward the Earth with exactly the same speed. This property of gravity was demonstrated by Galileo through his famous experiments at the Leaning Tower of Pisa. Thus, everything inside the spaceship starts floating. They experience zero gravity. For them, without the ability to look outside of their spaceship, there is no gravitational field and nothing is falling down.

Therefore, gravity is not absolute. While for some observers there is a gravitational field, for the free-falling observers inside the spaceship, there is none. If we want, we can therefore always consider some frame where there is no gravity at all! The gravitational force vanishes completely inside the free-falling spaceship. In contrast, an observer standing on Earth would describe the spaceship by taking the Earth's gravitational field into account. To such an observer everything falls down because of this gravitational field. However, for the astronauts inside the spaceship, nothing is falling.

This situation is exactly the reversed situation compared to our first scenario. In this first scenario, we considered a spaceship somewhere in the middle of nowhere. Then the spaceship was pulled by another spaceship, and suddenly the situation inside the original spaceship was as if they were immersed in a gravitational field. In our second scenario, we started with a spaceship immersed in a gravitational field. However, all effects of this gravitational field vanish immediately when the spaceship starts falling freely toward the Earth. Gravity has no absolute meaning. For some frames of reference there is gravity, for others, there isn't.

Now, what does all this have to do with the relative minus sign in front of the potential energy in the classical Lagrangian?

To understand this, we need to talk about a different kind of acceleration.

So far, we only considered linear acceleration. But we are not only dealing with acceleration when the absolute value of the velocity is non-constant ($\frac{d|\vec{v}|}{dt} \neq 0$) but also when the direction of the velocity vector changes in time.[56]

In this sense, when we are standing, for example, on a rotating disk, we are constantly accelerating too. Formulated differently, since the velocity of each point on the disk undergoes a change in direction at every instant, each such point is accelerating all the time.

Now for simplicity, let's assume that the disk rotates with a constant angular velocity Ω, which implies that each point on it moves with a constant absolute velocity

$$v \equiv \Omega r, \qquad (12.42)$$

where r is the distance from the center of the disk.

[56] In general, we are dealing with a non-zero acceleration when the velocity vector $\vec{v}$ changes in any way since $\vec{a} \equiv \frac{d}{dt}\vec{v}$.

We know from special relativity that clocks tick differently for different observers.[57] Moreover, we know that the time interval a particular clock records depends on its velocity. Therefore, if we put a clock at the center of the disk ($r = 0$ and thus $v = 0$), it will record something different from a clock located elsewhere on the disk ($r \neq 0$ and therefore $v \neq 0$).[58]

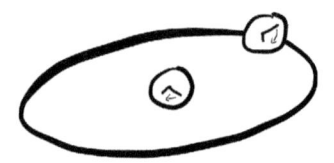

[57] We discussed this in Section 12.3.2.

[58] Take note that the time intervals recorded by these kinds of clocks are proper time intervals because they move with the points on the disk.

We can calculate

$$\Delta t(r) \stackrel{(12.23)}{=} \Delta t(0)\sqrt{1 - \frac{v^2(r)}{c^2}}$$

$$= \Delta t(0)\sqrt{1 - \frac{\Omega^2 r^2}{c^2}}. \qquad (12.43)$$

Eq. 12.42

If we now imagine that an observer sits in a closed cabin on a particular location r on the rotating disk, we can conclude that he will notice a centrifugal acceleration $a = \Omega^2 r$. However, according to the equivalence principle discussed above, it's impossible for such an observer to find out whether this acceleration is due to an acceleration of his cabin or due to a gravitational potential ϕ.

Specifically, if such a potential causes the force[59]

$$F \stackrel{(2.15)}{=} -m\frac{\partial \phi}{\partial r} = m\Omega^2 r, \qquad (12.44)$$

the effect inside the cabin would be exactly the same. We can

[59] For the first equality we use the usual relationship between a potential and the resulting force, c.f. Eq. 2.15.

see this because Newton's second law (Eq. 3.1) tells us

$$F = ma$$

$$\therefore m\Omega^2 r = ma \qquad \text{Eq. 12.44}$$

$$\Omega^2 r = a, \qquad \div m \qquad (12.45)$$

which is exactly the centrifugal acceleration due to the spinning of the disk.

By separating the variables in Eq. 12.44 and then integrating, we can determine the appropriate gravitational potential explicitly:

$$-m\frac{\partial \phi}{\partial r} = m\Omega^2 r$$

$$\therefore \frac{\partial \phi}{\partial r} = -\Omega^2 r \qquad \div m$$

$$\therefore \partial \phi = -\Omega^2 r \partial r \qquad \times \partial r$$

$$\therefore \phi = -\frac{\Omega^2 r^2}{2}. \qquad \int \qquad (12.46)$$

Next, using the principle of equivalence, we can substitute this into Eq. 12.43

$$\Delta t(\phi) \stackrel{(12.43)}{=} \Delta t(0)\sqrt{1 - \frac{\Omega^2 r^2}{c^2}}$$

$$= \Delta t(0)\sqrt{1 - \frac{(-2\phi)}{c^2}} \qquad \text{Eq. 12.46}$$

$$= \Delta t(0)\sqrt{1 + \frac{2\phi}{c^2}}. \qquad (12.47)$$

In words, this means that not only does a non-zero velocity change the flow of time, a gravitational potential ϕ does too.[60]

Therefore, we need to modify our time dilation formula (Eq. 12.23) a little bit to take this into account:

$$\Delta t \stackrel{(12.23)}{=} \Delta t(0)\sqrt{1 - \frac{v^2}{c^2}}$$

$$\rightarrow \quad \Delta t = \Delta t(0)\sqrt{1 - \frac{v^2}{c^2} + \frac{2\phi}{c^2}}. \qquad \text{Eq. 12.47} \qquad (12.48)$$

[60] Analogously, we can derive one of the most surprising consequences of the equivalence principle. By following the same steps as before, we can derive that a gravitational potential leads to length contraction and this, in turn, implies that gravity can be understood as curvature of spacetime.
To understand why, take note that each point on our spinning disk moves round and round, but not inward or outward. Thus, according to special relativity, there is length contraction along the circumference, but none along the radius. Thus, when we measure the circumference of a spinning disk, we measure a different value than an observer who sits on the spinning disk. For the observer sitting on the spinning disk, the disk is at rest, and hence no length contraction happens. But we agree with this observer on the diameter of the disk, since even for us there is no radial movement of the points on the disk. Now comes the punchline.
The formula we all learned in school for the relationship between the radius r and circumference C of a circle is $C = 2\pi r$. Therefore for an observer sitting on the disk for whom the disk appears at rest, we have $C/r = 2\pi$. However, if we look at the disk from the outside, the disk spins, and therefore there is length contraction along the circumference. Therefore, what we measure is not the same: $C/r \neq 2\pi$! We can understand this by noting that the formula $C = 2\pi r$ only holds in a flat space but not if the space in question is curved. Moreover, using the equivalence principle we can conclude that the same phenomena should hold for a disk at rest if there is a gravitational potential. Einstein connected these ideas and concluded that gravity = curvature of spacetime. This is the key idea at the heart of Einstein's theory of general relativity.

This formula describes that, in general, the flow of time can be modified by a non-zero velocity v and additionally by a gravitational potential.[61]

[61] Imagine that our disk not only rotates but also moves with a constant velocity in a specific direction.

Our discovery here implies that we also need to modify our proper time formula (Eq. 12.24)

$$\tau = \int \sqrt{1 - \frac{\dot{q}^2}{c^2}}\, dt$$

↻ Eq. 12.48

$$\rightarrow \quad \tau = \int \sqrt{1 - \frac{\dot{q}^2}{c^2} + \frac{2\phi}{c^2}}\, dt. \quad (12.49)$$

And therefore, we also need to modify our action functional (Eq. 12.35)

$$S[q(t)] = -mc^2 \int \sqrt{1 - \frac{\dot{q}^2}{c^2}}\, dt$$

↻ Eq. 12.49

$$\rightarrow \quad S[q(t)] = -mc^2 \int \sqrt{1 - \frac{\dot{q}^2}{c^2} + \frac{2\phi}{c^2}}\, dt. \quad (12.50)$$

If we now follow exactly the same steps that in Section 12.3.4 allowed us to derive the classical Lagrangian for a free object, we find[62]

[62] The steps here are analogous to what we did in Eq. 12.37. Moreover, take note what we do here only works if the gravitational potential ϕ is sufficiently weak $\frac{2\phi}{c^2} \ll 1$.

$$S[q(t)] \stackrel{(12.50)}{=} -mc^2 \int \sqrt{1 - \left(\frac{\dot{q}^2}{c^2} - \frac{2\phi}{c^2}\right)}\, dt$$

↻ Eq. 12.36 with $x = \frac{\dot{q}^2}{c^2} - \frac{2\phi}{c^2}$

$$\approx -mc^2 \int \left(1 - \frac{1}{2}\left(\frac{\dot{q}^2}{c^2} - \frac{2\phi}{c^2}\right)\right) dt$$

↻ rearranging terms

$$= -\int mc^2\, dt + \int \left(m\frac{1}{2}\dot{q}^2 - m\phi\right) dt. \quad (12.51)$$

So if we ignore once more the constant term $\int mc^2\, dt$, we find that (as long as all objects move slowly $v \ll c$ and the gravitational potential is sufficiently weak $\frac{2\phi}{c^2} \ll 1$) we can use the action functional

$$S[q(t)] = \int \left(m\frac{1}{2}\dot{q}^2 - m\phi\right) dt \equiv \int L\, dt, \qquad (12.52)$$

where
$$L = m\frac{1}{2}\dot{q}^2 - m\phi \equiv T - V \qquad (12.53)$$

is the Lagrangian.

We have therefore derived that the potential energy (here $V = m\phi$) indeed enters the Lagrangian with a relative minus sign.

Let's finish this chapter with a short comment on how Lagrangians are used generally in modern physics.

12.4 Lagrangians in Modern Physics

In Section 12.3, we've learned that we can derive the Lagrangian of special relativity by using that a Lagrangian must always respect the symmetries of the theory in question. And in fact, this trick seems to be working quite generally.

This means that we can start by identifying the symmetries of the theory, then we write down a Lagrangian which only includes terms which respect these symmetries, and if we finally restrict ourselves to the simplest, non-trivial of these terms, we end up with the correct Lagrangian.

As we've already seen above, not every Lagrangian is given by the difference between the kinetic and potential energy.[63] But the general algorithm outlined above works if we broaden our definition of what a Lagrangian is.[64] More generally, a Lagrangian is a function which allows us to calculate the action, and which is constructed such that nature chooses the trajectory for which the corresponding action is minimized.[65]

Our best model of elementary particles, known as the Standard Model, is defined in terms of a Lagrangian. And general relativity can be formulated using a Lagrangian too. Moreover, new models of fundamental physics are usually proposed using a Lagrangian.

So quite appropriately, Arthur Wightman once summarized the goal of most theoretical physicist as "*the hunt for the Green Lion, the ultimate Lagrangian of the world*". [66]

Now you properly wonder *why* all this works so well.

We already understand why the least action principle works. The economical behavior of nature turned out to be an illusion because really all paths are possible and it's just that the minimal action path is the most probable path.[67] But still, nature seems to prefer simple and beautiful Lagrangians because we

---

[63] For example, in Section 12.3.1, we learned that the Lagrangian in special relativity is given by the proper time (Eq. 12.35)

$$L = -mc^2\sqrt{1 - \frac{\dot{q}^2}{c^2}}.$$

[64] You can find a complete discussion in my book "*Physics from Symmetry*".

[65] As discussed in Section 12.2, we find the correct classical equations of motion by minimizing the action. In a quantum context, the action allows us to calculate the probability amplitude for each possible path.

[66] A. S. Wightman. The usefulness of a general theory of quantized fields. In *Conceptual foundations of quantum field theory. Proceedings, Symposium and Workshop, Boston, USA, March 1-3, 1996*, pages 41–46, 1996

[67] We discussed this in detail in Section 12.2.

only include the simplest non-trivial terms.

This is quite likely an illusion too.

Nowadays, most theoretical physicists are convinced that at higher energy scales (= smaller length scales) the correct Lagrangian contains lots of additional terms.[68] However, these higher-order terms become less and less important as we zoom out.[69] Therefore, no matter how complicated and ugly the Lagrangian is at high energy scales, at the low energy scales that we can probe using present-day technology, we are left with a simple Lagrangian.

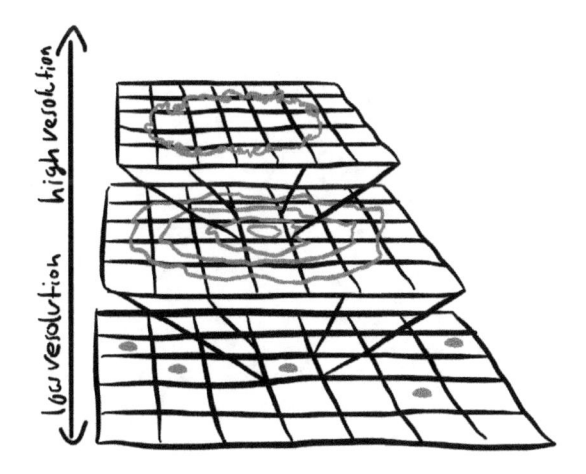

[68] When two particles collide at higher energies, they sometimes create heavier particles. In this sense, colliders are essentially large microscopes which allow us to look deeper. The higher the energy of the colliding particles, the deeper we can look. Moreover, take note that this is analogous to how nature really takes all possible paths into account, and the Lagrangian contains all possible terms if we look closely enough.

[69] Technically, this zooming out process is accomplished using a mathematical tool known as the **renormalization group**.

So the reason that simple and beautiful Lagrangians work so well is not that nature prefers simplicity for some magical reason, but instead, that if we look at something from a distance it usually appears simpler.

13

Further Reading Recommendations

Let's finish this book with a few comments on books you should consult if you want to read a second opinion on a specific topic or simply want to dive deeper.

Great books to learn more about the basic ideas of classical mechanics are

▷ **Analytical Mechanics** by Louis N. Hand and Janet D. Finch[1].

▷ **Introduction to Classical Mechanics** by David Morin[2].

▷ **Lagrangian and Hamiltonian Mechanics** by Melvin G. Calkin[3].

If you want to dive deeper, try

▷ **Classical Mechanics** by Herbert Goldstein[4].

[1] Louis Hand and Janet Finch. *Analytical Mechanics*. Cambridge University Press, Cambridge New York, 1998. ISBN 9780521573276

[2] David Morin. *Introduction to Classical Mechanics : With Problems and Solutions*. Cambridge University Press, Cambridge, UK New York, 2008. ISBN 9780511808951

[3] M. G. Calkin. *Lagrangian and Hamiltonian mechanics*. World Scientific, Singapore River Edge, NJ, 1996. ISBN 9810226721

[4] Herbert Goldstein. *Classical Mechanics*. Pearson, Essex, England, 2014. ISBN 9781292026558

[5] Alexei Deriglazov. *Classical Mechanics : Hamiltonian and Lagrangian Formalism.* Springer, Switzerland, 2016. ISBN 9783319441467

[6] V. I. Arnold. *Mathematical Methods of Classical Mechanics.* Springer New York, New York, NY, 1989. ISBN 978-1-4757-2063-1

[7] Jerrold Marsden. *Introduction to Mechanics and Symmetry : a Basic Exposition of Classical Mechanical Systems.* Springer, New York, 1999. ISBN 978-0-387-21792-5

[8] Jorge José and Eugene Saletan. *Classical Dynamics: A Contemporary Approach.* Cambridge University Press Textbooks, 1998. ISBN 9781299859579

[9] Jennifer Coopersmith. *The Lazy Universe : An Introduction to the Principle of Least Action.* Oxford University Press, Oxford New York, NY, 2017. ISBN 9780198743040

[10] Jakob Schwichtenberg. *No-Nonsense Electrodynamics.* No-Nonsense Books, Karlsruhe, Germany, 2018a. ISBN 978-1790842117

[11] Jakob Schwichtenberg. *No-Nonsense Quantum Mechanics.* No-Nonsense Books, Karlsruhe, Germany, 2018c. ISBN 978-1719838719

▷ **Classical Mechanics** by Alexei Deriglazov[5].

However, be warned that these books aren't as student-friendly as the books mentioned above.

To learn more about mathematical aspects related to classical mechanics consult

▷ **Mathematical Methods of Classical Mechanics** by Vladimir Arnold[6].

▷ **Introduction to Mechanics and Symmetry** by Jerrold E. Marsden and Tudor S. Ratiu[7].

▷ **Classical Dynamics** by Jorge V. José and Eugene J. Saletan[8].

And if you want to learn more about the history and philosophy of classical mechanics, a good starting place is

▷ **The Lazy Universe** by Jennifer Coopersmith[9].

Moreover, the two topics students usually study after classical mechanics are electrodynamics and quantum mechanics. So if you want to move on and learn more them, you might enjoy my books

▷ **No-Nonsense Electrodynamics**[10],

▷ **No-Nonsense Quantum Mechanics**[11].

In addition, you might want to learn more about special relativity and general relativity. Good books on special relativity are

▷ **Special Relativity** by Anthony French[12],

▷ **Special Relativity for Beginners** by Jürgen Freund[13],

▷ **Spacetime Physics** by Edwin F. Taylor and John A. Wheeler[14].

And if you want to understand Einstein's theory of general relativity try

▷ **Relativity, Gravitation and Cosmology** by Ta-Pei Cheng[15].

▷ **Einstein Gravity in a Nutshell** Anthony Zee[16].

[12] A. P. French. *Special relativity.* Norton, New York, 1968. ISBN 9780393097931

[13] Juergen Freund. *Special relativity for beginners : a textbook for undergraduates.* World Scientific, Singapore, 2008. ISBN 9789812771599

[14] Edwin Taylor. *Spacetime Physics : Introduction to Special Relativity.* W.H. Freeman, New York, 1992. ISBN 9780716723271

[15] Ta-Pei Cheng. *Relativity, Gravitation and Cosmology: A Basic Introduction.* Oxford University Press, 2nd edition, 1 2010. ISBN 9780199573646

[16] Anthony Zee. *Einstein Gravity in a Nutshell.* Princeton University Press, 1st edition, 5 2013. ISBN 9780691145587

One Last Thing

It's impossible to overstate how important reviews are for an author. Most book sales, at least for books without a marketing budget, come from people who find books through the recommendations on Amazon. Your review helps Amazon figure out what types of people would like my book and makes sure it's shown in the recommended products.

I'd never ask anyone to rate my book higher than they think it deserves, but if you like my book, please take the time to write a short review and rate it on Amazon. This is the biggest thing you can do to support me as a writer.

Each review has an impact on how many people will read my book and, of course, I'm always happy to learn about what people think about my writing.

PS: If you write a review, I would appreciate a short email with a link to it or a screenshot to JakobSchwich@gmail.com. This helps me to take note of new reviews. And, of course, feel free to add any comments or feedback that you don't want to share publicly.

Part IV
Appendices

A
Calculus

Of course, we can't discuss calculus in detail here. We will focus on a few key aspects which we need throughout the book. If you want to learn calculus, I highly recommend "Calculus Made Easy" by Silvanus P. Thompson which is available for free at http://calculusmadeeasy.org/.

In physics, we are mostly interested in how things change. So let's talk about derivatives.

In general, the derivative of a function $f(x)$ is defined by the **difference quotient**:

$$\frac{df(x)}{dx} = \lim_{h \to 0} \frac{f(x+h) - f(x)}{h} \quad (A.1)$$

Intuitively, the derivative of a function tells us how it changes as we move along the x-axis.

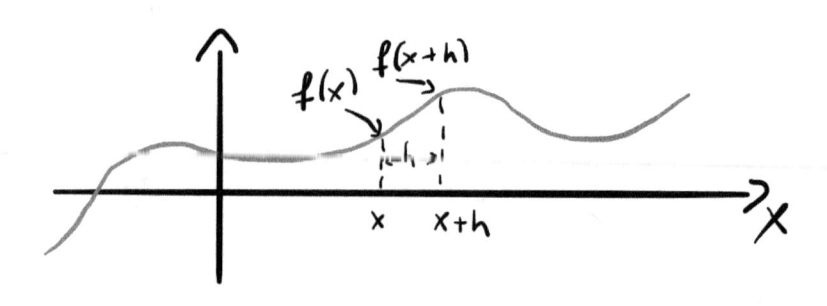

Imagine that we know the value of a function $f(x)$ at a specific location x_0. If we then want to know the value of the function as we move a little bit to the right on the x-axis $x_0 \to x_0 + \epsilon$, we can calculate it by using the derivative[1]

$$f(x_0 + \epsilon) = f(x_0) + \epsilon \frac{d}{dx} f(x). \tag{A.2}$$

This formula is correct because $\frac{df(x)}{dx}$ describes the rate of change of f, and therefore by multiplying it by the distance ϵ we are moving along the x axis, we get the total change in f.

One of the most important rules when it comes to derivatives is the product rule.

[1] Take note that technically this formula is only correct if ϵ is **infinitesimal**. Infinitesimal means as small as possible but still not zero. This implies that $\epsilon^2 = 0$, $\epsilon^3 = 0$ because for any positive number a smaller than 1, we have $a^2 < a$, e.g., $0.1^2 = 0.01 < 0.1$. But if, by assumption, ϵ is already as small as possible (but non-zero), the only way ϵ^2 and ϵ^3 can be even smaller is if they are exactly zero. If we want to find the value of f at a point which is not only infinitesimally away, the correct formula is given by the Taylor series which is discussed in Appendix F.

A.1 Product Rule

The product rule

$$\frac{d\big(f(x)g(x)\big)}{dx} = \left(\frac{df(x)}{dx}\right)g(x) + f(x)\left(\frac{dg(x)}{dx}\right) \equiv f'g + fg' \tag{A.3}$$

follows directly from the definition of the derivative

$$\begin{aligned}\frac{d}{dx}[f(x)g(x)] &= \lim_{h\to 0}\frac{f(x+h)g(x+h)-f(x)g(x)}{h}\\ &= \lim_{h\to 0}\frac{[f(x+h)g(x+h)-f(x+h)g(x)]+[f(x+h)g(x)-f(x)g(x)]}{h}\\ &= \lim_{h\to 0} f(x+h)\frac{g(x+h)-g(x)}{h}+g(x)\frac{f(x+h)-f(x)}{h}\\ &= f(x)g'(x)+g(x)f'(x).\end{aligned}$$

A.2 Integration by Parts

Using the product rule, we can derive immediately one of the most important tricks that we can use to simplify integrals. By integrating the product rule (Eq. A.3)[2]

[2] For the first term, we use the fundamental theorem of calculus, i.e., $\int_a^b dx\, h'(x) = h(b) - h(a)$.

$$\underbrace{\int_a^b dx\, \frac{d\big(f(x)g(x)\big)}{dx}}_{=f(x)g(x)\big|_a^b} = \int_a^b dx\left(\frac{df(x)}{dx}\right)g(x) + \int_a^b dx\, f(x)\left(\frac{dg(x)}{dx}\right) \quad (A.4)$$

and then rearranging the terms, we find

$$\int_a^b dx\left(\frac{df(x)}{dx}\right)g(x) = f(x)g(x)\Big|_a^b - \int_a^b dx\, f(x)\left(\frac{dg(x)}{dx}\right). \quad (A.5)$$

In classical mechanics, we often encounter functions (like the Lagrangian or Hamiltonian) which depend on multiple variables, e.g., $f = f(x,y,t)$. For such functions, there are different kinds of derivatives and this is what we will talk about next.

A.3 Total and Partial Derivatives

To understand the different kinds of derivatives, let's say we have a function $\rho\big(t, x(t), p(t)\big)$ which, in general, depends on

[3] The symbol ρ is typically used in physics to denote densities. We discuss the probability density $\rho(t, x(t), p(t))$ in a physical context in Section 11.2.

the location $x(t)$ and momentum $p(t)$ plus the time t.[3] A key observation is that the location $x(t)$ and momentum $p(t)$ are functions of t too. Therefore, we need to be extremely careful what we mean when we calculate the derivative with respect to the time t.

In such a situation, we use the symbol $\frac{d\rho}{dt}$ to denote the **total derivative** which is defined in terms of a difference quotient:

$$\frac{d\rho}{dt} = \lim_{\Delta t \to 0} \frac{\rho\big(t + \Delta t, x(t + \Delta t), p(t + \Delta t)\big) - \rho\big(t, x(t), p(t)\big)}{\Delta t} \tag{A.6}$$

The result is the total rate of change of ρ.

But it's also possible to consider to **partial derivative**, which is denoted by $\frac{\partial \rho}{\partial t}$ and defined as follows:

$$\frac{\partial \rho}{\partial t} = \lim_{\Delta t \to 0} \frac{\rho\big(t + \Delta t, x(t), p(t)\big) - \rho\big(t, x(t), p(t)\big)}{\Delta t} \tag{A.7}$$

The key difference is that we only vary t if it appears explicitly in ρ but not if it only appears implicitly because $x(t)$ and $p(t)$ also depend on t. Formulated differently, when we calculate partial derivatives we hold all other variables fixed. In contrast, if we calculate the total derivative we let changes in variables affect each other.

The total derivative of a function $\rho\big(t, x(t), p(t)\big)$ can be calculated by using the formula[4]

[4] We will derive this formula below.

$$\frac{d\rho}{dt} = \frac{\partial \rho}{\partial x}\frac{dx}{dt} + \frac{\partial \rho}{\partial p}\frac{dp}{dt} + \frac{\partial \rho}{\partial t}. \tag{A.8}$$

By using this formula we take the change of ρ in all "directions" (here x, p and t) into account.

We can see this more explicitly by multiplying our formula, formally, by dt:[5]

[5] The symbol d means "a little bit of". For example, dx means a little bit of x and dy means a little bit of y.

$$d\rho = \frac{\partial \rho}{\partial x}dx + \frac{\partial \rho}{\partial p}dp + \frac{\partial \rho}{\partial t}dt. \tag{A.9}$$

Now each term on the right-hand side is of the form "rate of change times distance" and therefore yields the total change of ρ in the corresponding direction. So the total change in ρ is the sum over the changes due to changes in position x, in momentum p, or in time t itself. We have here a sum, because the whole point of the total derivative is that when we move in the t direction, we automatically also move in the x and p direction because $x = x(t)$ and $p = p(t)$.

To understand what this means in intuitive terms, recall that one possible arena we can use to describe physics is phase space.[6] Our phase space coordinates are x and p. If we calculate the partial derivative $\frac{\partial \rho}{\partial t}$, we act as if phase space is a static background structure. This means that we hold x and p fixed and then only consider how our function $\rho\big(t, x(t), p(t)\big)$ changes in time.[7]

[6] Phase space is discussed in Section 2.3.

But if we calculate the total derivative $\frac{d\rho}{dt}$, we take into account that each point in phase space actually moves around as described by Hamilton's equations of motion (Eq. 5.15).[8]

In general, the total derivative of a function $f(x(t), y(t), z(t), t)$ reads

$$\frac{df}{dt} = \frac{\partial f}{\partial x}\frac{dx}{dt} + \frac{\partial f}{\partial y}\frac{dy}{dt} + \frac{\partial f}{\partial z}\frac{dz}{dt} + \frac{\partial f}{\partial t}. \quad \text{(A.10)}$$

[7] You might wonder why we want to do that. One important reason is that this partial change is part of the total change which we find when we calculate the total derivative. We can see in Eq. A.8 that the partial derivative is part of the total derivative. Moreover, take note that many important equations in physics like the Schrödinger equation or the Liouville equation contain solely a partial derivative with respect to time.

[8] This is discussed in more detail in Section 11.2.5.

Let's consider as a concrete example the function

$$\rho_s\big(t, x(t), p(t)\big) = x^2 + ct, \quad \text{(A.11)}$$

where $x(t) = at^2$ and c and a are constants. The partial derivative of this function with respect to t reads

$$\begin{aligned}\frac{\partial \rho_s}{\partial t} &= \frac{\partial (x^2 + ct)}{\partial t} \\ &= \frac{\partial x^2}{\partial t} + \frac{\partial (ct)}{\partial t} \\ &= c.\end{aligned} \quad \frac{\partial x}{\partial t} = 0 \text{ and } \frac{\partial t}{\partial t} = 1 \quad \text{(A.12)}$$

In contrast, the total derivative reads

$$\frac{d\rho_s}{dt} \stackrel{(A.8)}{=} \frac{\partial \rho_s}{\partial x}\frac{dx}{dt} + \frac{\partial \rho_s}{\partial p}\frac{dp}{dt} + \frac{\partial \rho_s}{\partial t}$$ ↷ Eq. A.11

$$= \frac{\partial(x^2+ct)}{\partial x}\frac{dx}{dt} + \frac{\partial(x^2+ct)}{\partial p}\frac{dp}{dt} + \frac{\partial(x^2+ct)}{\partial t}$$ ↷ Eq. A.12

$$= (2x)\frac{dx}{dt} + (0)\frac{dp}{dt} + (c)$$ ↷ $x(t) = at^2$

$$= 2(at^2)\frac{d(at^2)}{dt} + c$$

$$= 4a^2 t^3 + c.$$ ↷ $\frac{d(at^2)}{dt} = 2at$ (A.13)

Now, why is the general formula (Eq. A.8) for the total derivative correct?

To understand this, we first need to talk about the chain rule.

A.4 Chain Rule

The chain rule

$$\frac{df(g(x))}{dx} = \frac{\partial f}{\partial g}\frac{\partial g}{\partial x} \qquad (A.14)$$

follows directly from the definition of the derivative

$$\begin{aligned}\frac{d}{dx}f\big(g(x)\big) &= \lim_{h\to 0}\frac{f\big(g(x+h)\big)-f\big(g(x)\big)}{h}\\ &= \lim_{h\to 0}\frac{f\big(g(x+h)\big)-f\big(g(x)\big)}{h}\frac{g(x+h)-g(x)}{g(x+h)-g(x)}\\ &= \lim_{h\to 0}\frac{f\big(g(x+h)\big)-f\big(g(x)\big)}{g(x+h)-g(x)}\frac{g(x+h)-g(x)}{h}\\ &= \frac{\partial f}{\partial g}\frac{\partial g}{\partial x}.\end{aligned}$$

Completely analogously, we can derive that for a function f which depends on multiple functions $f = f\big(g(x), h(x)\big)$, we have

$$\frac{df\big(g(x),d(x)\big)}{dx} = \frac{\partial f}{\partial g}\frac{\partial g}{\partial x} + \frac{\partial f}{\partial d}\frac{\partial d}{\partial x}. \qquad (A.15)$$

To see this explicitly, we use once more the definition of the derivative

$$\begin{aligned}\frac{d}{dx}f\big(g(x),d(x)\big) &= \lim_{h\to 0}\frac{f\big(g(x+h),d(x+h)\big)-f\big(g(x),d(x)\big)}{h}\\ &= \lim_{h\to 0}\Bigg(\frac{f\big(g(x+h),d(x+h)\big)-f\big(g(x),d(x)\big)}{h}\\ &\quad + \frac{f\big(g(x+h),d(x)\big)-f\big(g(x+h),d(x)\big)}{h}\Bigg)\\ &= \lim_{h\to 0}\frac{f\big(g(x+h),d(x+h)\big)-f\big(g(x+h),d(x)\big)}{h}\\ &\quad + \lim_{h\to 0}\frac{f\big(g(x+h),d(x)\big)-f\big(g(x),d(x)\big)}{h}\\ &= \frac{\partial f}{\partial d}\frac{\partial d}{\partial x} + \frac{\partial f}{\partial g}\frac{\partial g}{\partial x},\end{aligned}$$

where in the last step, we used that the two terms are analogous to what we considered in the derivation of the chain rule above.

If instead, we consider a function which depends explicitly on x itself, i.e., $f = f(g(x), x)$, we have

$$\frac{df\big(g(x),x\big)}{dx} = \frac{\partial f}{\partial g}\frac{\partial g}{\partial x} + \frac{\partial f}{\partial x}, \qquad (A.16)$$

which follows directly from Eq. A.15 if we substitute $d(x)$ for x:

$$\frac{df(g(x),x)}{dx} = \frac{\partial f}{\partial g}\frac{\partial g}{\partial x} + \frac{\partial f}{\partial x}\frac{\partial x}{\partial x}.$$

$$\circlearrowright \frac{\partial x}{\partial x} = 1$$

$$= \frac{\partial f}{\partial g}\frac{\partial g}{\partial x} + \frac{\partial f}{\partial x} \qquad (A.17)$$

If we combine these puzzle pieces, we end up with the full formula for the total derivative of a general function $f = f(x(t), y(t), z(t), t)$:

$$\frac{df}{dt} = \frac{\partial f}{\partial x}\frac{dx}{dt} + \frac{\partial f}{\partial y}\frac{dy}{dt} + \frac{\partial f}{\partial z}\frac{dz}{dt} + \frac{\partial f}{\partial t}. \qquad (A.18)$$

B

The Legendre Transform

In general, a mathematical transformation is a method which allows us to express the information contained in a given function more conveniently.[1] The simplest kind of mathematical transformation is when we replace a given (invertible) function $y = f(x)$ with its inverse $x \equiv f^{-1}(y)$. Both functions contain exactly the same information but sometimes thinking about a problem in terms of the inverse helps us to move forward.[2]

The Legendre transform is just another method to express the information encoded in a given function a bit differently.

In classical mechanics, we use it to describe systems using the Legendre transform of the Lagrangian function. But the Legendre transform is also commonly used in thermodynamics to switch between descriptions in terms of internal energy, free energy, enthalpy and free enthalpy.

One reason why we might want to use it is that sometimes the differential equation for a given function becomes easier to solve once we rewrite it in terms of the corresponding Legendre transformed function. For example, in classical mechanics the behavior of the Lagrangian is governed by the Euler-Lagrange

[1] For example, the Fourier transform and Laplace transform $F(k)$ of a function $f(x)$ allow us to express the information contained in a function in terms of how much each basic building block (e^{ikx}) contributes. (k labels our basic building blocks.)

[2] To understand why $f(x)$ and its inverse contain the same information, recall that the graph of the inverse is the mirror image (mirrored at the line $y = x$) of the graph of $f(x)$. And we don't destroy any information by mirroring a graph.

[3] The Legendre transform sometimes helps but certainly not always.

[4] In general, the function we want to Legendre transform can depend on multiple variables, e.g., $L(q,v,\ldots)$. But this makes no difference for what we consider here, and everything works analogously. Moreover, a bit more context: L is the symbol that we typically use for the Lagrangian function (see Chapter 4). The Lagrangian is, in general, a function of the location q, velocities $\dot{q} \equiv v$ and can additionally depend on t: $L = L(q, \dot{q}, t)$. In this context the slope function $p \equiv \frac{\partial L(v)}{\partial v}$ is known as the conjugate momentum. We use the symbol $v \equiv \dot{q}$ for the velocity because the connection to q is not important.

[5] While this may seem awkward, we will see in a moment that it's actually helpful to write the function like this.

equation (Eq. 7.28) while the behavior of its Legendre transform (called the Hamiltonian) is governed by Hamilton's equations (Eq. 5.15). For some specific applications, Hamilton's equations are more convenient than the Euler-Lagrange equation.[3]

To understand how the Legendre transform works, let's say we start with a concrete function $L(v)$.[4] We can then calculate the corresponding slope function

$$p(v) = \frac{\partial L(v)}{\partial v}. \tag{B.1}$$

An overly complicated way to write our original function $L(v)$ is therefore[5]

$$L(v) = \int_0^v p(v')dv' \tag{B.2}$$

because integrating and differentiating are inverse procedures:

$$\frac{\partial L}{\partial v} \stackrel{B.2}{=} \frac{\partial}{\partial v} \int_0^v p(v')dv' = p \;\; \checkmark \tag{B.3}$$

We can understand this in geometrical terms by plotting the slope function $p(v)$ as a function of v:

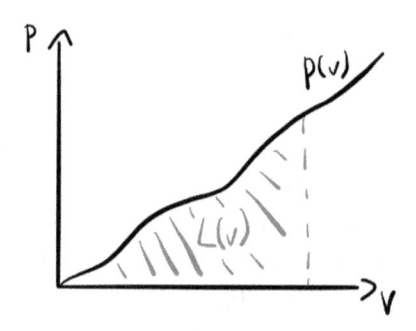

Eq. B.2 tells us that our function $L(v)$ is the area below the $p(v)$ curve.

So far nothing interesting has happened. But now let's switch perspectives.

THE LEGENDRE TRANSFORM 353

First of all, take note that we can equally talk about v as a function of p (i.e., $v(p)$), instead of about p as a function of v (i.e., $p(v)$). To understand this a bit better, we can simply rotate our graph:

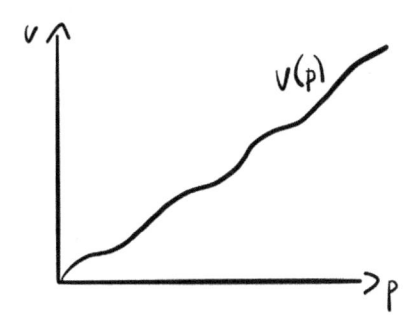

Now comes the first key idea. Analogous to how we defined our original function $L(v)$ as the area below the $p(v)$ curve (Eq. B.2), we can now introduce a new function $H(p)$ which describes the area below the $v(p)$ curve:

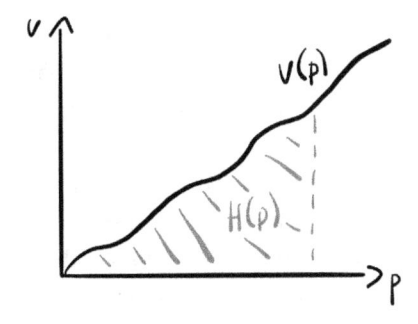

Mathematically, this means that

$$H(p) \equiv \int_0^p v(p')dp'. \tag{B.4}$$

We call $H(p)$ the **Legendre transform** of $L(v)$. This may seem like just another mathematical function. Of course, we can define it. But why should we care about it and how exactly is it related to our original function $L(v)$?

[6] Take note that in the figure here, the v and p axis are again back in their original positions. This means, the function $H(p)$ now describes the area *above* the curve.

To understand this, we need a second key idea: if we combine the two areas described by $L(v)$ (Eq. B.2) and $H(p)$ (Eq. B.4), we get a rectangle:[6]

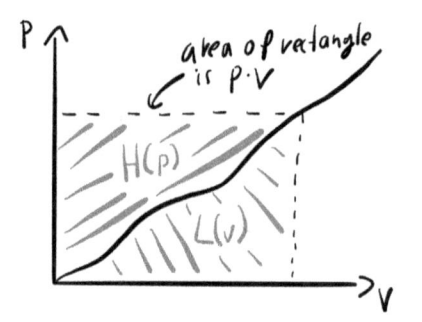

In mathematical terms this means that

$$L(v) + H(p) = pv. \quad (B.5)$$

By rearranging the terms, we find

$$\boxed{H(p) = pv - L(v).} \quad (B.6)$$

This is how our original function $L(v)$ is related to its Legendre transform $H(p)$. And the key observation is that $L(v)$ and $H(p)$ contain the same information but offer different perspectives.[7]

[7] Technically, this is only true if our original function is convex. A function $f(x)$ is convex if its second derivative $\frac{d^2 f(x)}{dx^2}$ is positive everywhere. Geometrically, this means that if we pick two points on the function, the straight line connecting them never goes below the function. Formulated differently, the first derivative $\frac{df(x)}{dx}$ (the slope) is strictly monotonic. This condition is necessary because otherwise the Legendre transform does not contain the same information as the original function. This assumption is fulfilled for all physical Lagrangians but, of course, not for any function we can write down.

The point of all this mathematical yoga is that we've switched from a function $L(v)$ which depends on v to a new function $H(p)$ which depends no longer on v but on the value of the slope function p at each location. And sometimes, our equations simplify if we write them in terms of $H(p)$ instead of $L(v)$.

Take note that the Legendre transform is really completely symmetric. If we calculate the derivative of $L(v)$ with respect to v we find the slope p (Eq. B.1), and if we calculate the derivative of $H(p)$ with respect to p, we find the original variable v:[8]

$$\frac{\partial L}{\partial v} = p$$
$$\frac{\partial H}{\partial p} = v \quad (B.7)$$

[8] This follows immediately when we use the definition of $H(p)$ as given in Eq. B.4. The steps are analogous to what we did in Eq. B.3.

To summarize: when we want to calculate the Legendre transform of a function $L(q, \dot{q}, t)$, we start by calculating the corresponding slope function

$$p \equiv \frac{\partial L}{\partial \dot{q}}. \tag{B.8}$$

Then, we calculate the Legendre transform of $L(q, \dot{q}, t)$ using the formula (Eq. B.6)

$$H(q, p, t) = p\dot{q}(p) - L(q, \dot{q}(p), t), \tag{B.9}$$

where we use the explicit relationship between $\dot{q}$ and p to get a function $H(q, p, t)$ which only depends on p but not on $\dot{q}$.

Take note that if we are dealing with a function that depends on multiple variables (like $L = L(q, \dot{q}, t)$), we can, in principle, perform a Legendre transformation with respect to each variable. In classical mechanics, we usually only consider the Legendre transform with respect to $\dot{q}$. But we can equally consider a Legendre transform with respect to q.

This means that we can consider the slope function[9]

$$f \equiv \frac{\partial L}{\partial q} \tag{B.10}$$

[9] We will talk about the physical meaning of this alternative slope function below.

instead of $p \equiv \frac{\partial L}{\partial \dot{q}}$ (Eq. B.8). The corresponding Legendre transform of the Lagrangian $L = L(q, \dot{q}, t)$ then reads

$$G(f, \dot{q}, t) \equiv fq - L(q(f), \dot{q}, t). \tag{B.11}$$

Therefore, we end up with a new function which no longer depends on q but on the new variable f.

In addition, we can even Legendre transform the Lagrangian

[10] The additional minus signs here are just conventions that make sure that we end up with the correct functions if we derive $K(f,p,t)$ by using $G(f,\dot{q},t)$ or by using $H(q,p,t)$. They have no deeper significance.

[11] The names "Kamiltonian" and "Gamiltonian" are jokes and, as mentioned below, no one uses them. Moreover, take note that $f = \frac{\partial L}{\partial q} = -\frac{\partial H}{\partial q}$. We will discuss this in more detail below.

twice such that we end up with a function:

$$K(f,p,t) \equiv p\dot{q} - \left(fq - L(q(f),\dot{q},t)\right)$$
$$\stackrel{B.11}{=} p\dot{q} - G(f,\dot{q},t).$$

Alternatively, we can calculate $K(f,p,t)$ by using $H(q,p,t)$:[10]

$$K(f,p,t) \equiv p\dot{q} - \left(fq - L(q(f),\dot{q},t)\right)$$
$$= (-f)q - \left((-p\dot{q}) - L(q(f),\dot{q},t)\right)$$
$$\stackrel{B.9}{=} (-f)q - H(q,-p,t)$$

To summarize:[11]

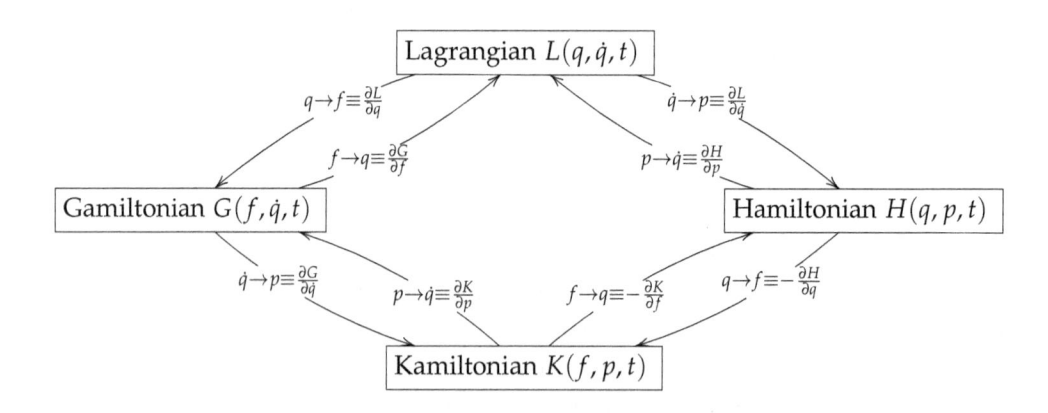

However, the only Legendre transform of the Lagrangian which is actually useful is the Hamiltonian $H(q,p,t)$, while the "Gamiltonian" $G(f,\dot{q},t)$ and the "Kamiltonian" $K(f,p,t)$ are rarely considered. To understand why, we need to understand what our new variable f represents.[12]

[12] Reminder: the slope function $p \equiv \frac{\partial L}{\partial \dot{q}}$ (Eq. B.8) we use to calculate the Hamiltonian represents the canonical momentum.

The Euler-Lagrange equation (Eq. 4.25) reads

$$\frac{d}{dt}\left(\frac{\partial L}{\partial \dot{q}}\right) = \frac{\partial L}{\partial q}. \tag{B.12}$$

And we discovered in Section 4.3.1, that, if we use $p \equiv \frac{\partial L}{\partial \dot{q}}$ and the definition of a force $F \equiv -\frac{\partial V}{\partial q}$, we can rewrite it as (Eq. 4.30):

$$\frac{d}{dt}p = F. \tag{B.13}$$

This suggests that our new slope function $f \equiv \frac{\partial L}{\partial q}$ describes the (generalized) force acting on the object in question.

We can also understand this by invoking Hamilton's equation (Eq. 5.15)

$$\frac{dp}{dt} = -\frac{\partial H}{\partial q} \tag{B.14}$$

and then recalling that in Section 5.1.1, we discovered that this equation essentially also states (Eq. 5.28)

$$\frac{d}{dt}p = F. \tag{B.15}$$

Therefore, we can again conclude that our slope function $f \equiv -\frac{\partial H}{\partial q}$ describes the (generalized) force.

This allows us to understand why the "Gamiltonian" $G(f, \dot{q}, t)$ and "Kamiltonian" $K(f, p, t)$ are almost never considered, while the Hamiltonian $H(q, p, t)$ can, in principle, always be used. The key observation here is that there is always a proper (bijective) relation between the velocity $\dot{q}$ and the momentum p.[13] But for most systems, there is no such relation between the location q and the force f. Mathematically, this implies that while the Hamiltonian is always well-defined, the Legendre transforms $G(f, \dot{q}, t)$ and $K(f, p, t)$ are ill-defined for most systems.[14]

[13] Bijective means one-to-one and onto. This means that each value of the velocity function $\dot{q}$ is mapped to *exactly* one unique momentum value p. In contrast, a formula for a force of the form $f(x) = bx^2$ is not bijective since $f(-x) = f(x)$, and therefore pairs of x values are mapped to the same value of f.

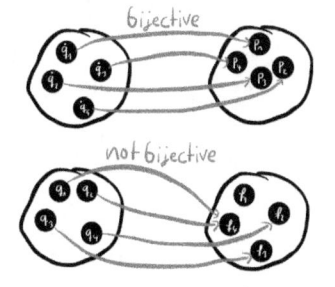

[14] An exception is an object attached to a spring (harmonic oscillator), for which $f = kx$ (Hooke's law). We discuss the harmonic oscillator in Chapter 8.

C
Lagrange Multipliers

The Lagrangian multiplier method is an extremely powerful tool which allows us to find the maxima or minima of functions which are subject to constraints.[1] In Chapter 4, we discuss that we can find the correct path in configuration space by minimizing the action. But for many applications, we are not just looking for any path that minimizes the action but for very specific ones. Quite often there are constraints which define which kind of paths are allowed. For example, the mass at the end of a pendulum cannot move around freely but needs to stay fixed to the string. This is a constraint.[2] And therefore, we need the Lagrangian multiplier method all the time in classical mechanics.

Let's start with a concrete example.

Suppose we have a marble which, for some unspecified reason, always remains attached to an inclined plane and moves on this plane only along a circle. What's the highest point the marble can reach?

[1] A typical example would be: find the path of minimum action under the condition that our path does not leave a fixed surface. But also: find a way to minimize the material costs to enclose a fixed volume.

[2] We discuss this application of Lagrange multipliers in more detail in Section 7.2.1. In particular we learn in this section that the Lagrangian multiplier method allows us to reduce a variational problem with constraints to a free variational problem without constraints. Moreover, we discuss the pendulum in detail in Chapter 9. In this appendix, we focus on the method as a general mathematical tool.

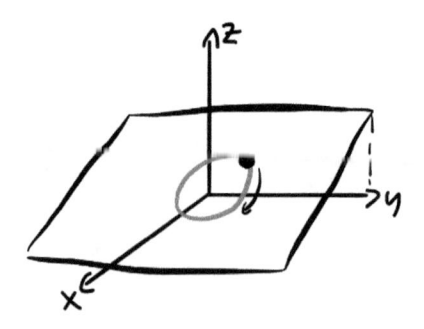

Let's say the plane is defined by

$$z(x,y) = -2x + y \qquad (C.1)$$

and the circle is defined by[3]

$$x^2 + y^2 - 1 = 0. \qquad (C.2)$$

Here Eq. C.1 tells us the height $z(x,y)$ at each point (x,y). In other words, Eq. C.1 defines a height function and our task is to find which point on the circle defined in Eq. C.2 is the highest.

Usually, we can find the maximum (or minimum) of a function by calculating the derivative and then determining the zeroes of the derivative. But here, this is not possible because additionally, we have to take the constraint in Eq. C.2 into account. One possibility is to solve the constraint equation (Eq. C.2) for one of the variables[4]

$$x = -\sqrt{1-y^2} \qquad (C.3)$$

and use this to eliminate it from the original equation:

$$z(y) \stackrel{(C.1)}{=} -2x(y) + y$$
$$= -2\left(-\sqrt{1-y^2}\right) + y. \qquad (C.4)$$

↪ Eq. C.3

Then afterwards, we can determine the maximum, as usual, by searching for the zeroes of the derivative with respect to the

---

[3] All values of x and y which fulfill the condition

$$x^2 + y^2 = 1$$

lie on a circle with radius 1 in the xy-plane.

[4] When we calculate the square root to get to Eq. C.3, we really find $x = \pm\sqrt{1-y^2}$. But by looking at our height function Eq. C.1, we can see that our maximum will correspond to a negative value of x. The minimum can be calculated using $x = +\sqrt{1-y^2}$.

remaining variable:

$$0 \stackrel{!}{=} \frac{dz(y)}{dy} \quad \circlearrowright \text{ Eq. C.4}$$

$$0 \stackrel{!}{=} \frac{d}{dy}\left(-2(-\sqrt{1-y^2}) + y\right) \quad \circlearrowright \text{ chain rule}$$

$$0 \stackrel{!}{=} -2\frac{y}{\sqrt{1-y^2}} + 1 \quad \circlearrowright \text{ rearranging}$$

$$\frac{1}{2} \stackrel{!}{=} \frac{y}{\sqrt{1-y^2}} \quad \circlearrowright \text{ rearranging}$$

$$\frac{\sqrt{1-y^2}}{2} \stackrel{!}{=} y \quad \circlearrowright \text{ squaring}$$

$$\frac{1-y^2}{4} \stackrel{!}{=} y^2 \quad \circlearrowright$$

$$\frac{1}{4} \stackrel{!}{=} \frac{5}{4}y^2 \quad \circlearrowright$$

$$\frac{1}{5} \stackrel{!}{=} y^2 \quad \circlearrowright \sqrt{}$$

$$\frac{1}{\sqrt{5}} \stackrel{!}{=} y. \tag{C.5}$$

This tells us that the maximum is located at $y_m = \frac{1}{\sqrt{5}}$.[5] We can then use Eq. C.3 to determine the x-coordinate of the maximum and minimum:

$$x_m \stackrel{(C.3)}{=} -\sqrt{1-y_m^2}$$

$$= -\sqrt{1-\left(\frac{1}{\sqrt{5}}\right)^2} \quad \circlearrowright \text{ Eq. C.5}$$

$$= -\sqrt{\frac{4}{5}}. \tag{C.6}$$

Using these coordinates, we can calculate the maximum height by plugging them into the height function (Eq. C.1)

$$z_m(x_m, y_m) = -2x_m + y_m$$

$$= -2\left(-\sqrt{\frac{4}{5}}\right) + \frac{1}{\sqrt{5}} \quad \circlearrowright \text{ Eq. C.5 and Eq. C.6}$$

$$= \sqrt{5}. \tag{C.7}$$

[5] The subscript m denotes the coordinates of our maximum. Moreover, again, when we calculate the square root in Eq. C.5, we really get $\pm\frac{1}{\sqrt{5}} \stackrel{!}{=} y$ since minima are characterized by a vanishing first derivative too. The maximum is at $y = +\frac{1}{\sqrt{5}}$ and the minimum at $y = -\frac{1}{\sqrt{5}}$.

While here this is possible without major problems, this approach quickly becomes cumbersome when we are dealing with a large set of variables. Eliminating all but one variable is often a lot of work.

The Lagrange multiplier method approaches this problem the other way around. Instead of eliminating variables, we introduce new ones.[6]

[6] Although this may sound stupid, we will see in a minute that this is actually a smart thing to do.

We replace the function that we want to maximize with a new function Λ which combines the original function and the constraint into a single function:

$$\Lambda(x, y, \lambda) = z(x, y) + \lambda(x^2 + y^2 - 1)$$

$$\quad \curvearrowright \text{Eq. C.1}$$

$$= -2x + y + \lambda(x^2 + y^2 - 1) \tag{C.8}$$

where we introduced a new variable λ. In general, λ is known as a **Lagrange multiplier**.

Now, let's see what happens when we try to find the maximum of this new function by determining its zeroes:

$$0 \stackrel{!}{=} \frac{d}{dx} \Lambda(x, y, \lambda)$$

$$\curvearrowright \text{Eq. C.8}$$

$$0 \stackrel{!}{=} \frac{d}{dx} \left(-2x + y + \lambda(x^2 + y^2 - 1) \right)$$

$$\curvearrowright \frac{dy}{dx} = 0, \frac{d\lambda}{dx} = 0$$

$$0 \stackrel{!}{=} -2 + 2\lambda x$$

$$\curvearrowright \text{rearranging}$$

$$\frac{1}{\lambda} \stackrel{!}{=} x. \tag{C.9}$$

Analogously, we can calculate the y-coordinate of the maximum

$$0 \stackrel{!}{=} \frac{d}{dy}\Lambda(x,y,\lambda)$$

⤷ Eq. C.8

$$0 \stackrel{!}{=} \frac{d}{dy}\left(-2x + y + \lambda(x^2 + y^2 - 1)\right)$$

⤷ $\frac{dx}{dy} = 0, \frac{d\lambda}{dy} = 0$

$$0 \stackrel{!}{=} 1 + 2\lambda y$$

⤷ rearranging

$$-\frac{1}{2\lambda} \stackrel{!}{=} y. \tag{C.10}$$

And for λ we find:

$$0 \stackrel{!}{=} \frac{d}{d\lambda}\Lambda(x,y,\lambda)$$

⤷ Eq. C.8

$$0 \stackrel{!}{=} \frac{d}{d\lambda}\left(-2x + y + \lambda(x^2 + y^2 - 1)\right)$$

⤷ $\frac{dx}{d\lambda} = 0, \frac{dy}{d\lambda} = 0$

$$0 \stackrel{!}{=} x^2 + y^2 - 1$$

⤷ rearranging

$$\tag{C.11}$$

The last line is exactly our constraint equation (Eq. C.2). This means that adding the new term $\lambda(x^2 + y^2 - 1)$ to our height function (as we did in Eq. C.8) allows us to describe the height and the constraint at the same time using a single function $\Lambda(x,y,\lambda)$.

The key observation is that the coordinates of the maximum (x_m, y_m) of our new function Λ depend on λ. This gives us the freedom to choose λ cleverly such that the constraint is fulfilled. And this particular value for λ allows us to find the correct maximum of our original function $z(x,y)$.

We solve the constraint equation for one of the variables (e.g., for x as we did in Eq. C.3) to solve Eq. C.9 for λ:[7]

[7] As before, the subscript m denotes the coordinates of our maximum. For your convenience: Eq. C.3 reads

$$x = -\sqrt{1 - y^2}.$$

$$x_m \stackrel{(C.9)}{=} \frac{1}{\lambda}$$

 Eq. C.3

$$-\sqrt{1 - y_m^2} = \frac{1}{\lambda}$$

 rearranging

$$\lambda = -\frac{1}{\sqrt{1 - y_m^2}} \qquad (C.12)$$

And this, in turn, allows us to calculate the y-coordinate of our maximum:

$$y_m \stackrel{(C.9)}{=} -\frac{1}{2\lambda}$$

 Eq. C.12

$$y_m = \frac{\sqrt{1 - y_m^2}}{2}$$

 squaring

$$y_m^2 = \frac{1 - y_m^2}{4}$$

 rearranging

$$y_m^2 = \frac{1}{5}$$

 ✓

$$y_m = \sqrt{\frac{1}{5}}. \qquad (C.13)$$

This is exactly the same result that we found previously (Eq. C.5). Moreover, we can again determine the x-coordinate of the maximum by using Eq. C.3.[8] Therefore, we find once more that the maximum height our marble can reach is $z_m = \sqrt{5}$.[9]

[8] We calculated this explicitly in Eq. C.6.

[9] We calculated this in Eq. C.7.

Let's summarize: We start with a function $f(x, y, \ldots)$, whose maximum (or minimum) we want to determine, and a constraint $g(x, y, \ldots) = 0$. We combine them into a single function

$$\Lambda(x, y, \ldots, \lambda) = f(x, y, \ldots) + \lambda g(x, y, \ldots), \qquad (C.14)$$

where λ is a new variable we call the Lagrange multiplier. If we now determine the maximum (or minimum) of this new function, we automatically take the constraint into account and, therefore, find the correct result.

So far, you might not be particularly impressed by the Lagrange multiplier method. But here's what makes it so useful. If there are multiple constraints $g_1(x,y,\ldots) = 0$, $g_2(x,y,\ldots) = 0, \ldots$, we simply have to use

$$\Lambda(x,y,\ldots,\lambda_1,\lambda_2,\ldots) = f(x,y,\ldots) + \lambda_1 g(x,y,\ldots) + \lambda_2 g(x,y,\ldots) + \ldots.$$

In words, this means that we introduce a new term and a Lagrange multiplier λ_i for each constraint. The rest of the method works completely analogously.[10]

[10] PS: If you want to understand in geometric terms why the Lagrange multiplier method works, have a look at this illuminating video by Grant Sanderson: https://www.youtube.com/watch?v=yuqB-d5MjZA .

D

Invariance, Covariance and Functional Form

First of all, it's essential to understand that when we switch coordinates $q \to q'$, in general, the functional form of a function which depends on q will be changed.

For example, the transformed Lagrangian is defined as

$$\tilde{L}(q',\dot{q}') \equiv L\big(q(q'),\dot{q}(q',\dot{q}')\big). \tag{D.1}$$

In words, this means that $\tilde{L}$ is the Lagrangian that we get when we replace in the old Lagrangian all coordinates with our new ones *using the explicit transformation rules*.

But in general, we have

$$\tilde{L}(q',\dot{q}') \neq L(q',\dot{q}')$$
$$L\big(q(q'),\dot{q}(q',\dot{q}')\big) \neq L(q',\dot{q}'). \tag{D.2}$$

↻ Eq. D.1

To understand this, let's assume our original Lagrangian reads

$$L(x,y) = x^2 - y^2, \tag{D.3}$$

and we switch to polar coordinates

$$x = r\cos(\phi), \quad y = r\sin(\phi). \tag{D.4}$$

Our Lagrangian becomes

$$\tilde{L}(r,\phi) = (r\cos(\phi))^2 - (r\sin(\phi))^2. \tag{D.5}$$

So the functional dependence on r, ϕ is clearly a different one. In other words, the tilde is necessary because, in general, the Lagrangian will look quite different in terms of the new coordinates.

We wouldn't need a tilde if

$$\tilde{L}(q',\dot{q}') = L(q',\dot{q}'), \tag{D.6}$$

which in words means that we can replace $q \to q'$ and $\dot{q} \to \dot{q}'$ and get the correct Lagrangian. For our example

$$L(r,\phi) = r^2 - \phi^2 \tag{D.7}$$

is wrong and therefore we need a tilde.

The special kind of transformations for which

$$\tilde{L}(q',\dot{q}') = L(q',\dot{q}') \tag{D.8}$$

holds are called **invariance transformations** and describe symmetries of the system.[1]

[1] Symmetries are discussed in detail in Section 10.

For example, let's consider the Lagrangian

$$L(x) = \left(\frac{dx}{dt}\right)^2 \tag{D.9}$$

and that we want to switch to a coordinate system which is shifted related to the old one

$$x' = x + d, \tag{D.10}$$

where d is the distance that we shift the new coordinate system relative to the old one.

Since d is a constant, we have

$$\frac{dx'}{dt} = \frac{d(x+d)}{dt} = \frac{dx}{dt} \quad (D.11)$$

and therefore the Lagrangian after the transformation in Eq. D.10 reads

$$\tilde{L}(x') = \left(\frac{dx(x')}{dt}\right)^2$$

$\circlearrowright$ Eq. D.11

$$= \left(\frac{dx'}{dt}\right)^2. \quad (D.12)$$

We can therefore conclude that our Lagrangian in Eq. D.9 is unchanged by such shifts. Mathematically, this means

$$\tilde{L}(x', \dot{x}') = L(x', \dot{x}'). \quad (D.13)$$

Not only can we talk about how functions change under coordinate transformations, but also about how equations change.

In Section 7.2.2, we check explicitly that no matter which coordinates we choose, the Euler-Lagrange equation (Eq. 4.25)

$$\frac{\partial L}{\partial q} = \frac{d}{dt}\left(\frac{\partial L}{\partial \dot{q}}\right) \quad (D.14)$$

always keeps its *form*. This means that if we switch coordinates, $q \to q'$ such that $L \to \tilde{L}$, we can still calculate the correct equation of motion using

$$\frac{\partial \tilde{L}}{\partial q'} = \frac{d}{dt}\left(\frac{\partial \tilde{L}}{\partial \dot{q}'}\right). \quad (D.15)$$

Whenever an equation has this property, we say it is **covariant**.

However, take note that while the Euler-Lagrange equation is covariant, the equations of motion usually aren't.

To understand this, let's consider the Lagrangian

$$L(x, \dot{x}) = \dot{x}^2 + x^2. \quad (D.16)$$

We can calculate the corresponding equation of motion by using the Euler-Lagrange equation (Eq. 4.25):

$$\frac{\partial L}{\partial x} = \frac{d}{dt}\left(\frac{\partial L}{\partial \dot{x}}\right)$$

↪ Eq. D.16

$$\frac{\partial(\dot{x}^2 + x^2)}{\partial x} = \frac{d}{dt}\left(\frac{\partial(\dot{x}^2 + x^2)}{\partial \dot{x}}\right)$$

↪

$$2x = 2\frac{d}{dt}\dot{x}. \tag{D.17}$$

If we switch to a shifted coordinate system

$$x' = x - d \tag{D.18}$$

the Lagrangian becomes

$$\tilde{L}(x', \dot{x}') \stackrel{(D.1)}{=} L\left(x(x'), \dot{x}(x', \dot{x}')\right)$$

↪ Eq. D.16

$$= \left(\dot{x}(x', \dot{x}')\right)^2 + \left(x(x')\right)^2$$

↪ Eq. D.18 using $\dot{x}' = \dot{x}$, c.f. Eq. D.11

$$= \dot{x}'^2 + (x' + d)^2$$

↪

$$= \dot{x}'^2 + x'^2 + 2x'd + d^2. \tag{D.19}$$

We can then calculate what the equation of motion looks like in terms of our new coordinate x' by using again the Euler-Lagrange equation, but this time for the new Lagrangian $\tilde{L}$:

$$\frac{\partial \tilde{L}}{\partial x'} = \frac{d}{dt}\left(\frac{\partial \tilde{L}}{\partial \dot{x}'}\right)$$

↪ Eq. D.19

$$\frac{\partial(\dot{x}'^2 + x'^2 + 2x'd + d^2)}{\partial x'} = \frac{d}{dt}\left(\frac{\partial(\dot{x}'^2 + x'^2 + 2x'd + d^2)}{\partial \dot{x}'}\right)$$

↪

$$2x' + 2d = 2\frac{d}{dt}\dot{x}'. \tag{D.20}$$

The form of this equation is clearly different from the form of the original equation of motion (Eq. D.17). Therefore, while the Euler-Lagrange is indeed always covariant, this is not true for the equations of motion.

To understand this even better, let's write the original equation of motion (Eq. D.17) in the following form

$$f(x, \dot{x}) = 0, \qquad (D.21)$$

where

$$f(x, \dot{x}) \equiv 2x - 2\frac{d}{dt}\dot{x}. \qquad (D.22)$$

The equation of motion would only be covariant if after the transformation it reads

$$f(x', \dot{x}') = 0. \qquad (D.23)$$

But we can calculate immediately

$$f(x', \dot{x}') = 0$$
$$\circlearrowright \text{ Eq. D.22}$$
$$2x' - 2\frac{d}{dt}\dot{x}' = 0 \qquad (D.24)$$

which is not the correct transformed equation of motion (Eq. D.20).

In contrast, if we start with the Lagrangian

$$L(x, \dot{x}) = \dot{x}^2 \qquad (D.25)$$

instead, we get equations of motion which are indeed covariant under shifts of the coordinate system. This follows because the Lagrangian is unchanged by such shifts (Eq. D.13) and, therefore, the equation of motion remains unaffected too.

E

Active vs. Passive Transformations and Symmetries vs. Redundancies

In Chapter 10, we only considered active transformations. For example, we discussed what happens when we rotate the ship. This is a real physical transformation.

However, there is also a different kind of transformation, called passive transformation. A passive transformation is a change in how we *describe* a given system.

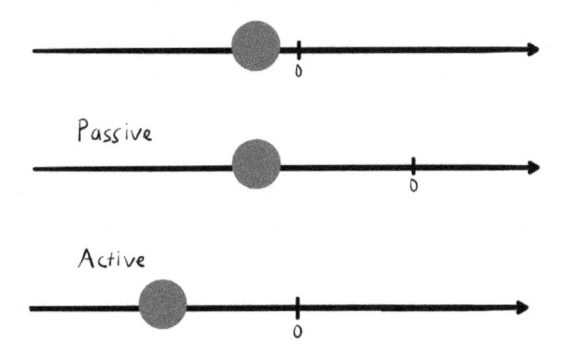

Passive

Active

For example, the physicist inside the boat can describe the experiment using curvilinear coordinates, a rotated or a shifted coordinate system. Such a change in how we describe a system has, of course, never any physical effect.

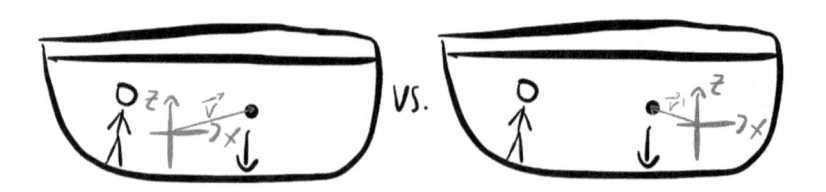

Passive transformations relate different *descriptions* of the same physical situation, while active transformations relate different physical situations.

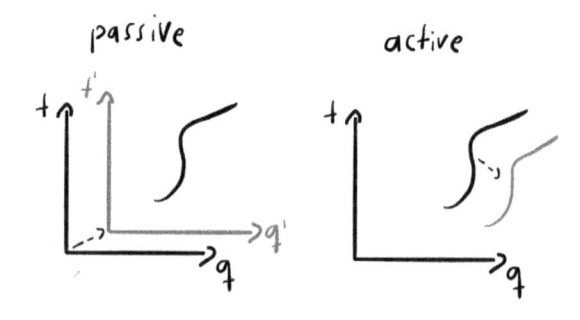

A key observation is that with enough mathematical shrewdness, we can make our description of a given experiment invariant under arbitrary coordinate transformations. But this does not mean that any experiment possesses all possible symmetries. Otherwise we wouldn't talk about symmetries at all because if all systems would possess a maximum amount of symmetry, symmetries wouldn't be useful to characterize individual systems.

A lot of confusion surrounding symmetries can be traced back to confusion about these two kinds of transformations.[1]

Since it is important to distinguish between passive and active transformations, we also need to distinguish between invari-

---

[1] Maybe it would be helpful to use "real transformation" and a "coordinate transformation" instead of "active transformation" and "passive transformation".

ance under active transformations and invariance under passive transformations. In particular, we call invariance under passive transformations a **redundancy** and invariance under active transformations a **symmetry**. This distinction is essential because a symmetry is a real feature of a system while a redundancy is only a feature of our description. It's a simple matter of consistency that we can make our description as redundant as we want. But we can't make a given system more symmetric by adding new mathematical symbols.[2]

[2] You can find a much more detailed discussion of this issue in

Jakob Schwichtenberg. Demystifying Gauge Symmetry. 2019a

F

Taylor Expansion

The Taylor expansion is one of the most useful mathematical tools, and we need it in physics all the time to simplify complicated systems and equations.

We can understand the basic idea as follows:

Imagine you sit in your car and wonder what your exact location $l(t)$ will be in 10 minutes: $l(t_0 + 10 \text{ minutes}) = ?$

▷ A first guess is that your location will be exactly your *current* location
$$l(t_0 + 10 \text{ minutes}) \approx l(t_0).$$
Given how large the universe is and thus how many possible locations there are, this is certainly not too bad.

▷ If you want to do a bit better than that, you can also include your *current* velocity $\dot{l}(t_0) \equiv \partial_t l(t)\big|_{t_0}$.[1] The total distance you will travel in 10 minutes if you continue to move at your current velocity is this velocity times 10 minutes: $\dot{l}(t_0) \times 10$ minutes. Therefore, your second estimate is your current location plus the velocity you are traveling times 10 minutes

$$l(t_0 + 10 \text{ minutes}) \approx l(t_0) + \dot{l}(t_0) \times 10 \text{ minutes}. \quad \text{(F.1)}$$

[1] Here ∂_t is a shorthand notation for $\frac{\partial}{\partial t}$, and $\partial_t l(t)$ yields the velocity (rate of change). After taking the derivative, we evaluate the velocity function $\dot{l}(t) \equiv \partial_t l(t)$ at t_0: $\dot{l}(t_0) = \partial_t l(t)\big|_{t_0}$.

[2] The factor $\frac{1}{2}$ and that we need to square the 10 minutes follows since, to get from an acceleration to a location, we have to integrate twice:

$$\int dt \int dt \ddot{x}(t_0) =$$
$$\int dt \ddot{x}(t_0) t =$$
$$\frac{1}{2}\ddot{x}(t_0)t^2$$

where $\ddot{x}(t_0)$ is the value of the acceleration at $t = t_0$ (= a constant).

[3] Here the superscript n denotes the n-th derivative. For example $f^{(0)} = f$ and $f^{(1)}$ is $\partial_x f$.

▷ If you want to get an even better estimate you need to take into account that your velocity can possibly change. The rate of change of the velocity $\ddot{l}(t_0) = \partial_t^2 l(t)\big|_{t_0}$ is what we call acceleration. So in this third step you additionally take your current acceleration into account[2]

$$l(t_0 + 10 \text{ minutes}) \approx l(t_0) + \dot{l}(t_0) \times 10 \text{ minutes}$$
$$+ \frac{1}{2}\ddot{l}(t_0) \times (10 \text{ minutes})^2.$$

▷ Our estimate will still not yield the perfect final location since, additionally, we need to take into account that our acceleration could change during the 10 minutes. We could therefore additionally take the current rate of change of our acceleration into account.

This game never ends and the only limiting factor is how precisely we want to estimate our future location. For many real-world purposes, our first order approximation (Eq. F.1) would already be perfectly sufficient.

The procedure described above is exactly the motivation behind the Taylor expansion. In general, we want to estimate the value of some function $f(x)$ at some value of x by using our knowledge of the function's value at some fixed point a. The **Taylor series** then reads[3]

$$f(x) = \sum_{n=0}^{\infty} \frac{f^{(n)}(a)(x-a)^n}{n!}$$
$$= \frac{f^{(0)}(a)(x-a)^0}{0!} + \frac{f^{(1)}(a)(x-a)^1}{1!} + \frac{f^{(2)}(a)(x-a)^2}{2!}$$
$$+ \frac{f^{(3)}(a)(x-a)^3}{3!} + \ldots, \tag{F.2}$$

where $f(a)$ is the value of the function at the point a we are expanding around. Moreover, $x - a$ is analogous to the 10-minute timespan we considered above. If we want to know the location at $x = 5{:}10$ pm by using our knowledge at $a = 5{:}00$ pm, we get $x - a = 5{:}10$ pm $- 5{:}00$ pm $= 10$ minutes. Therefore, this equation is completely analogous to our estimate of the future location we considered previously.

TAYLOR EXPANSION

To understand the Taylor expansion a bit better, it is helpful to look at concrete examples.

We start with one of the simplest but most important examples: the exponential function. Putting $f(x) = e^x$ into Eq. F.2 yields

$$e^x = \sum_{n=0}^{\infty} \frac{(e^0)^{(n)}(x-0)^n}{n!}.$$

The crucial puzzle pieces that we need are therefore $(e^x)' = e^x$ and $e^0 = 1$. Putting this into the general formula (Eq. F.2) yields

$$e^x = \sum_{n=0}^{\infty} \frac{e^0(x-0)^n}{n!} = \sum_{n=0}^{\infty} \frac{x^n}{n!}. \quad (F.3)$$

This result can be used as a definition of e^x.

Next, let's assume that the function we want to approximate is $\sin(x)$ and we want to expand it around $x = 0$. Putting $f(x) = \sin(x)$ into Eq. F.2 yields

$$\sin(x) = \sum_{n=0}^{\infty} \frac{\sin^{(n)}(0)(x-0)^n}{n!}.$$

The crucial information we therefore need is $(\sin(x))' = \cos(x)$, $(\cos(x))' = -\sin(x)$, $\cos(0) = 1$ and $\sin(0) = 0$. Because $\sin(0) = 0$, every term with even n vanishes, which we can use if we split the sum. Observe that

$$\sum_{n=0}^{\infty} n = \sum_{n=0}^{\infty} (2n+1) + \sum_{n=0}^{\infty} (2n)$$

$$1+2+3+4+5+6\ldots = 1+3+5+\ldots \quad +2+4+6+\ldots. \quad (F.4)$$

Therefore, splitting the sum into even and odd terms yields

$$\sin(x) = \sum_{n=0}^{\infty} \frac{\sin^{(2n+1)}(0)(x-0)^{2n+1}}{(2n+1)!}$$
$$+ \underbrace{\sum_{n=0}^{\infty} \frac{\sin^{(2n)}(0)(x-0)^{2n}}{(2n)!}}_{=0}$$
$$= \sum_{n=0}^{\infty} \frac{\sin^{(2n+1)}(0)(x-0)^{2n+1}}{(2n+1)!}. \quad (F.5)$$

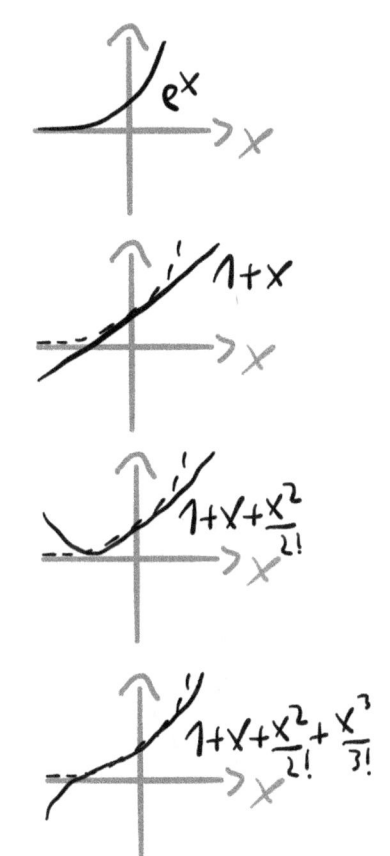

Moreover, every even derivative of $\sin(x)$ (i.e., $\sin^{(2n)}$) is again $\sin(x)$ or $-\sin(x)$. Therefore the second term vanishes since $\sin(0) = 0$. The remaining terms are odd derivatives of $\sin(x)$, which are all proportional to $\cos(x)$. We now use

$$\sin(x)^{(1)} = \cos(x)$$
$$\sin(x)^{(2)} = \cos'(x) = -\sin(x)$$
$$\sin(x)^{(3)} = -\sin'(x) = -\cos(x)$$
$$\sin(x)^{(4)} = -\cos'(x) = \sin(x)$$
$$\sin(x)^{(5)} = \sin'(x) = \cos(x).$$

The general pattern is $\sin^{(2n+1)}(x) = (-1)^n \cos(x)$, as you can check by putting some integer values for n into the formula[4].

[4] $\sin^{(1)}(x) = \sin^{(2\cdot 0+1)}(x) = (-1)^0 \cos(x) = \cos(x)$, $\sin^{(3)}(x) = \sin^{(2\cdot 1+1)}(x) = (-1)^1 \cos(x) = -\cos(x)$

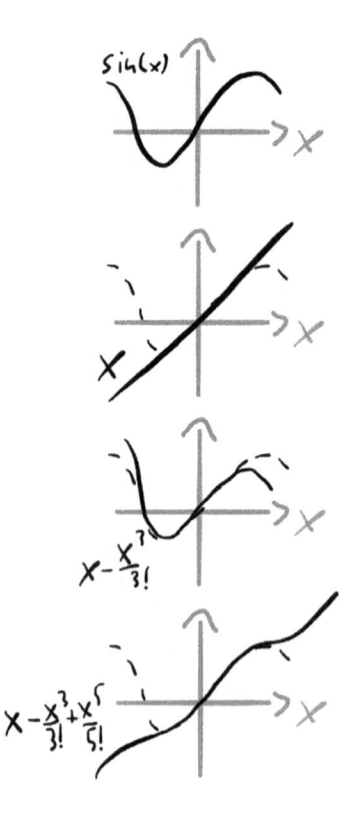

Thus, we can rewrite Eq. F.5 as

$$\sin(x) = \sum_{n=0}^{\infty} \frac{\sin^{(2n+1)}(0)(x-0)^{2n+1}}{(2n+1)!}$$
$$= \sum_{n=0}^{\infty} \frac{(-1)^n \cos(0)(x-0)^{2n+1}}{(2n+1)!}$$
$$\underbrace{=}_{\cos(0)=1} \sum_{n=0}^{\infty} \frac{(-1)^n (x)^{2n+1}}{(2n+1)!}. \qquad \text{(F.6)}$$

This is the Taylor expansion of $\sin(x)$, which we can also use as a definition of the sine function.

G

Vector Calculus

G.1 The Dot Product

The dot product allows us to combine two vectors $\vec{v}, \vec{w}$ in such a way that the result is a number[1]

$$\vec{v} \cdot \vec{w} = \begin{pmatrix} v_1 \\ v_2 \\ v_3 \end{pmatrix} \cdot \begin{pmatrix} w_1 \\ w_2 \\ w_3 \end{pmatrix} = v_1 w_1 + v_2 w_2 + v_3 w_3 . \quad (G.1)$$

In words, we can summarize the idea behind it as follows:

> The scalar product of two vectors $\vec{v} \cdot \vec{w}$ yields the projection of the first vector $\vec{v}$ onto the axis defined by the second vector $\vec{w}$ times the length of the second vector.

How does this interpretation fit together with the formula given in Eq. G.1?

To understand this, we need to talk about the projection of some vector $\vec{v}$ onto the axis defined by a second vector $\vec{w}$.[2]

[1] If we combine two vector functions, we get a scalar function.

[2] The easiest way to understand projections in general is to consider projections onto the coordinate axis $\vec{e}_x, \vec{e}_y, \vec{e}_z$. The projection of some vector $\vec{v}$ onto a coordinate axis like $\vec{e}_x$ is simply what we usually call the first component v_1. In words, the meaning of this component is how much our vector $\vec{v}$ spreads out in the x-direction. Analogously, the projection of $\vec{v}$ onto $\vec{e}_y$ is what we call the second component v_2 and it tells us how much $\vec{v}$ spreads out in the y-direction.

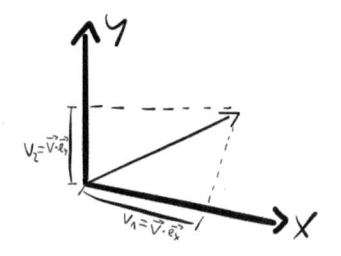

This allows us to write any vector in terms of basis vectors as follows:

$$\vec{v} = v_1 \vec{e}_x + v_2 \vec{e}_y + v_3 \vec{e}_z . \quad (G.2)$$

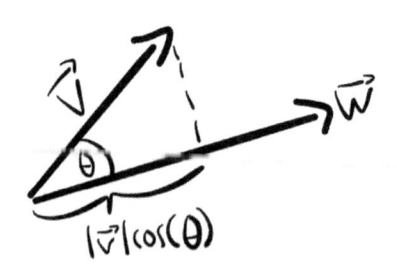

By looking at the figure above, we can see that the correct formula for the projection of $\vec{v}$ onto $\vec{w}$ is

$$\text{projection of } |\vec{v}| \text{ onto the axis defined by } \vec{w} = |\vec{v}| \cos \theta, \quad \text{(G.3)}$$

where θ denotes the angle between the two vectors. The statement from above in mathematical form therefore reads

$$\vec{v} \cdot \vec{w} = |\vec{v}| \cos \theta |\vec{w}|. \quad \text{(G.4)}$$

Therefore, the question we now need to answer is: How is this formula related to the usual formula in Eq. G.1?

To answer this question, we write our two general vectors in terms of our basis vectors:

$$\vec{v} = v_x \vec{e}_x + v_y \vec{e}_y + v_z \vec{e}_z$$
$$\vec{w} = w_x \vec{e}_x + w_y \vec{e}_y + w_z \vec{e}_z$$

We can then rewrite our dot product in terms of dot products of the basis vectors:

$$\vec{v} \cdot \vec{w} = |\vec{v}||\vec{w}| \cos(\theta)$$
$$= [v_x \vec{e}_x + v_y \vec{e}_y + v_z \vec{e}_z] \cdot [w_x \vec{e}_x + w_y \vec{e}_y + w_z \vec{e}_z]$$
$$= v_x w_x (\vec{e}_x \cdot \vec{e}_x) + v_x w_y (\vec{e}_x \cdot \vec{e}_y) + v_x w_z (\vec{e}_x \cdot \vec{e}_z)$$
$$+ v_y w_x (\vec{e}_y \cdot \vec{e}_x) + v_y w_y (\vec{e}_y \cdot \vec{e}_y) + v_y w_z (\vec{e}_y \cdot \vec{e}_z)$$
$$+ v_z w_x (\vec{e}_z \cdot \vec{e}_x) + v_z w_y (\vec{e}_z \cdot \vec{e}_y) + v_z w_z (\vec{e}_z \cdot \vec{e}_z).$$

Next we use that our basis vectors are normalized ($\vec{e}_x \cdot \vec{e}_x = 1$) and orthogonal ($\vec{e}_x \cdot \vec{e}_y = 0$):

$$\vec{v} \cdot \vec{w} = v_x w_x (1) + v_x w_y (0) + v_x w_z (0)$$
$$+ v_y w_x (0) + v_y w_y (1) + v_y w_z (0)$$
$$+ v_z w_x (0) + v_z w_y (0) + v_z w_z (1)$$
$$= v_x w_x + v_y w_y + v_z w_z \quad \checkmark.$$

We can visualize this calculation as follows:

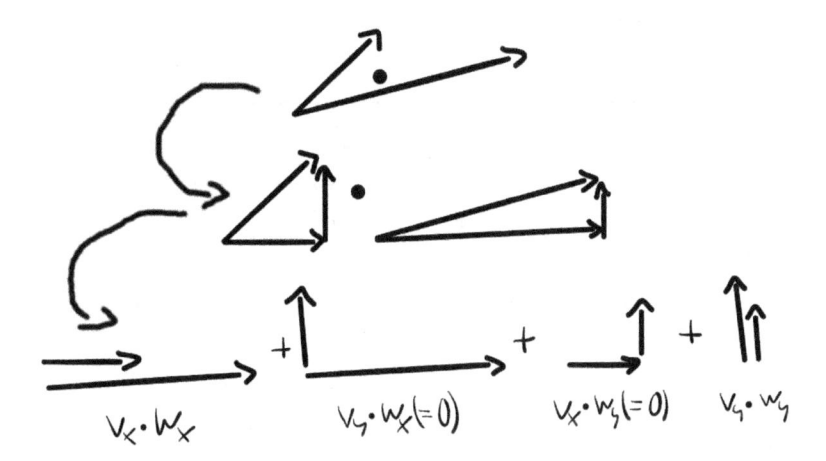

This tells us that we really can understand the result of the dot product as the projection of $\vec{v}$ onto $\vec{w}$ times the length of $\vec{w}$.

An important example is the dot product of a vector with itself $\vec{v} \cdot \vec{v}$. In words, the result is the projection of $\vec{v}$ onto itself times the length of $\vec{v}$. Since the projection of $\vec{v}$ onto itself yields the full vector length, we get the length of the vector squared

$$\vec{v} \cdot \vec{v} = |\vec{v}| \cos 0 |\vec{v}| = |\vec{v}|^2. \tag{G.5}$$

> **Example: dot product of two vectors**
>
> The dot product of
>
> $$\vec{v} = \begin{pmatrix} 1 \\ 4 \\ 9 \end{pmatrix} \quad \text{and} \quad \vec{w} = \begin{pmatrix} 2 \\ 2 \\ 1 \end{pmatrix} \tag{G.6}$$
>
> is given by
>
> $$\vec{v} \cdot \vec{w} = \begin{pmatrix} 1 \\ 4 \\ 9 \end{pmatrix} \cdot \begin{pmatrix} 2 \\ 2 \\ 1 \end{pmatrix} = 2 + 8 + 9 = 19. \tag{G.7}$$

G.2 The Cross Product

The cross product allows us to combine two vectors $\vec{A}$, $\vec{B}$ in such a way that the result is a vector

$$\vec{A} \times \vec{B} = \begin{pmatrix} A_1 \\ A_2 \\ A_3 \end{pmatrix} \times \begin{pmatrix} B_1 \\ B_2 \\ B_3 \end{pmatrix} = \begin{pmatrix} A_2 B_3 - A_3 B_2 \\ A_3 B_1 - A_1 B_3 \\ A_1 B_2 - A_2 B_1 \end{pmatrix}. \qquad (G.8)$$

In words, we can summarize the idea behind it as follows:[3]

[3] The direction in which the resulting vector points can be determined by the right-hand rule.

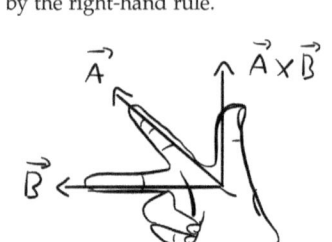

> The cross product of two vectors $\vec{A} \times \vec{B}$ yields a vector perpendicular to $\vec{A}$ and $\vec{B}$ whose magnitude is the area of the parallelogram spanned by $\vec{A}$ and $\vec{B}$.

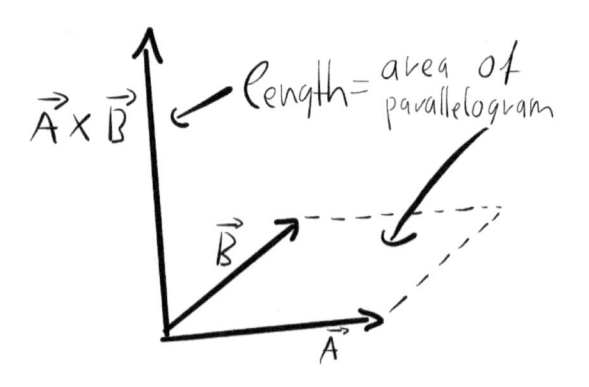

Now, how does this interpretation fit together with the formula in Eq. G.8?

To understand this, we first need to recall that the formula for the area of a parallelogram is base times height. Here, our base is given by the length of the vector $\vec{A}$ and the height by $\sin(\theta)|\vec{B}|$, where θ is the angle between $\vec{A}$ and $\vec{B}$.

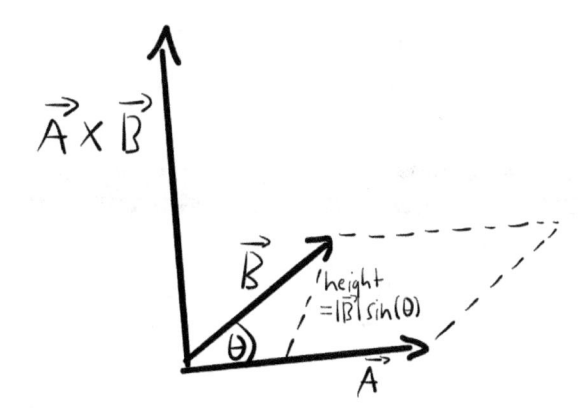

[Figure: hand-drawn diagram showing vectors $\vec{A}$, $\vec{B}$, and $\vec{A}\times\vec{B}$ with the parallelogram they span; height labeled $=|\vec{B}|\sin(\theta)$.]

The area of the parallelogram spanned by $\vec{A}$ and $\vec{B}$ is therefore

$$\text{area} = \vec{A}\sin(\theta)|\vec{B}|. \tag{G.9}$$

Therefore, the question we now need to answer is: How is this formula related to the usual formula in Eq. G.8?

To answer this question, we write our two general vectors in terms of our basis vectors:

$$\vec{v} = v_x\vec{e}_x + v_y\vec{e}_y + v_z\vec{e}_z$$
$$\vec{w} = w_x\vec{e}_x + w_y\vec{e}_y + w_z\vec{e}_z$$

We can then rewrite our cross product in terms of cross products of the basis vectors:

$$|\vec{v}\times\vec{w}| = \big|[v_x\vec{e}_x + v_y\vec{e}_y + v_z\vec{e}_z]\times[w_x\vec{e}_x + w_y\vec{e}_y + w_z\vec{e}_z]\big|$$
$$= \big|v_xw_x(\vec{e}_x\times\vec{e}_x) + v_xw_y(\vec{e}_x\times\vec{e}_y) + v_xw_z(\vec{e}_x\times\vec{e}_z)$$
$$+ v_yw_x(\vec{e}_y\times\vec{e}_x) + v_yw_y(\vec{e}_y\times\vec{e}_y) + v_yw_z(\vec{e}_y\times\vec{e}_z)$$
$$+ v_zw_x(\vec{e}_z\times\vec{e}_x) + v_zw_y(\vec{e}_z\times\vec{e}_y) + v_zw_z(\vec{e}_z\times\vec{e}_z)\big|.$$

Next we use that the cross product of a vector with itself yields zero ($\vec{e}_x\times\vec{e}_x = 0$) and that the cross product of two basis vectors yields the third basis vector ($\vec{e}_x\times\vec{e}_y = \vec{e}_z$):[4]

$$|\vec{v}\times\vec{w}| = \big|v_xw_x(0) + v_xw_y(\vec{e}_z) + v_xw_z(-\vec{e}_y)$$
$$+ v_yw_x(-\vec{e}_z) + v_yw_y(0) + v_yw_z(\vec{e}_x)$$
$$v_zw_x(\vec{e}_y) + v_zw_y(-\vec{e}_x) + v_zw_z(0)\big|$$
$$= \big|(v_yw_z - v_zw_y)\vec{e}_x + (v_zw_x - v_xw_z)\vec{e}_y + (v_xw_y - v_yw_x)\vec{e}_z\big|.$$

[4] This is necessarily the case since the resulting vector if we take the cross product of two vectors is orthogonal to them.

Therefore, we can conclude that the cross product really yields a vector whose length is the area of the parallelogram spanned by $\vec{A}$ and $\vec{B}$.

> ### Example: cross product of two vectors
>
> The cross product of
>
> $$ \vec{A} = \begin{pmatrix} 1 \\ 4 \\ 9 \end{pmatrix} \quad \text{and} \quad \vec{B} = \begin{pmatrix} 2 \\ 2 \\ 1 \end{pmatrix} \quad \quad (G.10)$$
>
> is given by
>
> $$ \vec{A} \times \vec{B} = \begin{pmatrix} 1 \\ 4 \\ 9 \end{pmatrix} \times \begin{pmatrix} 2 \\ 2 \\ 1 \end{pmatrix} = \begin{pmatrix} 4 - 18 \\ 18 - 1 \\ 2 - 8 \end{pmatrix} = \begin{pmatrix} -14 \\ 17 \\ -6 \end{pmatrix}. \quad (G.11)$$

Bibliography

V. I. Arnold. *Mathematical Methods of Classical Mechanics.* Springer New York, New York, NY, 1989. ISBN 978-1-4757-2063-1.

M. G. Calkin. *Lagrangian and Hamiltonian mechanics.* World Scientific, Singapore River Edge, NJ, 1996. ISBN 9810226721.

Ta-Pei Cheng. *Relativity, Gravitation and Cosmology: A Basic Introduction.* Oxford University Press, 2nd edition, 1 2010. ISBN 9780199573646.

Jennifer Coopersmith. *The Lazy Universe : An Introduction to the Principle of Least Action.* Oxford University Press, Oxford New York, NY, 2017. ISBN 9780198743040.

Alexei Deriglazov. *Classical Mechanics : Hamiltonian and Lagrangian Formalism.* Springer, Switzerland, 2016. ISBN 9783319441467.

Richard Feynman. *QED : the Strange Theory of Light and Matter.* Princeton University Press, Princeton, NJ, 2014. ISBN 978-0691164090.

A. P. French. *Special relativity.* Norton, New York, 1968. ISBN 9780393097931.

Juergen Freund. *Special relativity for beginners : a textbook for undergraduates.* World Scientific, Singapore, 2008. ISBN 9789812771599.

Herbert Goldstein. *Classical Mechanics*. Pearson, Essex, England, 2014. ISBN 9781292026558.

Louis Hand and Janet Finch. *Analytical Mechanics*. Cambridge University Press, Cambridge New York, 1998. ISBN 9780521573276.

Jorge José and Eugene Saletan. *Classical Dynamics: A Contemporary Approach*. Cambridge University Press Textbooks, 1998. ISBN 9781299859579.

Jerrold Marsden. *Introduction to Mechanics and Symmetry : a Basic Exposition of Classical Mechanical Systems*. Springer, New York, 1999. ISBN 978-0-387-21792-5.

David Morin. *Introduction to Classical Mechanics : With Problems and Solutions*. Cambridge University Press, Cambridge, UK New York, 2008. ISBN 9780511808951.

Roger Penrose. *The Emperor's New Mind : Concerning Computers, Minds and the Laws of Physics*. Oxford University Press, Oxford, 2016. ISBN 9780198784920.

Jakob Schwichtenberg. *No-Nonsense Electrodynamics*. No-Nonsense Books, Karlsruhe, Germany, 2018a. ISBN 978-1790842117.

Jakob Schwichtenberg. *Physics from Symmetry*. Springer, Cham, Switzerland, 2018b. ISBN 978-3319666303.

Jakob Schwichtenberg. *No-Nonsense Quantum Mechanics*. No-Nonsense Books, Karlsruhe, Germany, 2018c. ISBN 978-1719838719.

Jakob Schwichtenberg. Demystifying Gauge Symmetry. 2019a.

Jakob Schwichtenberg. *Physics from Finance*. No-Nonsense Books, Karlsruhe, Germany, 2019b. ISBN 978-1795882415.

Z. K. Silagadze. Gauge transformations are canonical transformations, redux, 2014.

Edwin Taylor. *Spacetime Physics : Introduction to Special Relativity*. W.H. Freeman, New York, 1992. ISBN 9780716723271.

Steven Weinberg. What is quantum field theory, and what did we think it is? In *Conceptual Foundations of Quantum Field Theory. Proceedings, Symposium and Workshop, Boston, USA, March 1-3, 1996*, pages 241–251, 1996.

A. S. Wightman. The usefulness of a general theory of quantized fields. In *Conceptual foundations of quantum field theory. Proceedings, Symposium and Workshop, Boston, USA, March 1-3, 1996*, pages 41–46, 1996.

Anthony Zee. *Einstein Gravity in a Nutshell*. Princeton University Press, 1st edition, 5 2013. ISBN 9780691145587.

Index

acceleration, 28
action, 73
action functional, 86
angular momentum, 30, 31

canonical coordinates, 153
canonical momentum, 91
canonical Poisson bracket relation, 156
canonical transformation, 122, 134, 153
   infinitesimal, 169
classical limit, 304, 306, 310
classical mechanics, 298
classical path, 304
commutator bracket, 110
configuration space, 18, 50, 52
conjugate momentum, 91
conservative forces, 34
constraint, 130
constructive interference, 305
contact transformation, 153
coordinate transformation, 122
covariant, 135, 369
cross product, 384
cube of physics, 297

destructive interference, 305
difference quotient, 343
dot product, 381

energy, 38

equation of motion, 41
Euler-Lagrange equation, 20, 89, 114
   meaning, 91

formulation, 18
functional, 78

Galilean relativity, 325
gauge transformation, 124, 141, 142
   canonical transformation, 160
general relativity, 298
generalized coordinates, 135
generalized force, 92
generalized momentum, 91
generating function, 172
generator, 170

Hamilton's equation, 95
   meaning, 105
Hamilton's equations, 20, 99, 108, 115
Hamilton's general equation, 107
Hamilton's principal function, 253
Hamilton-Jacobi equation, 254
Hamilton-Jacobi Mechanics, 252
Hamiltonian, 20, 98
   meaning, 105
Hamiltonian formalism, 115
   algorithm, 148
Hamiltonian formulation, 18, 57

Hamiltonian mechanics, 93
harmonic oscillator, 181
   Hamiltonian description, 188
   Lagrangian description, 187
   Newtonian description, 184
Heisenberg equation, 110
Hilbert space, 18, 289
holonomic constraint, 131
Hooke's law, 184

inertial observer, 63
infinitesimal, 344
initial conditions, 44
invariance transformations, 242, 368

kinetic energy, 33, 38
Koopman-von Neumann equation, 291
Koopman-von Neumann formulation, 18, 252

Lagrange multiplier, 131, 362
Lagrangian, 19
Lagrangian formalism
   algorithm, 127
Lagrangian formulation, 18, 57, 114
Legendre transform, 20, 353
length contraction, 317
Liouville equation, 275
Liouville operator, 291
Liouville's theorem, 280

mass, 28
mathematical arenas, 49
momentum, 30

Newton's second law, 60
Newtonian formalism
   algorithm, 125
Newtonian formulation, 18, 57, 113
Newtonian mechanics, 59
Noether charge, 244
   bare, 246

full, 246
Noether theorem
   converse, 242
   extended, 241
   generalized, 245
   intuitively, 228
   Lagrangian, 244
Noether's theorem
   Hamiltonian, 235
non-holonomic constraint, 131

partial derivative, 346
path integral, 301
pendulum, 201
   Hamiltonian description, 211
   Lagrangian description, 207
   Newtonian description, 203
phase space, 18, 55
   filamentation, 286
   flow, 263
physical space, 18, 49
Planck constant, 297, 300
   reduced, 297
point transformation, 122, 134
   point transformation, 160
Poisson bracket, 108
position, 26
potential energy, 38
principle of equivalence, 325
principle of relativity, 311
probability amplitude, 300
probability distribution, 264
product space, 51
proper time, 318

quantum field theory, 298
quantum mechanics, 298
   phase space, 293
   pilot wave formulation, 293, 310
quantum potential, 310
quasi-symmetries, 240

redundancy, 375
renormalization group, 333
rest energy, 321
rheonomic, 131

Schwarz's theorem, 145
scleronomic, 131
separation of variables, 44
special relativity, 298, 314
state vector, 289
subsystem, 225
symmetry, 224, 375

Taylor expansion, 377
Taylor series, 378
theory of everything, 298

time dilation, 317
total derivative, 346
total energy, 33
trajectory, 45

variation, 84
variational calculus, 81
velocity, 27

wave function, 308
work, 34